HOMOPLASY

The Recurrence of Similarity in Evolution

HOMOPLASY

The Recurrence of Similarity in Evolution

Michael J. Sanderson

Section of Evolution and Ecology
University of California, Davis
Davis, California

Larry Hufford

Department of Botany
Washington State University
Pullman, Washington

Academic Press

San Diego   London   Boston   New York   Sydney   Tokyo   Toronto

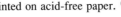
Copyright © 1996 by ACADEMIC PRESS

Academic Press, Inc.
525 B Street, Suite 1900, San Diego, California 92101-4495, USA
http://www.apnet.com

Academic Press Limited
24-28 Oval Road, London NW1 7DX, UK
http://www.hbuk.co.uk/ap/

Library of Congress Cataloging-in-Publication Data

Homoplasy / edited by Michael J. Sanderson, Larry Hufford
        p.   cm.
    Includes bibliographical references and index.
    ISBN 0-12-618030-X (alk. paper)
    1. Homoplasy.   I. Sanderson, Michael J.   II. Hufford, Larry,
date.
QH372.5.H65   1996
575--dc20   96-28243
                                                 CIP

PRINTED IN THE UNITED STATES OF AMERICA
96   97   98   99   00   01   BB   9   8   7   6   5   4   3   2   1

CONTENTS

Introduction

David B. Wake

PART I IMPLICATIONS OF HOMOPLASY

Explanations of Homoplasy at Different Levels of Biological Organization

Daniel R. Brooks

Homoplasy Connections and Disconnections: Genes and Species, Molecules and Morphology

Jeff J. Doyle

The Relationship between Homoplasy and Confidence in a Phylogenetic Tree

Michael J. Sanderson and Michael J. Donoghue

Nonfloral Homoplasy and Evolutionary Scenarios in Living and Fossil Land Plants
Richard M. Bateman

Behavioral Characters and Homoplasy: Perception versus Practice
Heather C. Proctor

PART II MEASURES OF HOMOPLASY

Measures of Homoplasy

James W. Archie

The Measurement of Homoplasy: A Stochastic View
Joseph T. Chang and Junhyong Kim

PART III   GENERATION OF HOMOPLASY

Complexity and Homoplasy
Daniel W. McShea

Exaptation, Adaptation, and Homoplasy: Evolution of Ecological Traits in Dalechampia Vines
W. Scott Armbruster

Patterns of Homoplasy in Behavioral Evolution

Susan A. Foster, William A. Cresko, Kevin P. Johnson, Michael U. Tlusty,
and Harleigh E. Willmott

Ontogenetic Evolution, Clade Diversification, and Homoplasy

Larry Hufford

Homoplasy in Angiosperm Flowers

Peter K. Endress

Homoplasy and the Evolutionary Process: An Afterword

Michael J. Sanderson and Larry Hufford   327

CONTRIBUTORS

Numbers in parentheses indicate the pages on which the authors' contributions begin.

James W. Archie (153) Department of Biological Sciences, California State University, Long Beach, California 90840.

W. Scott Armbruster (227) Department of Biology and Wildlife and Institute of Arctic Biology, University of Alaska, Fairbanks, Alaska 99775.

Richard M. Bateman (91) Department of Geology, Royal Botanic Garden and Royal Museum of Scotland, Edinburgh EH1 1JF, Scotland, United Kingdom.

Daniel R. Brooks (3) Department of Zoology, University of Toronto, Toronto, Ontario M5S 3G5 Canada.

Joseph T. Chang (189) Department of Statistics, Yale University, New Haven, Connecticut 06520.

William A. Cresko (245) Department of Biology, Clark University, Worcester, Massachusetts 01610.

Michael J. Donoghue (67) Department of Organismic and Evolutionary Biology, Harvard University, Cambridge, Massachusetts 02138.

Jeff J. Doyle (37) L. H. Bailey Hortorium, Cornell University, Ithaca, New York 14853.

Peter K. Endress (303) Institute of Systematic Botany, University of Zurich, 8008 Zurich, Switzerland.

Susan A. Foster (245) Department of Biology, Clark University, Worcester, Massachusetts 01610.

Larry Hufford (271, 327) Department of Botany, Washington State University, Pullman, Washington 99164.

Kevin P. Johnson (245) Department of Ecology, Evolution, and Behavior, University of Minnesota, St. Paul, Minnesota 55108.

Junhyong Kim (189) Department of Biology, Yale University, New Haven, Connecticut 06511.

Daniel W. McShea (207) Department of Zoology, Duke University, Durham, North Carolina 27708.

Heather C. Proctor (131) Department of Biology, Queen's University, Kingston, Ontario K7L 3N6, Canada.

Michael J. Sanderson (67, 327) Section of Evolution and Ecology, University of California, Davis, Davis, California 95616.

Michael U. Tlusty (245) Department of Biology, Clark University, Worcester, Massachusetts 01610.

David B. Wake (xvii) Department of Integrative Biology and Museum of Vertebrate Zoology, University of California, Berkeley, Berkeley, California 94720.

Harleigh E. Willmott (245) Department of Biological Sciences, University of Arkansas, Fayetteville, Arkansas 72701.

PREFACE

More than one hundred years after E. Ray Lankester coined the term "homoplasy" and Darwin explored its evolutionary implications, homoplasy remains a concept that evokes strong reactions among evolutionists and systematists. It is variously and inconsistently seen as fascinating, inexplicable, or downright annoying. That homoplasy could elicit such strong responses seems curious at first, but homoplasy is tied to long-standing unresolved disagreements about the origin of similarity among organisms. These are the same themes that have underlain discussions of homology, another subject that elicits strong responses. However, perhaps because of the widespread acceptance of the idea that homoplasy is just nonhomology, and therefore that homoplasy is somehow subsumed by discussions of homology, homoplasy has not received the attention that it deserves.

Homoplasy was first discussed by comparative anatomists, such as Richard Owen, and natural historians, such as Charles Darwin, who marveled at the convergence of forms possessed by organisms in similar habitats. The famous homoplasies, like wings of birds and bats or succulence in cacti and euphorbs, around which elaborate narratives of natural history have been woven, have become icons of modern evolutionary biology. Phylogenetic systematics has added a wealth of knowledge about empirical patterns of homoplasy, beginning with the discovery that parallelisms and reversals pervade phylogenies of every clade. Discussions on the nature of homoplasy have been heavily influenced by the view of some phylogeneticists that homoplasy is merely a mistake, an error made in homology assessment and character coding.

Given this long, rich, and now controversial history, it is surprising that homoplasy has never formed the basis for a book-length treatment. The explosive growth in phylogenetic information in recent years, the growing recognition of the need to integrate evolutionary/comparative biology and phylogenetics, and the recent publication of major works on homology all suggest to us that the time is finally right for a detailed examination of the issue of homoplasy. Our aims are several. Mainly we seek some consensus about whether the concept of homoplasy can stand alone in evolutionary biology and phylogenetics or whether it is inextricably linked to homology. At the same time, by eliciting contributions from ecologists, evolutionists, morphologists, and developmental and molecular biologists, we hope to evaluate the notion that homoplasy does indeed reflect an evolutionary process (or set of processes) in its own right and perhaps provide some initial circumscription of the nature of those processes. Finally, we examine the relationship between homoplasy and phylogenetic systematics, which have become so intertwined in recent years and which can finally be explored in the context of broader evolutionary issues.

We thank Chuck Crumly at Academic Press for suggesting that the time was right for this project and all the contributors to this volume for their patience and diligence.

Michael J. Sanderson
Larry Hufford

INTRODUCTION

DAVID B. WAKE

University of California
Berkeley, California

Homology and homoplasy are terms that travel together; homoplasy being close to, but not quite, the inverse of homology. If homology is "the same thing" (persistence of traits in their various transformed states), homoplasy is the *appearance* of "sameness" that results from independent evolution. Homoplasy is *derived similarity* that is *not* synapomorphy (Archie, in his chapter, usefully differentiates between evolutionary and phylogenetic homoplasy). To my knowledge, this is the first book devoted to the subject, and the authors have assumed the challenge to define the field of investigation and to lay the foundation for future research.

Similarity, and the reasons for it, are central issues in studies of homoplasy, as in controversies over homology. Virtually all of the problems associated with the study of homology (well displayed in the diverse contributions in Hall 1994; see also Wake 1994) also relate to homoplasy. However, some issues relate more importantly to homoplasy than to homology. I believe that a full exposition of these issues has long been needed, and this volume explores homoplasy in broad perspective, while at the same time giving attention to many detailed aspects of homoplasy.

There are two general categories of issues that relate to homoplasy. One is the means of detection and the other is the reason that apparently similar features evolve independently. Detection was made objective, but not simplified, by the development of a well-formulated, logical philosophy of phylogenetics and of associated analytical methods. The seminal work of Hennig (1966) accelerated progress in this area, and thanks to contributors such as Farris (Hennig-86), Felsenstein (PHYLIP), Swofford (PAUP), and Maddison and Maddison (MacClade), to mention only the most frequently cited, a wide variety of tools is now available for phylogenetic analyses. Every contributor to the present volume explicitly (and appropriately) discusses this area of research on homoplasy.

The second category concerns the biological basis of homoplasy, that is, the biological processes and mechanisms responsible for the production of homoplastic traits. This is a venerable area of inquiry that was largely shunted aside during the past 25 years as interest focused on phylogenetics *per se*. But there has always been an interest in the underlying causes of homoplasy, and our vastly improved means of detection has made the study of causes more sophisticated. Recent efforts exploring the biological basis of homology (Hall 1995; Wagner 1989, 1995) have relevance to this topic, and I have examined the causes of homoplasy in some recent papers (Wake 1991; Shubin et al. 1995). Most of the authors in this book discuss at least some of the proximal biological reasons for homoplasy and some (e.g., McShea, Armbruster, and Hufford) emphasize it.

For one focused on phylogenetic analysis it matters little whether a trait subject to homoplasy is a convergence, a parallelism, or a reversal, but instead what is critical is that the phylogenetic signal (based on "true" synapomorphies in classical and mainly morphological cladistics) be stronger than the homoplastic noise. As data bases grow, with concomitant analytical challenges, there is less interest in analysis of individual traits. In molecular systematics it is not unusual to encounter hundreds of equally parsimonious trees for a large data set (Sanderson and Donoghue, this volume, find morphological data sets to be no better in this respect). In such circumstances homoplasy is rampant and one does not worry about determining synapomorphies (but if one is using a coding sequence there will probably be more interest in specific second-position transversions than in third-position transitions). For those focused on trait evolution and evolutionary mechanisms, however, the phylogenetic analysis is usually a precursor to a study of the reasons for particular homoplasies.

At no hierarchical level is homoplasy more common, and arguably better understood, than at the most reduced—the nucleotide level. Possibilities are severely constrained by the availability of only four bases, and in parts of the genome the rate of substitution is relatively high. At the level of detection of a "G" in a particular nucleotide position, homoplasy may be common, although homology (in the sense of chemically being "the same thing") may be precise. By reducing the homoplasy problem to the lowest hierarchical level that is practical, all statements of homology can be seen to be provisional and cladogram dependent. Homoplasy may increase as the number of taxa studied increases, but it also increases when there is low trait persistence or when there are few alternatives and transitions between them are possible.

Problems with homoplasy parallel those related to homology, and they are unlikely to be resolved any more readily. Homologous traits provide insight into the connectedness of life, its genealogy. For some researchers, mainly phylogeneticists (or "phylogenists"), history and homology are essentially congruent, while for others, mainly morphologists and developmentalists, history and structural identity need to be teased apart. Wagner (1989, 1995) argued that homologous characters are parts of the genotype that become individualized because of the local regulation of development, which acts as a constraint on further change; consequently, the individuated parts tend to persist. Wagner's (1995) "building block" hypothesis is a recent and I believe heuristic attempt to explain how characters become individuated (possibly as a result of selection "taming" spontaneously produced morphological variants) and then play important roles in adaptive processes by virtue of their replaceability and combinability. This conceptualization is relevant for homoplasy of characters that on face value appear too complicated or novel to evolve repeatedly. Perhaps characters reappear because small changes in regulation of local developmental systems trigger latent building blocks. This connects to a long recognized area of discussion in character homology. Latent homology is the situation in which the developmental precursor of a structure persists and is triggered in different lineages, sometimes by different means (de Beer 1971; Hall 1995). There is a long history of interest in the phenomenon, expressed variously in such concepts as homologous series (Vavilov, 1922), canalized evolutionary potential (Saether 1983), and apomorphic tendencies (Rasmussen 1983) (reviewed by Sanderson 1991). This is a "levels problem," so that the precursor itself is the appropriate level for homology but the phenotypic outcome of phylogenetically independent triggerings is the level at which homoplasy occurs. One interested in trait evolution might then classify the different triggers, thus sorting different reasons for parallel evolution which could lead to improved character coding or additional characters to be coded.

It is not possible to state a hard and fast distinction between parallel and convergent evolution, but categorization can help sort morphological homoplasies. In general, parallelism is the production of apparently identical traits by the same generative system and convergence involves the production of similar traits by different generative systems. Elongation is a common homoplasy in salamanders: parallelism is the independent increase in numbers of segments, and ultimately vertebrae, during early embryogenesis in different taxa; convergence is the much rarer increase in vertebral length in a few taxa. This example illustrates the importance of levels of analysis in homology and homoplasy research. "Elongation" can be explained by "added segments" and "increased segment length," each of which in turn generates homoplasy.

Organisms are integrated systems showing complicated couplings that limit and bias the kinds and direction of trait evolution (Roth and Wake 1989; Schwenk 1995). The homoplastic evolution of freely projectile tongues in salamanders has been explored in some detail with respect to these ideas (Lombard and Wake 1977, 1986; Wake 1982, 1991; Roth and Wake 1985). The hyolingual system is used for both filling lungs and propelling the tongue. Lung reduction and loss is a common homoplasy in salamanders and appears to be a necessary, but not sufficient, condition for the evolution of extreme tongue projection by releasing the constraint of lung filling and permitting specialization. Specialization takes specific, alternative forms. For example, the tongue skeleton must fold for long-distance projection, and only two geometrical arrangements are possible, both of which have been used by different clades to accomplish the same end, which produces the same homoplasy at the level of tongue projection, but is revealed to be convergence at the level of biomechanics.

Paedomorphic homoplasy may be the easiest to understand and one of the most common kinds. The pattern of some taxa having "backed down" plesiomorphic ontogenetic trajectories in relation to others can produce confusingly similar morphologies in separate lineages within a clade. Perennibranchiate species of salamanders, forms in which sexual maturity is attained by gilled larvae and metamorphosis is eliminated, come to resemble each other in most morphological details. This is because differences among taxa are most commonly terminal ontogenetic additions and the larval stage is relatively undifferentiated. The perennibranchiate forms represent evolutionary reversals to more simple, and hence less variable, early ontogenetic stages. Less obvious kinds of ontogenetic transformations are probably associated with many homoplastic trait transformations.

Certain homoplasies appear commonly in some clades but not others, and hierarchical approaches may be heuristic. Parthenogenesis has evolved repeatedly in lizards, and Moritz et al. (1992) hypothesized that a general ("phylogenetic") constraint in vertebrates had been overcome in some manner in squamate vertebrates, representing a necessary condition for the homoplasy. Schwenk (1995) offered what he called an internalist (or structuralist) perspective as an alternative, suggesting that there is no prior constraint, but rather that squamates have evolved novel genetic or developmental conditions which bias them toward the homoplastic evolution of parthenogenesis. In this case the parallelism is taken as evidence of a synapomorphic evolutionary constraint. I offer this example to show that homoplasy may give insight into underlying traits that are of both phylogenetic and evolutionary significance. Despite widespread interest in homoplastic tendencies such as parthenogenesis in lizards, Sanderson's (1991) statistical analysis failed to find compelling evidence for their existence. However, it may be that homoplastic characters of this sort are routinely excluded from phylogenetic analyses by systematists (the "file-drawer" problem identified by Sanderson and Donoghue, this volume).

Since I began biological research I have been fascinated with morphological homoplasies, especially the biological basis for their independent generation. Perhaps this fascination developed because I chose to pursue evolutionary morphological and systematic studies of a difficult group, relatively featureless salamanders. The most featureless were the most difficult—clades that contained miniaturized species, clades that displayed general uniformity despite being speciose and in which the few derived traits were distributed in bewildering arrays, and clades that contained species displaying varying degrees of paedomorphosis. It was my studies of salamanders that first made clear to me that the study of the causes of homoplasy requires a hierarchical approach.

That homoplasy detection is accomplished through a genealogical hierarchical (phylogenetic) analysis is now widely understood. Less attention has been given to hierarchically based, biological explanations for homoplasy; I am pleased that several authors in this book (e.g., Armbruster, Bateman, Brooks, Hufford, and McShea) emphasize such perspectives, because I believe they are key to our understanding of the phenomena involved in the production of biological similarity. Hierarchical approaches have been used with respect to homology by several researchers (e.g., Roth 1991) with excellent results.

Important issues in evolutionary biology involve the reasons that morphological change takes specific forms (e.g., Wagner 1989; Wake 1991), and hierarchical investigations of the biological bases of homoplasy have provided insight. Homoplasy alerts the researcher to the possibility of limits on character production and spurs inquiry into the mechanistic foundation for change. If we demand an explanation for specific homoplasies without taking into account the full extent of homoplasy and the degree of intercorrelation, a hierarchical error can result. For example, researchers might devise a research program that demands a selective explanation for a specific homoplastic trait, when this is an inappropriate explanatory level. The trait in question might be part of a more general phenomenon. An example which illustrates many points related to homoplasy is miniaturization.

Miniaturization results from dynamics at the level of populations and communities, but often has evolutionary consequences quite independent of factors that led to size decrease (Hanken and Wake 1993). A frequent outcome is ontogenetic truncation, or progenesis (Gould 1977), and the homoplasy thus generated is failure of plesiomorphic characters to appear, recorded as a loss. One of the best understood in vertebrates is the repeated loss of a toe in miniaturized amphibians (Alberch and Gale 1985). In frogs it is the first and in salamanders it is the fifth toe that invariably is lost. This can be understood, and even predicted, from the different, but standard, patterns of morphogenesis in the two taxa. Frogs display postaxial-to-preaxial digital morphogenesis, a pattern shared with amniotes, while salamanders display a preaxial-to-postaxial pattern. The first toe of frogs and the fifth of salamanders fail to appear in four-digited taxa. However, the last formed digit is often larger than expected from out-group comparisons. Occasional individuals of five-toed species of salamanders are found in which only four external digits are present but in which there are five skeletons; the full skeleton of the last digit may be imperfectly duplicated, so that there are two skeletal but only one integumentary digit (Wake 1991). The last digit of four-toed salamanders then is not number four but some undifferentiated combination of four and five. Here another hierarchical issue becomes relevant. Amphibians have large to very large genomes, relative to other tetrapods, and genome size is a relatively persistent trait, less likely to change during phylogenesis than body size, for example (Sessions and Larson 1987). Genome size translates directly to cell size, so miniaturized amphibians are caught in a hierarchical crunch, organismal form being simultaneously affected by downward causation from factors at the population level leading to small body size and by upward causation from persistent genome size. Limb buds of amphibians have large cells, so miniaturized taxa with smaller limb buds will have fewer cells. There are important allometric considerations. Reduction in limb bud size in a large celled species will have greater consequences than in a small celled species, because cell number is critically important in morphogenetic processes such as condensation, segmentation, and bifurcation (Shubin and Alberch 1986). The large-celled species may encounter developmental thresholds that they fail to cross because they have too few cells and digital reduction may result. The combination of few cells and reduced cell division rate (a consequence of large genome size; Sessions and Larson 1987) causes truncation of developmental pathways and digital loss occurs. In contrast, miniaturized lizards, derived from small-genomed ancestors, are capable of producing five small digits. Some miniaturized lizards are nearly as small as tiny frogs and salamanders, but the amphibians are effectively (i.e., with respect to developmental mechanics and some other organismal-level phenomena) much smaller because of their much larger cells (Hanken and Wake 1993).

One might expect the hierarchical factors outlined above to have general effects beyond digit loss and such effects do occur. Homoplasy at the level of tissue histogenesis in the brains of large-genomed lungfish, frogs, and salamanders has been documented (Roth et al. 1994); the proximal reasons for such homoplasy are relatively well understood (e.g., Roth et al. 1995). It would be a mistake to try to explain on a point-by-point basis the reasons for detailed neuroanatomical similarity between lungfishes and plethodontid salamanders when there is a more general explanation available at a different hierarchical level (this of course assumes that evidence is good that the two taxa do not form a monophyletic group). There are many examples of paedomorphic homoplasy in amphibians, and unless one uses an hierarchical approach, traits that are correlated are treated as separate. A large problem with ontogenetically based homoplasy is that there may be different thresholds for different characters. Thus, some four-toed salamanders also lack prefrontal bones and have large cranial foramina (e.g., *Batrachoseps*), but others (e.g., *Hemidactylium*) do not (Wake 1966).

Genome size, patterns of limb morphogenesis, and tissue histogenesis in the brain can all be viewed as factors that produce bias in the production of variation and make some outcomes far more likely than others (see also discussions in this book by Brooks, Hufford, and McShea). Traits characteristically are not "free" to vary in any direction; they vary predictably (Alberch 1989; Wake 1991). Variant traits within a population frequently duplicate conditions fixed in other taxa, illustrating why study of factors responsible for production of homoplastic traits is likely to be a fruitful area of inquiry (Shubin et al. 1995).

Once one initiates a hierarchical approach to homoplasy, unanticipated insights emerge. Surprisingly, retinotectal projections of paedomorphic bolitoglossine salamanders with large genomes share a homoplastic relationship with primates and "megabats." This is an example of ontogenetic repatterning in the salamanders (Wake and Roth 1989), in which paedomorphosis has led to a derived morphology that in turn serves as a point of departure for a derived ontogenetic trajectory. In this case, one level of homoplasy that has predictable results (in essence, backing down a persistent ontogenetic trajectory, using the formalism of Alberch et al. 1979; see also Hufford 1996) leads to another, unanticipated and less-predictable level of homoplasy. In all three cases of homoplasy stereoscopic vision is enhanced, but whether there is a common ontogenetic phenomenon is unclear. Ontogenetic repatterning of the kind found in the salamanders is a far more profound event than a simpler, phase- or stage-specific ontogenetic transformation, which, however, might also have major consequences (again, see Hufford 1996).

The study of homoplasy offers a rich and diverse array of opportunities. The chapters in this book represent the collective focus on homoplasy by a wide array of biologists, mainly systematists, but also organismal and developmental biologists. The rigor and discipline represented in these contributions establish a solid foundation on which to build future studies of homoplasy. I look forward to a unified approach that incorporates functionalist (selectionist or externalist), structuralist (mechanistic or internalist), and phylogenetic perspectives. One can view homoplasy as the major problem in phylogenetic inference; alternatively, once we have some assurance that homoplasy exists it becomes a challenge to explain its origins. The pursuit of such questions will contribute greatly to our understanding of how diversity evolves.

REFERENCES

Alberch, P. 1989. The logic of monsters: Evidence for internal constraint in development and evolution. Geobios, mém. spec. 12:21–57.

Alberch, P. and E. Gale. 1985. A developmental analysis of an evolutionary trend: Digital reductions in amphibians. Evolution 39:8–23.

Alberch, P., S. J. Gould, G. F. Oster, and D. B. Wake. 1979. Size and shape in ontogeny and phylogeny. Paleobiology 5:296–317.

Archie, J. 1996. Measures of homoplasy. Pp. 153–187 *in* M. J. Sanderson and L. Hufford, eds. Homoplasy: The recurrence of similarity in evolution. Academic Press, San Diego.

Armbruster, W. S. 1996. Exaptation, adaptation, and homoplasy: Evolution of ecological traits in *Dalechampia*. Pp. 227–243 *in* M. J. Sanderson and L. Hufford, eds. Homoplasy: The recurrence of similarity in evolution. Academic Press, San Diego.

Bateman, R. M. 1996. Nonfloral homoplasy and evolutionary scenarios in living and fossil land plants. Pp. 91–130 *in* M. J. Sanderson and L. Hufford, eds. Homoplasy: The recurrence of similarity in evolution. Academic Press, San Diego.

Brooks, D. R. 1996. Explanations of homoplasy at different levels of biological organization. Pp. 3–36 *in* M. J. Sanderson and L. Hufford, eds. Homoplasy: The recurrence of similarity in evolution. Academic Press, San Diego.

de Beer, G. 1971. Homology, an unsolved problem. Oxford Univ. Press, Oxford.

Gould, S. J. 1977. Ontogeny and phylogeny. Harvard Univ. Press, Cambridge.

Hall, B. K. 1994. Homology: The hierarchical basis of comparative biology. Academic Press, New York.

Hall, B. K. 1995. Homology and embryonic development. Evolutionary Biology 28:1–37.

Hanken, J. and D. B. Wake. 1993. Miniaturization of body size: Organismal consequences and evolutionary significance. Annual Review of Ecology and Systematics 24:501–521.

Hennig, W. 1966. Phylogenetic systematics. Univ. Illinois Press, Urbana.

Hufford, L. 1996. Ontogenetic evolution, clade diversification, and homoplasy. Pp. 271–301 *in* M. J. Sanderson and L. Hufford, eds. Homoplasy: The recurrence of similarity in evolution. Academic Press, San Diego.

Lombard, R. E. and D. B. Wake. 1977. Tongue evolution in the lungless salamanders, family Plethodontidae. II. Function and evolutionary diversity. Journal of Morphology 153: 39–80.

Lombard, R. E. and D. B. Wake. 1986. Tongue evolution in the lungless salamanders, family Plethodontidae. IV. Phylogeny of plethodontid salamanders and the evolution of feeding dynamics. Systematic Zoology 35:532–551.

McShea, D. W. 1996. Complexity and homoplasy. Pp. 207–225 *in* M. J. Sanderson and L. Hufford, eds. Homoplasy: The recurrence of similarity in evolution. Academic Press, San Diego.

Moritz, C.J., J. W. Wright, and W. M. Brown. 1992. Mitochondrial DNA analyses and the origin and relative age of parthenogenetic *Cnemidophorus*: Phylogenetic constraints on hybrid origins. Evolution 46:184–192.

Rasmussen, F. N. 1983. On "apomorphic tendencies" and phylogenetic inference. Systematic Botany 8:334–337.

Roth, G. and D. B. Wake. 1985. Trends in the functional morphology and sensorimotor control of feeding behavior in salamanders: An example of the role of internal dynamics in evolution. Acta Biotheoretica 34:175–192.

Roth, G. and D. B. Wake. 1989. Conservatism and innovation in the evolution of feeding in vertebrates. Pp. 7–21 *in* D. B. Wake and G. Roth, eds. Complex organismal functions: Integration and evolution in vertebrates. John Wiley & Sons, Chichester.

Roth, G., J. Blanke and M. Ohle. 1995. Brain size and morphology in miniaturized plethodontid salamanders. Brain, Behavior and Evolution 45:84–95.

Roth, G., J. Blanke and D. B. Wake. 1994. Cell size predicts morphological complexity in the brains of frogs and salamanders. Proceedings of the National Academy of Sciences, U.S.A. 91:4796–4800.

Roth, V. L. 1991. Homology and hierarchies: Problems solved and unsolved. Journal of Evolutionary Biology 4:167–194.

Saether, O. 1983. The canalized evolutionary potential: In consistencies in phylogenetic reasoning. Systematic Zoology 32:343–359.

Sanderson, M. J. 1991. In search of homoplastic tendencies: Statistical inference of topological patterns in homoplasy. Evolution 45:351–358.

Sanderson, M. J. and M. J. Donoghue. 1996. The relationship between homoplasy and confidence in a phylogenetic tree. Pp. 67–90 *in* M. J. Sanderson and L. Hufford, eds. Homoplasy: Academic Press, San Diego.

Schwenk, K. 1995. A utilitarian approach to evolutionary constraint. Zoology 98:251–262.

Sessions, S. K. and A. Larson. 1987. Developmental correlates of genome size in plethodontid salamanders and their implications for genome evolution. Evolution 41:1239–1251.

Shubin, N. and P. Alberch. 1986. A morphogenetic approach to the origin and basic organization of the tetrapod limb. Evolutionary Biology 20:319–387.

Shubin, N., D. B. Wake, and A. Crawford. 1995. Morphological variation in the limbs of *Taricha granulosa* (Caudata:Salamandridae): Evolutionary and phylogenetic implications. Evolution 49:874–884.

Vavilou, N. 1922. The law of homologous series in variation. J. Genetics 12:47–89.

Wagner, G. P. 1989. The origin of morphological characters and the biological basis of homology. Evolution 43:1157–1171.

Wagner, G. P. 1995. The biological role of homologues: A building block hypothesis. Neue Jahrbucher Geologie Paläontologie Abhandlungen 195:279–288.

Wake, D. B. 1966. Comparative osteology and evolution of the lungless salamanders, family Plethodontidae. Memoirs of the Southern California Academy of Sciences 4:vii, 1–111.

Wake, D. B. 1982. Functional and developmental constraints and opportunities in the evolution of feeding systems in vertebrates. Pp. 51–66 *in* D. Mossakowski and G. Roth, eds. Environmental adaptation and evolution. Gustav Fischer, Stuttgart.

Wake, D. B. 1991. Homoplasy: The result of natural selection, or evidence of design limitations. American Naturalist 138:543–567.

Wake, D. B. 1994. Comparative terminology. Review of Hall, B. K. (ed.), Homology. Science 265:268–269.

Wake, D. B. and G. Roth. 1989. The linkage between ontogeny and phylogeny in the evolution of complex systems. Pp. 361–377 *in* D. B. Wake and G. Roth, eds. Complex organismal functions: Integration and evolution in the vertebrates. John Wiley & Sons, Chichester.

IMPLICATIONS OF HOMOPLASY

EXPLANATIONS OF HOMOPLASY AT DIFFERENT LEVELS OF BIOLOGICAL ORGANIZATION

DANIEL R. BROOKS

University of Toronto
Ontario, Canada

Comparative evolutionary biologists have a love/hate relationship with homoplasy. On the one hand, contemporary comparative studies require robust estimates of phylogenetic relationships and, because phylogenetic relationships are denoted by homology, homoplasy confounds efforts to recover phylogeny. On the other hand, many comparative studies aimed at uncovering evolutionary mechanisms utilize particular patterns of homoplasy as important, or even primary evidence, and this requires explicit enumeration of as many homoplasious changes as possible. *Phylogenetic systematics* (Hennig 1950, 1966) represented a breakthrough in a long-standing problem with the relationship between homology and homoplasy that permits workers to highlight homoplasies explicitly within a phylogenetic framework.

Traditional systematic methods defined homologous traits by phylogeny, and then used the traits thus identified to reconstruct phylogeny, making phylogenetic reconstruction irreducibly circular. Hennig believed that homologous characters outnumber *covarying* homoplasious characters within any given group if an energetic search for characters is made. He suggested that researchers assume that all characters which conform to nonphylogenetic criteria for homology (e.g., Remane 1956) are evolutionarily homolo-

HOMOPLASY: *The Recurrence of Similarity in Evolution,* M. J. Sanderson and L. Hufford, eds.

gous. This means that homoplasious traits will initially be labeled incorrectly as homologs. When a phylogeny is reconstructed by grouping taxa according to their shared presumed homologies, such misidentifications will be revealed, which then can be recognized using the phylogenetic hypothesis. Phylogenetic systematics is not circular because homologies, which indicate phylogenetic relationships, are determined without a priori reference to a phylogeny, while homoplasies, which are inconsistent with phylogeny, are determined a posteriori as such by reference to the phylogeny. Thus, while phylogenetic systematic studies represent a search for sister group relationships indicated by homology, they also provide a strong test of homoplasy because homoplasies are highlighted explicitly against a background of homology. In this way, the method satisfies both needs of comparative evolutionary biologists.

The principles of phylogenetic systematic reasoning serve to document homology and homoplasy at all levels of biological organization that potentially show phylogenetic effects. Homology is indicated by *patterns* of congruence resulting from application of phylogenetic systematic methods. There are three patterns of occurrence for homologous traits on the phylogenetic tree for a given clade, *plesiomorphy, synapomorphy,* and *autapomorphy* (Fig. 1). These differ only in the extent of common history that is implied, so the *explanation* for homology, common history, is the same for all of them. Likewise, analysis of interspecific biological organization that show phylogenetic effects produces same three *patterns* of congruence, differing only in their extent of common history of interclade associations. The *explanation* for congruence, common history, therefore, is the same for all levels of biological organization, whether it is a common history of character evolution or of clade association and speciation.

The *patterns* of homoplasious characters are only slightly more complicated than those for homologies (once a particular phylogenetic hypothesis has been established—see chapters on methodologies). Homoplasies may

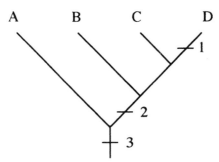

FIGURE 1   Hypothetical phylogenetic tree showing the three patterns of homology: autapomorphy (1), synapomorphy (2), and symplesiomorphy (3).

occur as separate appearances of an apomorphic trait or the reappearance(s) of plesiomorphic traits (reversals). In addition, homoplasies can occur in paraphyletic taxa (parallelism) or in polyphyletic taxa (convergence) (Fig. 2). By contrast with homology, however, the *explanations* for homoplasy invoke different processes within and between different levels of biological organization. Some types of data, such as nucleotide sequences, are so limited in their options for evolutionary change that high levels of homoplasy are an inevitable result of simple mutation processes. By contrast, when we speak of homoplasious traits for individual taxa, we might postulate multiple origins of traits mediated by the effects of a variety of selection processes or common developmental pathways or a combination of more than one process. In analyses of geographical or ecological associations among different clades, patterns of homoplasy indicate episodes of colonization, called dispersal in biogeography, and resource- or host-switching in coevolutionary studies, or of extinction. Hence, the analytical method produces patterns of phylogenetic congruence for those elements sharing a common history and phylogenetic incongruence for those having independent histories; calling the elements in the analysis characters or taxa notwithstanding. The same methodological principles should apply to any biological systems potentially showing phylogenetic effects. The question, then, is one of process (explanation) and not simply of pattern (analysis).

As suggested above, homoplasy can be viewed as error or noise in a data set that makes delineating phylogenetic relationships difficult. It can also be viewed as a source of opportunity for a variety of evolutionary studies. In the following discussion I discuss representative, but by no means exhaustive, types of explanations for homoplasy occurring at different levels of

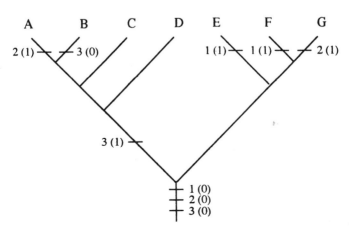

FIGURE 2 Hypothetical phylogenetic tree showing the three patterns of homoplasy: parallelism (1), convergence (2), and reversal (3).

biological organization. This is intended to show how broadly the fundamental principles of phylogenetic reasoning can be applied to exciting studies
of evolutionary mechanisms.

"BUILT-IN" HOMOPLASY

A common metaphorical perception among biologists is that the "closer to
the genome" one samples characters, the closer to the true phylogeny will be
the results of phylogenetic analysis on those data. Many believe that data
"closer to the genome" should be closer to the actual underpinnings of evolution and should not be affected by homoplasy produced by selection processes. Thus, for example, we might expect nucleotide sequences to perform
best at recovering phylogeny, not as well at recovering morphology, and even
worse at recovering behavior. Phylogenetic systematics provides a means of
testing such perceptions. Results to date have been surprising, highlighting
several sources of "built-in" homoplasy arising from the basic nature of the
type of data or type of process affecting the system.

Felsenstein (1982) was among the first to recognize that the inherently
limited range of options for change in nucleotides should produce extraor-
dinarily large amounts of homoplasy as a result of simple mutation processes. He showed that in certain cases, particularly in rapid multiple lineage
splitting occurring over a relatively short time but long ago, nucleotoide sequence evolution could produce covarying homoplasies outnumbering homologies, resulting in multiple confusing equally parsimonious trees or a
single most parsimonious tree postulating wildly improbable phylogenetic
relationships. This effect has even become known popularly as the "Felsenstein Zone." Phylogenetic analyses of nucleotide data routinely encounter
high levels of homoplasy, making determination of phylogenetic relationships difficult. As a consequence, molecular systematists have been forced to
consider ways of screening particular types of data for use in particular types
of phylogenetic studies, and are often forced to consider methods other than
phylogenetic systematics to try to produce a robust hypothesis of evolutionary history (e.g., Hillis and Moritz 1990 and references therein). This in
turn has led to a highly productive investigation of models of genetic evolution, most notably those embodied in maximum likelihood estimates of
phylogeny.

Other types of "genetic" data may also carry a high propensity for homoplasy. Some recent studies have suggested that viral transmission of genetic elements across species may occur; this would be an additional source
of homoplasy at the molecular level. Likewise, any clade experiencing gene
duplication and subsequent loss of one or the other of the duplicates will
exhibit homoplasy in single copy occurrences of the genes that have been

duplicated and the differentially lost. Although molecular and "genetic" data have been found to show higher levels of homoplasy than some originally expected, in some cases morphology and behavior have been found to be less prone to homoplasious changes than expected. These are cases in which the expectation of high levels of homoplasy are derived not from empirical study but from theoretical postulates.

There is a perception that becoming parasitic carries with it homoplasious evolutionary consequences (Price 1980; Brooks and McLennan 1993a and references therein). One of these is extreme secondary evolutionary simplification. For example, Rogers (1962) stated that "the loss of sense organs is a common feature of parasitism." Rohde (1989), however, discovered in a single species of endoparasitic flatworm eight types of sense receptors not reported for their free-living relatives, opening the door for a reconsideration of our beliefs about parasite evolution. Brooks and McLennan (1993a,b) extended Rohde's line of investigation, examining a data base comprising 1882 morphological character changes for various groups of parasitic platyhelminths. They found that only 10.6% of the character changes involved secondary loss. They also found that about half of the character loss was due to homoplasious losses in a small number of characters. The second commonly held view is that the evolutionary success of parasites is due, in part, to their ability to adapt rapidly and in a similar manner to similar challenges posed by the "environment," often equated with "host." If parasites are adaptively plastic in this sense, then homoplasious changes should be numerous. Brooks and McLennan (1993a,b) examined levels of homoplasy for the parasitic platyhelminth data base. The homoplasy slope ratio (Meier et al. 1991) value for the total data base was 0.002, corresponding roughly to 0.2% of the homoplasy expected for a randomly generated data base of the same number of taxa and characters. In this case, examination of levels of homoplasy leads us to question seriously, if not abandon entirely, some strongly held traditional beliefs about the evolution of parasites.

Perhaps even more startling has been recent analysis of behavioral traits, beginning with McLennan et al. (1988), who produced a fully resolved phylogenetic tree of gasterosteid fishes using a suite of 26 characters drawn from courtship behavior. McLennan (1993) updated that data base to include a total of 50 behavioral characters, producing a phylogenetic tree congruent with morphological, karyological, and allozyme data and exhibiting only three instances of homoplasy. DeQueiroz and Wimberger (1993) analyzed 22 data sets for which there were both morphological and behavioral data and concluded that levels of homoplasy were about the same in each case.

Finally, speciation by hybridization may produce high levels of homoplasy; in fact, the widespread occurrence of speciation by hybridization among flowering plants is of special concern to botanical systematists. Funk (1985) produced the first general method for trying to identify species of hybrid origin from a phylogenetic tree. She reasoned that particular patterns of homoplasy on phylogenetic trees might serve to identify species carrying genetic information of two different parental species. In the simplest case, consider a species resulting from hybridization of two sister species, each possessing an autapomorphy. If the species of hybrid origin expresses the apomorphic traits of both parental species, the result would be the particular pattern of homoplasy shown in Fig. 3. Funk (1985) showed the efficacy of her method in a series of examples involving progressively more distantly related parental species. For distantly related parental species, she suggested that a phylogenetic systematic analysis would tend to produce alternate equally parsimonious trees differing primarily in the placement of the species of hybrid origin. Thus, removal of species of hybrid origin from a phylogenetic systematic analysis should reduce the number of equally parsimonious trees. McDade (1990, 1992) pointed out that Funk's method works in direct proportion to number of apomorphies of each parental species expressed by the species of hybrid origin. This requires that each apomorphy be genetically dominant to each plesiomorphy and that the apomorphies of each parental species be genetically and developmentally independent of each other. McDade showed that hybrids produced under laboratory settings often express only some of the apomorphies of each parent; in some cases, only the apomorphies of one parent are expressed.

HOMOPLASY WITHIN CLADES: CONSTRAINTS AND SELECTION

Sæther (e.g., 1977, 1979a,b,c, 1983) listed a number of general processes by which homoplasious character changes might occur within particular clades. These included homoplasies stemming from (1) fixation, mediated by selec-

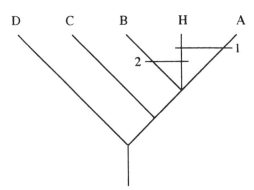

FIGURE 3   Phylogenetic tree depicting hybrid species H expressing the autapomorphies (1 and 2) of each of its parents, which in this case are sister-species.

tion processes, of apomorphic traits in paraphyletic taxa derived from ancestors that were polymorphic for the plesiomorphic and apomorphic condition; (2) developmental constraints, or developmental processes common to paraphyletic or polyphyletic species, producing convergences, secondary reductions, and evolutionary reversals; and (3) what he termed *underlying synapomorphies,* or canalized evolutionary potential producing the same apomorphic condition repeatedly.

HOMOPLASY POINTING TO SELECTION

Brooks and McLennan (1991 and references therein) echoed the sentiments of many authors by pointing out that selection processes can potentially be evolutionarily significant for homologous or homoplasious traits. Homoplasious traits, however, are the best data for studies of evolutionary mechanisms because they provide degrees of freedom on explanations, whereas singularly occurring homologies do not. By this logic, the more times the same or similar apomorphic traits appear under the same or similar selection regimes, the more robust are our explanations of the mechanisms producing the changes.

Schaeffer and Rosen (1961) stated: "In the main stream of actinopterygian evolution from palaeoniscoid to acanthopterygian, there has been a progressive improvement in a fundamentally predaceous feeding mechanism." The oldest members of the Actinopterygii are predators, so predation appears to be plesiomorphic and herbivory apomorphic within the group. Winterbottom and McLennan (1993; see also McLennan 1994) optimized diet preferences onto the phylogenetic tree for the Acanthuroidei, a group including the surgeon and unicorn fishes. Their analysis corroborated Schaeffer and Rosen's hypothesis: predation on benthic invertebrates is plesiomorphic for the group, while herbivory on macrophytic algae is apomorphic, originating in the ancestor of the Siganidae + Luvaridae + Zanclidae + Acanthuridae clade (Fig. 4). The origin of herbivory in this ancestor is associated with morphological changes reducing the protrusibility of the upper jaw and limiting its movement primarily to rotation (Tyler et al. 1989). A protrusible premaxilla is hypothesized to increase a fish's ability to seize moving prey items (Schaeffer and Rosen 1961), so this change to a more rigid jaw may be causally associated with the appearance of herbivory. Once herbivory appeared it was retained through the majority of speciation events in the clade, although there have been at least six subsequent evolutionary changes in diet: from macrophytic algae to filamentous algae in the ancestor of the acanthurins, from herbivory to predation on jellyfish in *Luvarus,* from herbivory to omnivory by adding sponges to the diet in *Zanclus,* and from

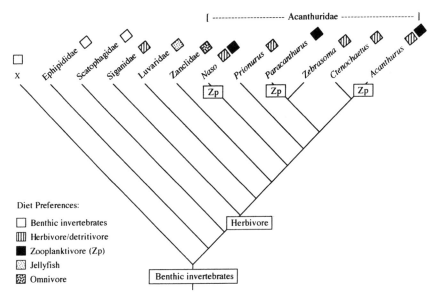

FIGURE 4   Phylogenetic diversity of diet preferences in the Acanthuroidei (Surgeonfishes). *Acanthurus* and *Naso* contain herbivorous and zooplanktivorous species. *Drepane* is the out-group (X) (from McLennan 1994).

herbivory to zooplankton picking in *Paracanthurus* and in some species of both *Naso* and *Acanthurus*.

Having identified the homoplasious origins of zooplankton picking, Winterbottom and McLennan searched for correlated changes in morphology and/or the environment that would provide an adaptive explanation for those switches. Twelve species of *Naso* and four species of *Acanthurus* are zooplankton pickers. Given the genealogical relationships and the foraging mode of the outgroup, the most parsimonious explanation for the origin of zooplanktivory in the Acanthuridae is that it has evolved independently from an herbivorous ancestor in *Naso, Acanthurus,* and *Paracanthurus* (Fig. 5). If this explanation is correct, the basal members of the polytypic acanthurid genera should be herbivores, something that will be corroborated or refuted by species-level phylogenetic analyses in the future.

The evolution of zooplanktivory in *Naso, Acanthurus,* and *Paracanthurus* involved a shift from medium-sized food items to extremely small prey. In two of the three genera, it also involved the movement from foraging close to the cover provided by coral and rocky outcroppings to foraging on the perimeter of the reefs in more open waters close to the drop-off. Two environmental problems are posed by these ecological shifts: (1) find and consume small prey items, and (2) avoid being eaten yourself while foraging in the more open mid-waters. Are there any morphological character changes

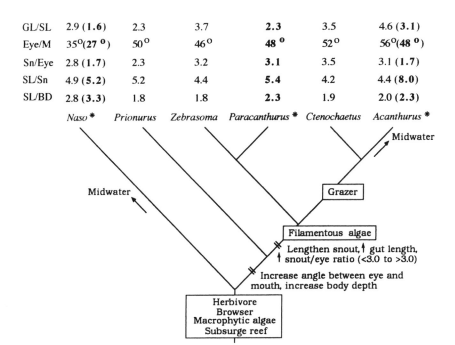

	Naso *	Prionurus	Zebrasoma	Paracanthurus *	Ctenochaetus	Acanthurus *
GL/SL	2.9 (**1.6**)	2.3	3.7	**2.3**	3.5	4.6 (**3.1**)
Eye/M	35°(**27** °)	50°	46°	**48 °**	52°	56°(**48** °)
Sn/Eye	2.8 (**1.7**)	2.3	3.2	**3.1**	3.5	3.1 (**1.7**)
SL/Sn	4.9 (**5.2**)	5.2	4.4	**5.4**	4.2	4.4 (**8.0**)
SL/BD	2.8 (**3.3**)	1.8	1.8	**2.3**	1.9	2.0 (**2.3**)

FIGURE 5 Phylogenetic diversification in ecology and morphology in the Acanthuridae. Ecological traits are optimized on the phylogenetic tree. Morphological traits are listed above each terminal taxon; morphometric trends are indicated by double bars on the phylogenetic tree. Boldface = data for zooplanktivorous species (*); BD = body depth; Eye/M = angle between eye and mouth; GL = gut length; Sn = snout length; SL = standard length; ↑ = increased value (from McLennan 1994).

correlated with the new ecology that might enhance performance in these respects? It is possible to open interesting avenues for future research by examining morphological characters that have been postulated to influence a fish's ability to detect and engulf small prey items (snout length, eye size relative to head size, and the angle between the snout and eye), to digest food (relative intestine length), and to avoid predation (the degree of streamlining).

Winterbottom and McLennan discovered that the change toward a deep-bodied fish with a mouth well below eye level began in the ancestor of the *Prionurus* + Acanthurinae (*Zebrasoma* + *Paracanthurus* + *Ctenochaetus* + *Acanthurus*) clade, and the change toward a longish snout, a smaller eye relative to snout length, and a long intestine began in the ancestor of the acanthurins. Change in intestine length is associated with a dietary shift from leafy/fleshy algae to filamentous algae. The zooplanktivores in *Paracanthurus* and *Acanthurus*, therefore, evolved from a different morphological background than the zooplanktivores in *Naso*. Despite that, the similarities in the direction of morphological change exhibited by zooplanktivores in all three genera are quite striking. Movement of new species off the reef in pursuit of small zooplankters have been associated with changes hypothesized to increase the efficiency of foraging and predator avoidance in coral reef zooplanktivores (Jones 1968; Hobson 1991): body elongation, decreasing gut and snout length, increasing eye size, and decreasing angle between the eye and mouth. The phylogenetic analysis demonstrated that the ability of different organisms to respond to similar selection pressures is constrained by their underlying history. In this case, zooplanktivorous species of *Naso* and *Acanthurus* demonstrate similarities in the direction of evolutionary change for certain characters, but not in the magnitude of those changes.

Some butterflies lay their eggs in clusters rather than singly, as is the behavior for most species. Once hatched, larvae from egg clusters tend to remain aggregated, which should leave them highly susceptible to predation, and thus at a strong selective disadvantage. Three observations appear to explain the apparent paradox: (1) larvae of many butterfly species are aposematically colored; (2) there is a correlation between larval clustering (gregariousness) and aposematism; and (3) there is a correlation between larval warning coloration and distastefulness. Fisher (1930) proposed that the development of prey unpalatability required the involvement of kin selection, while Turner (1971) and Harvey et al. (1982) extended the kin selection hypothesis to include the development of aposematic color patterns. Some bad-tasting mutants with bright coloration would be eaten, predators would associate bad taste with the mutant's color, and those predators would avoid brightly colored caterpillars. Under this hypothesis, gregariousness should either evolve first, providing a context in which kin selection could work, followed by the evolution of distastefulness/aposematic coloration, or appear at the same time as warning coloration/distastefulness. The first evolutionary sequence provides the strongest corroboration for the hypothesis that kin selection produced the link between gregariousness and the appearance of "taste bad/look bad" traits. Sillen-Tullberg (1988) examined this hypothesis for a number of butterfly lineages, compiling data from the literature concerning larval aggregation habits (solitary versus gregarious) and color (cryptic versus aposematic). She asked if gregariousness evolved before, with, or after warning coloration?

In the Papilionini, the plesiomorphic state is "larvae solitary and cryptic" (Fig. 6). Within the clade gregariousness evolved in three separate lineages and aposematic coloration in two lineages. All four possible combinations of the grouping and color traits exist in these butterflies, but in no case did gregariousness evolve before aposematism. This example is representative of Sillen-Tullberg's overall results: of the 23 independent evolutions of gregariousness, 3 evolved in conjunction with the appearance of aposematism, 15 evolved after aposematic coloration evolved, and in 5 cases aposematism

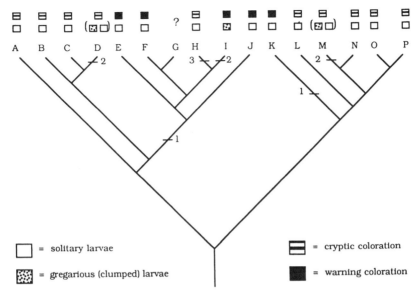

FIGURE 6 Temporal sequence of the evolution of larval clustering and warning coloration mapped onto a phylogenetic tree for the butterfly tribe Papilionini. A = *Pterourus* (5 species groups); B = *Heraclides thoas*; C = *H. torquatus*; D = *H. anchisiades*; E = *Chilasa elwesi*; F = *C. clytia*; G = *C. veiovis*; H = *C. agestor*; I = *C. laglaizei*; J = *Eleppone anactus*; K = *Papilio machaon*; L = *Princeps xuthus*; M = *P. demolion*; N = *Princeps* (9 species groups); O = *Princeps* (6 species groups); P = *Princeps* (10 species groups). Distribution of larval aggregation and color characters mapped above the taxa and optimized onto the tree. 1 = Warning coloration present; 2 = gregariousness present; 3 = reversal to crypsis (from Brooks and McLennan 1991).

never evolved. Maddison (1990) showed that the overall distribution of gregariousness and the distribution of aposematic coloration on the trees used by Sillen-Tullberg did not depart significantly from a model of random character evolution. He thus suggested that there might be no causal relationship between aposematism and gregariousness. Sillen-Tullberg (1993) responded correctly that the critical issue was not the overall distribution of characters on the trees, but the observation that in no instance did gregariousness evolve before warning coloration. Maddison's study suggested that aposematism and gregariousness evolve independently; Sillen-Tullberg's study indicated that when gregariousness evolves in an aposematic species, the combination is selectively advantageous. Both Maddison's and Sillen-Tullberg's observations are supported by the phylogenetic patterns, but for the hypothesis that Sillen-Tullberg wishes to test, it is the unambiguous sequence of evolutionary change in each lineage and not the overall distribution of the traits on the tree as a whole that is critical.

Sanderson (1991) found little evidence that homoplasies are clustered nonrandomly on phylogenetic trees based on real data sets. He suggested that if homoplasy is distributed nonrandomly across large clades, we need to examine either more inclusive clades than in previous studies or the distribution of traits which we believe are functionally related evolutionarily. Sillen-Tullberg's study shows that analysis of character evolution on a branch by branch basis may provide better insights than analysis of overall patterns.

HOMOPLASY POINTING TO DEVELOPMENTAL CONSTRAINTS

The developmental phenomena covered by the term *heterochrony*, or changes in the timing of development that produce changes in morphology, have been the focus of considerable recent interest by some evolutionary biologists (Gould 1977; Alberch et al. 1979; Bonner 1982; Fink 1982; Raff and Raff 1987; McKinney 1988). Alberch et al. (1979) developed a model based upon the idea that the development of any part of an organism can be represented by a positive trajectory with a starting point, α (growth starts), an endpoint, β (growth stops), and a rate of growth, k, for changes in shape, γ, or size, S. Development could thus be tracked graphically by plotting α and β on the x-axis (time), γ on the y-axis (increasing complexity), and by varying the onset of growth, rate of growth, and ending of growth (Fig. 7). They recognized two general categories of heterochrony, *paedomorphosis* and *peramorphosis*.

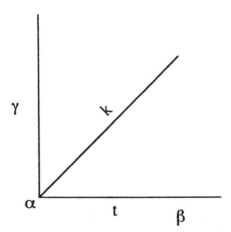

FIGURE 7   Schematic representation of the elements of the Heterochrony Model of Alberch et al. (1979). α = Starting point; β = ending point; γ = degree of morphological complexity of the trait; k = rate of change; t = time.

Paedomorphosis results in the production of adult descendant morphologies that are less complex than those of their immediate ancestor. This does not necessarily imply that the descendant will have a morphology comparable to that of a juvenile or even a larval ancestor; it may resemble a less developed ancestral adult (McKinney 1988). Paedomorphosis can be accomplished in three ways: growth onset can be delayed (*postdisplacement*), growth can be terminated earlier (*progenesis*), or development can proceed at a slower rate (*neoteny*). Paedomorphic phenomena can in some cases produce morphological changes in apomorphic traits that hark back to their plesiomorphic roots, manifested on phylogenetic trees as putative evolutionary reversals. Peramorphosis, on the other hand, results in the production of adult descendant morphologies that are more complex than those of their immediate ancestor. Since this produces a morphological trait in an organism that passes beyond the condition found in its ancestor, the result will be a recapitulation of the ancestral ontogeny during development. Not surprisingly, there are three ways in which peramorphosis can happen: growth onset can begin earlier (*predisplacement*), growth can continue for a longer period of time (*hypermorphosis*), and development can proceed at a faster rate (*acceleration*: see also McNamara 1986). Recurring peramorphic phenomena can produce homoplasious apomorphic changes. McNamara (1986) suggested that any modification of a developmental sequence, whether by the addition or deletion of stages, may therefore be interpreted as an outcome of heterochrony (but see Alberch 1985). Based upon this, McNamara extended paedomorphosis to include a descendant passing through fewer ontogenetic stages than the ancestor and peramorphosis to include a descendant passing through more ontogenetic stages than the ancestor.

Heterochronic processes might be an important explanation for some of the homoplasy we see in phylogenies. For example, let us consider some characters of larval digeneans (a group of platyhelminth parasites). The plesiomorphic digenean ontogeny includes three larval stages; the miracidium, sporocyst, and cercaria. Early in the phylogeny of digeneans a fourth stage, the redia, was intercalated between the sporocyst and cercaria. Not all descendants of the ancestral digenean in which the redia arose are characterized by the presence of a redial stage. Many exhibit a stage termed a daughter sporocyst that occurs, like the redia, between the sporocyst (now called a mother sporocyst) and the cercarial stages. The presence of daughter sporocysts coincides with the absence of the redia and appears convergently among digeneans (Brooks and McLennan 1993b).

Rediae and daughter sporocysts are both derived from mother sporocyst germinal tissue, both have a birth pore and both give rise to cercariae having pharynges and guts. The difference between a redia and a daughter sporocyst is morphological; a redia has a pharynx and saccate gut while a daughter sporocyst has neither. This suggests the presence of paedomorphic phenomena leading to the later-than-expected expression of the pharynx and gut

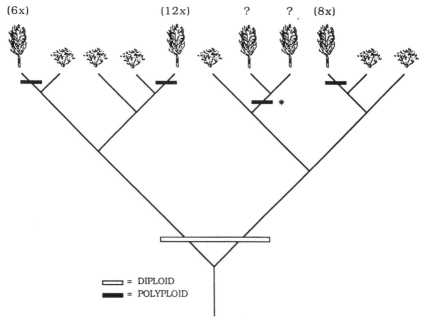

FIGURE 8   Simplified phylogenetic tree for *Montanoa* (daisy trees), showing distribution of shrub and tree habits. Ploidy levels of tree forms listed above the taxa. * = hypothesized polyploidy (from Brooks and McLennan 1991).

during ontogeny in those species having "daughter sporocysts." In the absence of experimental studies, we cannot tell which of the three categories of paedomorphosis may have been involved, but we can say that daughter sporocysts appear to be paedomorphic rediae. Having established a mechanistic basis for the appearance of "daughter sporocysts" more than one time during digenean phylogeny, workers may now proceed to an investigation of whether or not there is a selective component to their subsequent evolutionary persistence.

HOMOPLASY POINTING TO UNDERLYING SYNAPOMORPHIES

Sæther (1977), following Brundin (1972), suggested that parallelisms could be produced by "noninherited" factors (i.e., various forms of selection) and by "inherited" factors collectively called *canalized evolutionary potential*. Sæther linked this discussion with the formulations by Tuomikoski (1967) who defined the concept of *underlying synapomorphy* as an agreement in capacity to develop parallel similarity (see also Crampton 1929). Simply put, this means that among groups within a larger clade, and within that clade only, certain homoplasious changes occur repeatedly. The appearance of any of these traits conforms to a phylogenetic pattern of homoplasy, but the evolutionary capacity to produce it actually evolved only one time, in the common ancestor of the larger clade containing all the species exhibiting the trait. Underlying synapomorphies are intriguing because there are systems providing *patterns* of character changes supportive of the concept, but the nature of, and *process(es)* producing, canalized evolutionary potential remain areas of interesting theoretical discourse (e.g., Salthe 1993).

The Compositae genus *Montanoa* comprises approximately 30 taxa living throughout Central America, extending as far north as central Mexico and as far south as central Colombia. Twenty-one of these species are shrubs, 5 are "daisy trees" reaching 20 m in height, and 4 are vines. Funk (1982) performed a phylogenetic analysis of the genus, discovering that the shrub-like habit is plesiomorphic (ancestral) in *Montanoa*. Seven of 47 branches on her phylogenetic tree are associated with a change in habit (to tree forms 4 times and to vines 3 times). All five tree species have a number of similar morphological and physiological characters permitting them to survive in cloud forests and none has ever been found outside a cloud forest. They are members of four different clades, their sister-species being shrubs living at adjacent lower elevations in each case. They represent cases of convergent adaptation (Brooks and McLennan 1991) because they exhibit a convergent trait correlated with the same environmental variable. The explanation could be that natural selection favored the evolution of the same kind of strategy for surviving in cloud forests, but we do not know if the correlation between recurring environment and homoplasious character evolution is causal, or is an emergent result of the production of the character by some other mechanism. Examination of the distribution of the *Montanoa* habit and ploidy level on the phylogenetic tree reveals a recurring pattern of association between the appearance of tree forms and increased polyploidy (Fig. 8). Diploidy is the ancestral condition in *Montanoa*, but all the tree species, and only the tree species, are high-level polyploids (Funk and Raven 1980; Funk 1982; Funk and Brooks 1990). It is common for diploids to produce polyploid seeds in the Compositae under a variety of environmental conditions and for polyploids to be viable, so the homoplasious appearance of polyploids is not a surprising phylogenetic pattern. It is also common for polyploids to be larger than diploids, so the tree habit may be an emergent product of polyploidization. The apparent adaptive nature of the ecological relationship between tree forms of *Montanoa* and cloud forests could be one evolutionary consequence of this developmentally mediated production of homoplasies. It is thus possible that the repeated convergent evolution of tree-like species is a result of an underlying developmental propensity to

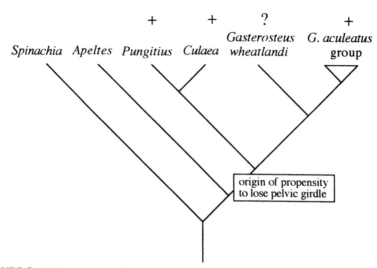

FIGURE 9   Phylogenetic tree for major taxa of Gasterosteidae (Stickleback fishes), showing distribution of populations exhibiting secondary pelvic girdle loss.

produce polyploid seeds in Asteraceae, expressed in a clade (*Montanoa*) in which shrubby forms can give rise to tree forms, the kind of phenomenon the underlying synapomorphy concept was formulated to explain.

The pelvic girdle of stickleback fishes provides an evolutionary system for integrating concepts of underlying synapomorphies, developmental constraints, and environmental selection. Within the clade comprising the genera *Culaea, Pungitius,* and *Gasterosteus,* populations of at least four species [*Culaea inconstans, Pungitius pungitius, P. occidentalis,* and *Gasterosteus aculeatus (sensu lato)*] contain members lacking pelvic girdles (Fig. 9). The appearances of this trait in each population of each species are thought to represent indpendent (homoplasious) occurrences. No other members of the Gasterosteidae lack pelvic girdles. The combination of the restriction of this trait to one clade, coupled with its repeated appearance within the clade, corresponds to the phylogenetic pattern expected for an underlying synapomorphy. Studies of *P. occidentalis* (listed as *P. pungitius*) have indicated that the trait is heritable, at least in part (Blouw and Boyd 1992). Developmental studies indicate two different pathways by which the loss of the pelvic girdle is manifested, at least one of which involves heterochronic processes (Bell and Foster 1994).

Given the apparent evolutionary propensity for loss of the pelvic girdle to occur, one is led to ask why the plesiomorphic presence of a pelvic girdle persists in all species within the clade exhibiting the underlying synapomorphy. Answering this question leads to studies of environmental selection. Bell et al. (1993) summarized studies of various populations in the *G. aculeatus* species complex. Those studies have shown statistically significant correlations between loss of pelvic girdle and living in lakes with (1) few or no predatory fish and (2) low calcium concentrations. In their overview, Bell et al. were able to show that lakes with predatory fish and low calcium concentrations did not contain significant numbers of sticklebacks lacking pelvic girdles. This suggests that the primary factor influencing the success of the apomorphic "girdle-less" condition is the presence or absence of predators. Strong support for this idea comes from a report by Nelson and Atton (1971) of a lake lacking predatory fish in which populations of both *P. occidentalis* (reported as *P. pungitius*) and *C. inconstans* showed significant numbers of individuals lacking pelvic girdles. Reist (1980) further suggested that the absence of a pelvic girdle in *Culaea* made them more susceptible to predation by dityscid beetles.

It is important to remember that selection explanations and underlying synapomorphy explanations need not be mutually exclusive. If the recurring production of tree forms of *Montanoa*, or of girdleless sticklebacks, results from biased developmental evolution, it is still possible that the ultimate survival and persistence of these forms is mediated by local selection processes.

HOMOPLASY AMONG CLADES

Comparative evolutionary studies of interspecific associations attempt to uncover the historical patterns of geographical and ecological associations among clades and to provide explanations for them (Brooks and McLennan 1991, 1993a, and references therein). The range of historical patterns between and among geographically and ecologically associated species encompasses the complete range of phylogenetic patterns.

Species may be geographically or ecologically associated today because their ancestors were associated with each other in the past. In those cases, the contemporaneous relationship is a persistent ancestral component of the biotic structure within which the species reside. Each of the species has "inherited" being part of the association. In those cases of "association by descent" (Brooks and McLennan 1991), the species act like "homologies" of the geographic areas or the ecological associations of which they are a part. Alternatively, two or more species may be associated because at least one of the species originated in a different geographical area or ecological context and subsequently became involved in the present-day association by dispersal, host switching, or some other form of "association by colonization" (Brooks and McLennan 1991). Finally, there may be geographical areas or ecological associations from which some species historically present have now become extinct. In the latter two categories of associations, species would act as homoplasies. We expect contemporary biotas to contain species representing a variety of different historical influences, so the challenge has been to provide a means for documenting the patterns as explicitly as possible and then to provide an explanation for each species in each of its geographical and ecological contexts.

HOW DO WE FIND
THE PATTERNS?

The most basic case is one in which we have a phylogenetic tree for a group of species, the "clade of interest," coupled with information about the geographical distributions of the species in the clade, or about species with which the members of the clade of interest are associated ecologically. The areas in which each species resides, or the species with which each member of the study clade is associated ecologically, can be treated as characters and optimized onto the phylogenetic tree for the clade of interest. The result is a historical "picture" of the geographical or ecological context during the diversification of the clade of interest. Such patterns have little explanatory value, however, unless we make the assumption that all of the geographical or ecological associations represent instances of one particular kind of phenomenon. And if we do make such assumptions, our explanations become highly circular.

If we have information about the historical relationships among the geographical areas, or the phylogenetic relationships among the associated spe-

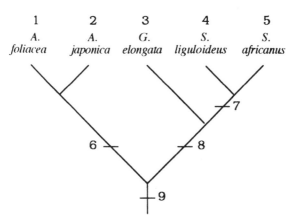

FIGURE 10   Phylogenetic tree for five species of amphilinid flatworms with terminal and internal branches numbered for phylogenetic analysis. A = *Amphilina*; G = *Gigantolina*; S = *Schizochoerus* (from Brooks and McLennan 1991).

cies, in addition to a phylogenetic tree for the clade of interest, the power of phylogenetic studies increases dramatically. In such cases, we can compare the pattern of relationships among the geographical areas or among the associated species with the clade of interest. This can be done by considering the phylogenetic tree for the clade of interest as completely polarized and ordered multistate transformation series and optimizing that transformation series onto the the cladogram representing the pattern of relationships among the geographical areas or among the ecologically associated species. Points of congruence conform to association by descent, points of incongruence conform to association by colonization or, in the case of reversals, extinctions.

Alternatively, we may use phylogenetic systematic methods to construct a cladogram of areas or of ecologically associated species derived solely from the relationships indicated by the members of the clade of interest treated as a multistate transformation series. In such a case, we explain points of congruence between the two cladograms as instances of association by descent and points of incongruence as instances of association by colonization or extinction. As an example, consider the Amphilinideans, the sister-group of the species-rich true tapeworms, comprising eight known species living in the body cavities of freshwater and estuarine fishes and one species of freshwater turtle.

The first requirement is a robust hypothesis of the phylogenetic relationships for the group. To illustrate this approach, I will begin with five of the eight species (Fig. 10). Treating the phylogenetic relationships of the five amphilinid species as if they were a completely polarized multistate transformation series, construct an area cladogram based on the phylogenetic relationships of the species. This produces a picture of the areas' historical involvement in the parasites' speciation (Fig. 11). The area cladogram in this case is identical to the historical relationships among the areas as indicated

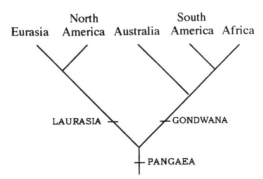

FIGURE 11 Area cladogram for five areas based on the phylogenetic relationships of five species of amphilinid flatworms that inhabit those areas (from Brooks and McLennan 1991).

by geological evidence, so we can hypothesize that the occurrence of the study species in the study areas is a result of a long history of association between amphilinids and the areas in which they now occur.

The complete phylogenetic tree for the Amphilinidea includes *Gigantolina magna* (taxon 6) and *Schizochoerus paragonopora* (taxon 7) in Indo-Malaysia, and *Schizochoerus janickii* (taxon 8) in South America (Fig. 12). Taxon 10 appears twice on the area cladogram constructed using all eight species (Fig. 13). This indicates that the common ancestor (species 10) of *Gigantolina elongata* (species 3) and *G. magna* (species 6) occurred in both Australia and Indo-Malaysia. Its occurrence in Australia coincides with the geological history of the areas, so we explain this by saying that *G. elongata* evolved in the same place as its ancestor. On the other hand, the occurrence of 10 in Indo-Malaysia does not coincide with the geological history of the areas. We explain this by hypothesizing that at least some members of ancestor 10 dispersed to Indo-Malaysia, where the population evolved into *G. magna*. Hence, the occurrence of *S. paragonopora* (species 7) in Indo-Malaysia is due to common history, whereas the occurrence of *G. magna* in Indo-Malaysia is due to the dispersal of its ancestor into that area. If this is true, then what we have called "Indo-Malaysia" is, from a historical perspective, two different areas for those species.

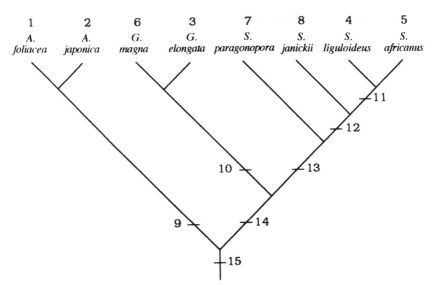

FIGURE 12 Phylogenetic tree for eight species of amphilinid flatworms with internal branches numbered for phylogenetic analysis. A = *Amphilina*; G = *Gigantolina*; S = *Schizochoerus* (from Brooks and McLennan 1991).

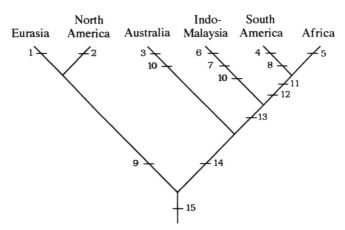

FIGURE 13 Area cladogram based on phylogenetic relationships among eight species of am-philinid flatworms. A = Eurasia; B = North America; C = Australia; D = South America; E = Africa; F = Indo-Malaysia. "Characters," represented by numbers accompanying slash marks, are species. 1 = *Amphilina foliacea*; 2 = *A. japonica*; 3 = *Gigantolina elongata*; 4 = *Schizochoerus liguloideus*; 5 = *S. africanus*; 6 = *Gigantolina magna*; 7 = *Schizochoerus paragonopora*; and 8 = *Schizochoerus janickii*. 9–15 = Ancestral species (from Brooks and McLennan 1991).

We can further examine the question of ancestor 10's dispersal by listing *G. magna* and *S. paragonopora* in different subsections of Indo-Malaysia. When we perform a phylogenetic analysis using this new designation, we obtain the area cladogram depicted in Fig. 14. We now find the two Indo-Malaysian areas in different parts of the geographic cladogram, with IM_1 connected to Australia and IM_2 associated with South America and Africa. The placement of IM_2 is in accordance with the patterns of continental drift, but the placement of IM_1 is not. This strengthens our hypothesis that ancestor 10 dispersed from Australia to Indo-Malaysia. Interestingly, this dispersal involved a movement into a different habitat than that occupied by *S. paragonopora*; F_1 encompasses estuarine Indo-Malaysian habitats, while F_2 represents freshwater, nuclear Indian subcontinent habitat.

On the ecological side, amphilinids live in the body cavities of freshwater and estuarine ray-finned fishes and in one species of freshwater turtle. By replacing geographical areas with hosts, we can obtain a picture of the history of involvement of various types of hosts in the evolutionary diversification of amphilinideans. Phylogenetic analysis using the hosts as taxa and the amphilinid phylogenetic tree as a multistate transformation series produces the host cladogram shown in Fig. 15. The species of parasites inhabiting acipenseriforms are a monophyletic group, as are those inhabiting osteoglossiforms. However, the turtle *Chelodina longicollis* is not the sister-group of perciform teleostean fishes, so *G. elongata* must inhabit this turtle as the result of a host switch. In addition, contrary to the current phylogenetic

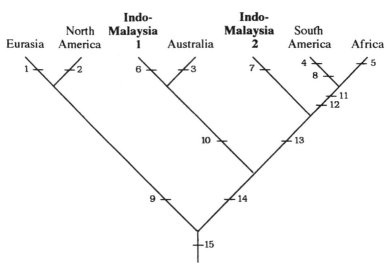

FIGURE 14   Area cladogram based on phylogenetic relationships of eight species of amphilinid flatworms, listing Indo-Malaysia (F) as two separate areas, depicting the two separate historical biogeographical influences in the area (from Brooks and McLennan 1991).

analysis of the actinopterygians, this cladogram places siluriform fishes with the osteoglossiforms rather than with the perciforms; therefore, *S. paragonopora* presumably lives in a siluriform host as the result of a host switch as well.

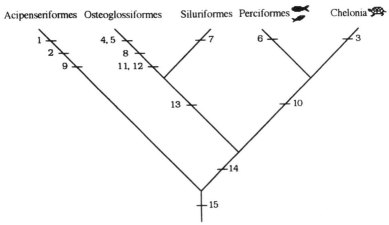

FIGURE 15   Host cladogram based on phylogenetic relationships of amphilinid flatworm parasites. *Note:* Turtles are not the sister-group of perciform fishes (from Brooks and McLennan 1991).

The approach described above refers to existing hypotheses about the relationships among areas or among ecologically associated species which can be compared with the relationships among the members of the clade of interest. By applying the principles of phylogenetic systematics fully, we may use phylogenetic information to produce the general patterns from which we can determine particular cases of association by descent and association by colonization or extinction. When we use phylogenetic systematics to produce a general pattern of phylogenetic relationships among a group of species, we expect that the robustness of the phylogenetic hypothesis will be proportional to the number of different characters used to construct it. No self-respecting phylogeneticist would use only a single transformation series to produce a phylogenetic tree, but rather would rely on the most parsimonious hypothesis resulting from the analysis of as many characters as possible. In the same way, by treating phylogenies as transformation series as above, we can generate general patterns of historical relationships among areas or among ecologically associated species by examining multiple clades having species occurring in the same areas or associated ecologically with the same species (see Brooks 1981, 1985, 1990; Brooks and McLennan 1991, 1993b). Although this approach to studying coexisting clades has been called Brooks Parsimony Analysis (Wiley 1988a,b) and Component Compatibility Analysis (Zandee and Roos 1987), it is important to remember that the methodological underpinnings are those of phylogenetic systematics.

HOW DO WE EXPLAIN
THE PATTERNS?

Biogeographical homoplasies indicate four different phenomena. The first of these is postspeciation dispersal. In such cases, one species occurs in more than one of the areas designated in a biogeographical analysis. This indicates that a species has expanded its range from its area of origin into one or more other areas. This class of evolutionary phenomena has been called by some the "widespread species problem" (e.g., Nelson and Platnick 1981). The converse of postspeciation dispersal is the situation in which two or more species representing a given clade are endemic to only one of the areas designated in a biogeographical analysis. This can occur in two general ways. First, one species occurs in the area as a result of vicariance and the other(s) occur in the area as a result of some form of speciation by colonization. Second, all the species occur in the area as a result of vicariance, indicating that what we recognize as a single geographical area today is actually a hybrid historically, at least with respect to the species living in it. Finally, geographical homoplasy can result from unexpected absence, in which an area lacks a member of a clade that should have a representative in it. This can indicate extinction or it can suggest sampling error, indicating areas in which we should do more collecting.

When discussing historical biogeography, the possibility of shared phylogenetic histories among geographically co-occurring species is investigated without making assumptions about the type or extent of their ecological interactions. Beyond occupying the same geographical areas, the study species need not interact at all. Phylogenetic systematic principles can be used to analyze putatively coevolving interspecific ecological associations. This is conceptually and analytically analogous to historical biogeographical studies, except that in this case the group of taxa designated as the "host" group is used to symbolize the ecological association.

Three major classes of coevolutionary models have emerged from 30 years of study initiated by the landmark publication of Ehrlich and Raven (1964). Each model implies a range of phylogenetic patterns characteristic for it. Allopatric cospeciation (Brooks 1979), or the "California Model," is the null model for comparative coevolutionary studies. It is based on the assumption that hosts and associates are simply sharing space and energy. It predicts congruence between host and associate phylogenies based solely upon simultaneous allopatric speciation in associate and host lineages. Like any null model, support for the hypothesis of cospeciation offers relatively weak explanatory power. For example, discovering congruence between host and associate phylogenies does not allow us to distinguish the effects of a coincidental historical correlation based on common geographical occurrence from the effects of various forms of mutual adaptive interactions. Brooks and McLennan (1991, 1993b) showed that various outcomes of all three classes of coevolutionary models could result in congruent host and associate phylogenies. Thus, it is the discovery of homoplasy, in this case incongruence between host and associate phylogenies, that provides the strongest means of differentiating the effects of different coevolutionary phenomena.

Resource tracking or colonization models are based on the concept that hosts represent patches of necessary resources which associates have tracked through evolutionary time (Kethley and Johnston 1975). The phylogenetic distribution of the resource among potential hosts, coupled with the evolutionary opportunity for colonization (Farrell and Mitter 1994), determines which macroevolutionary patterns will be produced. For example, the sequential colonization model (Jermy 1976, 1984) proposes that the diversification of phytophagous insects took place after the radiation of their host plants. The insects are hypothesized to have colonized new host plants many times during their evolution. In each case the colonization was the result of the evolution of insects responding to a particular biotic resource that already existed in at least one plant species. That resource, in turn, is postulated to have been either plesiomorphically or convergently widespread, so the predicted macroevolutionary pattern is that host and associate phylogenies will show no congruence.

The classical coevolution model is sometimes termed the exclusion or evolutionary arms race model (Mode 1958; Ehrlich and Raven 1964; Feeny 1976; Berenbaum 1983). The primary assumption in arms race models of coevolution is that coevolving ecological associations are maintained by mutual adaptive responses. For example, it is possible that during the course of evolution novel traits arise that protect the host from the effects of the associate. It is also possible that traits countering such defense mechanisms may evolve in the associate lineage. The macroevolutionary patterns that result depend upon the time scale on which the adaptive responses occur.

Evolutionary arms race models generally assume that, in many cases, the time scale on which the defense and counterdefense traits originate in response to reciprocal selection pressure is longer than the time between speciation events. Given this, we might expect to find macroevolutionary patterns in which the associate group is missing from most members of the host clade characterized by possession of the defense trait. A second pattern results when one or more relatively plesiomorphic members of a host clade are colonized by more recently derived members of the associate group bearing the counterdefense trait. In this case, host and associate phylogenies demonstrate some degree of incongruence and we would expect to find evidence that some associates have back colonized hosts in the clade diagnosed by the presence of the defense trait. Finally, Ehrlich and Raven (1964) postulated that coevolutionary patterns similar to ones expected for the sequential colonization resource-tracking model would result when host shifts by insects with a counterdefense trait occurred between plants that had convergently evolved similar secondary metabolites in response to insect attack. In both cases, there is a departure from phylogenetic congruence between hosts and associates; however, the resource-tracking model requires only that the associates be opportunistic. The host resource can be widespread due to either plesiomorphic occurrence or convergence, but its evolutionary patterns are not affected by the presence or absence of the associates. In contrast, the mechanism by which the host resource evolves in the Ehrlich and Raven case requires a high degree of convergent mutual modification on the part of the host and associate groups. Differentiation between the two models requires information about the evolutionary elaboration of putative defense and counter defense traits.

EXPLAINING ECOLOGICAL HOMOPLASY: COMMUNITY STRUCTURE

Comparative studies of community evolution need not be based on the assumption that strong and often highly specialized ecological interactions

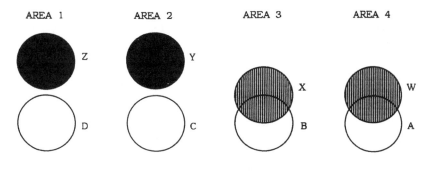

(a) DISTRIBUTION OF FISH SPECIES IN WATER COLUMN

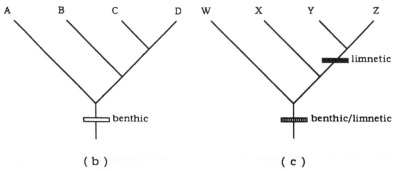

(b)   (c)

FIGURE 16   The influence of history at the community level. (a) Distribution of foraging preferences. White circle = species foraging in the benthos; striped circle = species foraging in both the benthos and the limnos; black circle = species foraging in the limnos. (b) Phylogenetic relationships for clade A–D based on nonecological data. (c) Phylogenetic relationships for clade W–Z based on nonecological data (from Brooks and McLennan 1991).

occur between species, although this may be true. Communities do not evolve in the same way species evolve; they are assembled through time. If they are more than arbitrary units, randomly dispersed through time and space, we should be able to document the effects of both macro- and micro-evolutionary processes on their evolutionary assemblage.

Consider the following hypothetical example of the interaction between past and present at the community level (Fig. 16). Imagine discovering that two fish species in a large lake (Fig. 16, area 1) do not overlap ecologically; say, for example, one is a benthic forager (species D) and one is limnetic (species Z). A possible explanation for such habitat separation is that it represents the effects of competition between these two species in the past. Without a phylogeny for the fishes and a record of their relatives' interactions with each other, however, it is impossible to ascertain whether the association in our research lake is a result of interactions between Z and D or a historical legacy of interactions between their ancestors. Let us now consider an appropriate comparative data base: species C (benthic) and Y (limnetic) in area 2, species X (demonstrates both foraging modes) and B (benthic) in area 3, and species W (demonstrates both foraging modes) and A (benthic) in area 4 (Fig. 16a). When the foraging modes are optimized on the phylogenetic trees for these two groups, we discover that foraging on the benthos is plesiomorphic for all members of the A + B + C + D clade. These species have not changed their foraging habits, interactions with members of the other clade notwithstanding. Conversely, foraging on both benthic and limnetic prey was primitive for the W + X + Y + Z clade but something happened during the interaction between the ancestor of Y + Z and the ancestor of C + D and the former moved out of the benthic, into the limnetic, realm. So, while this does rule out a role for interspecific competition between past populations of species D and Z in shaping the current foraging modes in these fishes, it does not rule out the possibilities that competition was either involved in the habitat shift in the appropriate ancestors or is maintaining the divergent foraging habits today.

There is nothing less interesting to a community ecologist than communities in which the only variables that show evolutionary changes are patterns of species occurrence and co-occurrence. What community ecologists wish to find is evidence of particular evolutionary changes in the ways species go about their business associated with particular environments or community structures. Thus, homoplasy is essential for an interesting research program. Many community ecologists have substituted geographic distance for phylogeny, asserting that intercontinental similarities must be due to convergence while intracontinental similarities might well be due to common history. If there is any common theme associated with phylogenetic systematic studies in biogeography and coevolution it is that most contemporary biotas are much older than we thought previously. Since many areas were much closer together in the past than they are today, using today's proximity or distance as an indicator of homology or homoplasy in ecological structure is problematical at best.

Losos (1994) recently summarized a series of elegant integrative studies detailing a comparative approach to studying the evolution of community structure. Using the anoline lizards of the islands of the Caribbean Sea as a model system, Losos was able to distinguish the relative sequence of addition of species to the lizard communities of different islands and to document the evolutionary diversification of microhabitat preference. He discovered significant similarity in the anoline community structures on the islands, achieved in each case by an historically unique, but repeating, mixture of colonization and diversification of microhabitat preference. Losos interpreted this homoplasy in current community structure, achieved through a series of historically unique circumstances, as evidence of the convergent evolution of community structure mediated by similar interspecific competitive interactions on each island.

AN ALTERNATIVE TO PHYLOGENETIC SYSTEMATIC APPROACHES

There is an alternative approach to historical analyses of geographically or ecologically associated clades. The first methods stemming from this perspective originated in historical biogeographic studies. Rosen (1975) presented an approach derived from theoretical discussions first articulated by Nelson (1969) in which distributional elements not common to all clades were eliminated from the database, resulting in a simplified area cladogram depicting the components of geographic distribution patterns common to all clades. Platnick and Nelson (1978) and Nelson and Platnick (1981) refined this approach, calling it component analysis (see also Humphries and Parenti, 1986), relying on consensus trees to summarize the common biogeographic elements. Elements that depart from the general pattern are depicted as ambiguities resulting from mistakes in phylogenetic analysis, extinctions, or sampling error. The plausibility of such interpretations is investigated by invoking one of two assumptions (assumptions 1 and 2 of Nelson and Platnick 1981). Component analysis approaches have been criticized by ecologists and systematists for the ways in which homoplasy is treated (e.g., Endler 1982; Simberloff 1987, 1988; Wiley 1988a,b; Zandee and Roos 1987; Brooks 1990).

Recent discussions centered around phylogenetic systematic methods and component methods have been based on a perception that all authors have the same worldview of the subject, demanding a common explanatory domain. If this is true, both classes of methods should be compared directly and one should be chosen over the other for general use because it best represents the explanatory domain. However, what if there are actually two different worldviews involved, each pointing to a different explanatory domain, thus requiring two different methods of analysis? If this is true, their methods and explanatory domains overlap to a considerable degree, and this may be a source of confusion. There seems to be agreement on three conceptual points: (1) significant aspects of the diversification of life can be explained by examining the association between the phylogeny of any given clade and the history of abiotic (e.g., geography) or biotic (e.g., other species with which they have a close and intimate ecological association) components; (2) association by descent occurs; (3) association by descent (absence of dispersal/host switching) is the explanation of first choice. Furthermore, both groups agree on two major *methodological* points: (1) association by descent is signified by phylogenetic congruence and (2) alternatives to association by descent are signified by homoplasy and require ad hoc explanations. Hoberg et al. (in press) and Zandee et al. (in preparation) and Klassen (in preparation) have discovered that both methods agree when there is no departure from association by descent, or when extinction is unambiguously more parsimonious than horizontal transfer. Furthermore, each method differs when association by colonization is either the most parsimonious explanation or is equally parsimonious with an explanation of extinction. Both approaches behave in a predictable and consistent manner; consequently, if the methods map the desires of the research group onto the data, they correspond to two different explanatory domains.

Finally, both approaches are (1) *testable* empirically, and both rely on the testability of cladograms by characters; (2) *parsimonious* operations within their explanatory domains. Neither one multiplies entities beyond necessity, but they differ with respect to the definition of "entities" (single areas or host taxa in the case of phylogenetic systematic approaches or the entire solution tree in the case of component approaches), and why they choose those particular kinds of entities; (3) *bold* philosophically, but in different ways. Phylogenetic systematic methods are bolder with respect to prohibited explanations of the results, whereas component methods are bolder with respect to a priori prohibitions (i.e., association by colonization is a prohibited class of events within the explanatory framework). Component approaches are weaker than phylogenetic systematic approaches is terms of *empirical content* by not permitting association by colonization to be identified and confronted *within* the explanatory domain. Phylogenetic systematic approaches do not suffer from this problem because the falsification of association by colonization is phylogenetic congruence and phylogenetic systematics (a) finds congruence when the data warrant it and (b) can be modified to increase estimates of congruence if new characters emerge and change the cladograms in such a way that congruence is increased.

What is important from the standpoint of this volume is that the crucial aspect of evaluating these divergent approaches is the way they provide explanations for homoplasies, which are test cases for the worldviews. In the component explanatory domain, reversals are the result of extinction or sampling error, and parallelisms and convergences are the result of poor taxonomy (poor discrimination of taxa) or poor systematics (having the wrong cladogram). The result (areagram, host cladogram) is both the solution tree (the geological hypothesis or the coevolutionary hypothesis) and the determinant of the "correctness" of the phylogenetic hypotheses for each clade. In the phylogenetic systematic explanatory domain, by contrast, reversals are the result of extinction or sampling error, and parallelisms and convergences are the result of a myriad of phenomena (dispersal, host switching, peripheral isolates speciation, colonization, widespread taxa). Characters determine the trees for each clade. The combined analysis of taxa, using phylogenetic reasoning, determines the solution tree, and explanations are based on current knowledge. All components can be wrong, and all are subject to being changed by subsequent evidence brought to bear on the issue of sister-group relationships within particular clades.

When component approaches remove association by colonization to the assumptions without confronting them within the explanatory framework, at least one form of falsification is prohibited. Dispersal and host switching are not denied, but neither are they confronted. The phylogenetic systematic worldview allows more possibilities a priori to be incorporated directly into explanations, thus not prejudging or restricting the range of results.

A recent computer program produced by Rod Page, called TREEMAP, moves component studies closer to phylogenetic systematic ones by permitting host-switching or dispersal by terminal taxa, though not by ancestral taxa. This approach is conceptually aligned with phylogenetic systematic approaches, but still does not permit the full range of possibilities for homoplasy to be explored within an analysis. True to its roots in the component philosophy, TREEMAP does not permit the simultaneous analysis of multiple phylogenetic trees—each tree must be evaluated against an a priori hypothesis of geographic history or host phylogeny. If TREEMAP were modified to permit simultaneous evaluation of multiple trees to produce the general patterns and to permit the possibility of host-switching or dispersal by ancestors, it would achieve the status of phylogenetic methods.

SUMMARY

I began by suggesting that homoplasy can be viewed as an obstacle or as an opportunity. What this means, not surprisingly, is that the way in which we deal with homoplasy is not theory-free. Worldviews, which determine the kinds of research to be done and the methods of analysis to be used, have everything to do with theories, in this case theories about the way in which biodiversity has come to be arrayed vertically and horizontally and at different levels of organization. For some, homoplasy requires complex explanations which introduces an unwelcome inference of evolutionary mechanisms. Homoplasy is simply a pattern phenomenon, and those observations that obscure the general patterns and which can be made to disappear by appropriate methodological manipulations. For others, homoplasy is more than a feature of a particular way of assessing phylogenetic relationships, it is an indication of evolutionary processes themselves. To them, a purely pattern approach robs evolutionary biology of the ability to link description and explanation. Clearly, my sympathies lie with the evolutionary biologists, and it is my hope that this brief excursion through biological diversity has exposed a "tangled bank" of exciting research that can be done when homoplasy is treated as an opportunity rather than a mistake.

REFERENCES

Alberch, P. 1985. Problems with the interpretation of developmental sequences. Systematic Zoology 34:46–58.

Alberch, P., S. J. Gould, G. F. Oster, and D. B. Wake. 1979. Size and shape in ontogeny and phylogeny. Paleobiology 5:296–315.

Bell, M. A., G. Orti, J. A. Walker, and J. P. Koenings. 1993. Evolution of pelvic reduction in threespine stickleback fish: A test of competing hypotheses. Evolution 47:906–914.

Bell, M. A., and S. A. Foster. 1994. The evolutionary biology of the threespine stickleback. Oxford Univ. Press, Oxford.

Berenbaum, M. R. 1983. Coumarins and caterpillars: A case for coevolution. Evolution 37: 163–179.

Blouw, D. M., and G. J. Boyd. 1992. Inheritance of reduction, loss, and asymmetry of the pelvis of Pungitius pungitius (ninespine stickleback). Heredity 68:33–42.

Bonner, J. T. (Ed.). 1982. Heterochrony in evolution: A multidisciplinary approach. Plenum, New York.

Brooks, D. R. 1979. Testing the context and extent of host-parasite coevolution. Systematic Zoology 28:299–307.

Brooks, D. R. 1981. Hennig's parasitological method: A proposed solution. Systematic Zoology 30:229–249.

Brooks, D. R. 1985. Historical ecology: A new approach to studying the evolution of ecological associations. Annals of the Missouri Botanical Garden 72:660–680.

Brooks, D. R. 1990. Parsimony analysis in historical biogeography and coevolution: Methodological and theoretical update. Systematic Zoology 39:14–30.

Brooks, D. R., and D. A. McLennan. 1991. Phylogeny, ecology and behavior: A research program in comparative biology. Univ. of Chicago Press, Chicago.

Brooks, D. R., and D. A. McLennan. 1993a. Parascript: Parasites and the language of evolution. Smithsonian Institution Press, Washington, D.C.

Brooks, D. R., and D. A. McLennan. 1993b. Macroevolutionary trends in the morphological diversification among the parasitic flatworms (Platyhelminthes: Cercomeria). Evolution 47:495–509.

Brundin, L. 1972. Evolution, causal biology and classification. Zoologica Scripta 1:107– 120.

Crampton, G. C. 1929. The terminal abdominal structures of female insects compared throughout the orders from the standpoint of phylogeny. Journal of the New York Entomological Society 37:453–496.

De Queiroz, A., and P. H. Wimberger. 1993. The usefulness of behavior for phylogeny estimation: Levels of homoplasy in behavioral and morphological characters. Evolution 47: 46–60.

Ehrlich, P. R. and P. H. Raven. 1964. Butterflies and plants: A study in coevolution. Evolution 18:586–608.

Endler, J. A. 1982. Problems in distinguishing historical from ecological factors in biogeography. American Zoologist 22:441–452.

Farrell, B. D., and C. Mitter. 1994. Adaptive radiation in insects and plants: Time and opportunity. American Zoologist 34:57–69.

Feeny, P. P. 1976. Plant apparency and chemical defense. Recent Advances in Phytochemistry, Biochemical Interactions between Plants and Insects 10:1–40.

Felsenstein, J. 1982. Numerical methods for inferring phylogenetic trees. Quarterly Review of Biology 57:379–404.

Fink, W. L. 1982. The conceptual relationships between ontogeny and phylogeny. Paleobiology 8:254–264.

Fisher, R. 1930. The genetical theory of natural selection. Clarendon Press, Oxford.

Funk, V. A. 1982. Systematics of Montanoa (Asteraceae: Heliantheae). Memoirs of the New York Botanical Garden 36:1–135.

Funk, V. A. 1985. Phylogenetic patterns and hybridization. Annals of the Missouri Botanical Garden 72:681–715.

Funk, V. A., and D. R. Brooks. 1990. Phylogenetic systematics as the basis of comparative biology. Smithsonian Contributions in Botany 73:1–45.

Funk, V. A., and P. H. Raven. 1980. Ploidy in Montanoa (Cerv.) (Compositae, Heliantheae). Taxon 29:417–419.

Gould, S. J. 1977. Ontogeny and phylogeny. Harvard Univ. Press, Cambridge, Massachusetts.

Harvey, P., J. J. Bull, M. Pemberton, and R. J. Paxton. 1982. The evolution of aposematic coloration in distasteful prey: A family model. American Naturalist 119:710–719.

Hennig, W. 1950. Grundzüge einer Theorie der phylogenetischen Systematik. Deutscher Zentralverlag, Berlin.

Hennig, W. 1966. Phylogenetic systematics. Univ. of Illinois Press, Urbana.

Hillis, D. M., and C. Moritz. 1990. Molecular systematics. Sinauer, Sunderland, MA.

Hoberg, E. P., D. R. Brooks, and D. Siegel-Causey. In press. Host-parasite phylogenies and cospeciation. In D. Clayton and J. Moore, eds. Coevolutionary ecology of birds and their parasites. Oxford Univ. Press, Oxford.

Hobson, E. S. 1991. Trophic relationships of fishes specialized to feed on zooplankters above coral reefs. Pp. 69–95 in Sale, P.F., ed. The Ecology of Fishes on Coral Reefs. Academic Press, New York.

Humphries, C. J., and L. Parenti. 1986. Cladistic biogeography. Academic Press, London.

Jermy, T. 1976. Insect-host plant relationships- coevolution or sequential evolution? Symposiae Biologicae Hungarica 16:109–113.

Jermy, T. 1984. Evolution of insect/host plant relationships. American Naturalist 124:609–630.

Jones, R. S. 1968. Ecological relationships in Hawaiian and Johnston Island Acanthuridae (Surgeonfishes). Micronesia 4:310–361.

Kethley, J. B., and D. E. Johnston. 1975. Resource tracking patterns in bird and mammal ectoparasites. Miscellaneous Publications of the Entomological Society of America 9: 231–236.

Losos, J. B. 1994. Integrative approaches to evolutionary ecology: Anolis lizards as model systems. Annual Review of Ecology and Systematics 25:467–493.

Maddison, W. P. 1990. A method for testing the correlated evolution of two binary characters: Are gains and losses concentrated on certain branches of a phylogenetic tree? Evolution 44:539–557.

McDade, L. A. 1990. Hybrids and phylogenetic systematics I. Patterns of character expression in hybrids and their implications for cladistic analysis. Evolution 44:1685–1700.

McDade, L. A. 1992. Hybrids and phylogenetic systematics II. The impact of hybrids on cladistic analysis. Evolution 46:1329–1346.

McKinney, M. L. (Ed.). 1988. Evolution and development. Dahlem Conference. Springer Verlag, New York.

McLennan, D. A. 1993. Phylogenetic relationships in the Gasterosteidae: An updated tree based on behavioral characters with a discussion of homoplasy. Copeia 1993:318–326.

McLennan, D. A. 1994. A phylogenetic approach to the evolution of fish behaviour. Fish Biology and Fisheries 4:430–460.

McLennan, D. A., D. R. Brooks, and J. D. McPhail. 1988. The benefits of communication between comparative ethology and phylogenetic systematics: A case study using gasterosteid fishes. Canadian Journal of Zoology 66:2177–2190.

McNamara, K. J. 1986. A guide to the nomenclature of heterochrony. Journal of Paleontology 60:4–13.

Meier, R., P. Kores, and S. Darwin. 1991. Homoplasy slope ratio: a better measurement of observed homoplasy in cladistic analyses. Systematic Zoology 40:74–88.

Mode, C. J. 1958. A mathematical model for the co-evolution of obligate parasites and their hosts. Evolution 12:158–165.

Nelson, G. 1969. The problem of historical biogeography. Systematic Zoology 18:243–246.

Nelson, G., and N. Platnick. 1981. Systematics and biogeography: Cladistics and vicariance. Columbia Univ. Press, New York.

Nelson, J. S. and F. M. Atton. 1971. Geographical and morphological variation in the presence and absence of the pelvic skeleton in the brooks stickleback, Culaea inconstans (Kirkland), in Alberta and Saskatchewan. Canadian Journal of Zoology 49:343–352.

Platnick, N. I., and G. Nelson. 1978. A method of analysis for historical biogeography. Systematic Zoology 27:1–16.

Price, P. W. 1980. Evolutionary biology of parasites. Princeton Univ. Press, Princeton, NJ.

Raff, R. A., and E. C. Raff., eds. 1987. Development as an evolutionary process. Proceedings of a meeting held at the Marine Biological Laboratory in Woods Hole, Massachusetts, August 23 and 24, 1985. A. R. Liss, New York.

Reist, J. D. 1980. Selective predation upon pelvic phenotypes of brooks stickleback, Culaea inconstans, by selected invertebrates. Canadian Journal of Zoology 58:1253–1258.

Remane, A. 1956. Die Gründlagen des naturlichen System der vergleichenden Anatomie und Phylogenetik. 2. Geest und Portig, K.G, Leipzig.

Rogers, W. P. 1962. The nature of parasitism: The relation of some metazoan parasites to their hosts. Academic Press, New York.

Rohde, K. 1989. At least eight types of sense receptors in an endoparasitic flatworm: A countertrend to sacculinization. Naturwissenschaften 76:383–385.

Rosen, D. E. 1975. A vicariance model of Caribbean biogeography. Systematic Zoology 24:431–464.

Sæther, O. A. 1977. Female genitalia in Chironomidae and other Nematocera:Morphology, phylogenies, keys. Bulletin of the Fisheries Research Board of Canada 197:1–210.

Sæther, O. A. 1979a. Hierarchy of the Chironomidae with special emphasis on the female genitalia. Pp. 17–26 in O.A. Sæther, ed. Recent developments in chironimid studies (Diptera: Chironomidae). Entomol. Scand. Suppl. 10.

Sæther, O. A. 1979b. Underlying synapomorphies and anagenetic analysis. Zoologica Scripta 8:305–312.

Sæther, O. A. 1979c. Underliggende synapomorfi enestênde innvendig parallellisme belyst ved eksempler fra Chironomidae og Chaoboridae (Diptera). Entomologiskes Tidskrift 100:173–180.

Sæther, O. A. 1983. The canalized evolutionary potential: Inconsistencies in phylogenetic reasoning. Systematic Zoology 32:343–359.

Salthe, S. N. 1993. Development and evolution: Complexity and change in biology. MIT Press, Cambridge, MA.

Sanderson, M. J. 1991. In search of homoplastic tendencies: Statistical inference of topological patterns in homoplasy. Evolution 45:351–358.

Schaeffer, B., and D.E. Rosen. 1961. Major adaptive levels in the evolution of the actinopterygian feeding mechanism. American Zoologist 1:187–204.

Sillen-Tullberg, B. 1988. Evolution of gregariousness in aposematic butterfly larvae: A phylogenetic analysis. Evolution 2:293–305.

Sillen-Tullberg, B. 1993. The effect of biased inclusion of taxa on the correlation between discrete characters in phylogenetic trees. Evolution 47:1182–1191.

Simberloff, D. 1987. Calculating probabilities that cladograms match: A method of biogeographical inference. Systematic Zoology 36:175–195.

Simberloff, D. 1988. Effects of drift and selection on detecting similarities between large cladograms. Systematic Zoology 37:56–59.

Tuomikoski, R. 1967. Notes on some principles of phylogenetic systematics. Annales Entomologicae Fennica 33:137–147.

Turner, J. R. G. 1971. Studies of Müllerian mimicry and its evolution in burnet moths and heliconid butterflies. Pp. 224–260 in R. Creed, ed. Ecological genetics and evolution. Blackwell, Oxford.

Tyler, J. C., G. D. Johnson, I. Nakamura, and B. B. Colette. 1989. Morphology of Luvarus imperialis (Luvaridae), with a phylogenetic analysis of the Acanthuroidei (Pisces). Smithsonian Contributions in Zoology 485:1–78.

Wiley, E. O. 1988a. Vicariance biogeography. Annual Review of Ecology and Systematics 19: 513–542.

Wiley, E. O. 1988b. Parsimony analysis and vicariance biogeography. Systematic Zoology 37: 271–290.

Winterbottom, R., and D. A. McLennan. 1993. Cladogram versatility—evolution and biogeography of acanthuroid fishes. Evolution 47:1557–1571.

Zandee, M., and M. C. Roos. 1987. Component-compatibility in historical biogeography. Cladistics 3: 305–332.

HOMOPLASY CONNECTIONS AND DISCONNECTIONS: GENES AND SPECIES, MOLECULES AND MORPHOLOGY

JEFF J. DOYLE
Cornell University
Ithaca, New York

INTRODUCTION

The tremendous advances of modern biology have placed systematists in a position where we are able to analyze genotypic variation directly and draw on it as a source of characters for phylogeny reconstruction. This has resulted in what is today called "molecular" systematics. Far from providing a panacea, however, these new sources of data have raised novel issues and engendered new controversies. Among the most prominent of these is the problem of inferring organismal relationships from gene phylogenies—the "gene tree vs species tree" problem (e.g., Nei 1987; reviewed in Doyle 1992). Meanwhile, systematists still rely on more traditional characters, particularly morphological ones, not only for phylogeny reconstruction, but for day to day taxonomic work involving identification and classification. Both the "molecules vs morphology" controversy and the more recent debate over approaches to data combination are symptomatic of the tension between the desire to incorporate new methods and sources of data in

HOMOPLASY: *The Recurrence of Similarity in Evolution*, M. J. Sanderson and L. Hufford, eds.

systematics and the need to meet traditional taxonomic goals. At the heart of these controversies is homoplasy, primarily because of its connection to the reliability of different types of data as sources of characters for phylogenetic inference.

To a considerable degree, the inspiration for this chapter comes from Louise Roth's 1991 paper entitled "Homology and Hierarchies: Problems Solved and Unresolved." A thesis of that paper was that different organizational levels are "screened off"—disconnected—from one another, making it difficult to apply information gained from one hierarchical level to problems at other levels. For the most part, she dealt with issues of ontogeny, where the complexity of developmental networks is responsible for screening off molecular change from morphological variation. In this chapter, I explore some practical consequences of disconnections between organizational and hierarchical levels when molecular data are used to reconstruct phylogeny. One conclusion is that crossing from a lower to a higher level—from genes to species or from genotype to phenotype—can simultaneously introduce new kinds of homoplasy and mask homoplasy from the lower level. The concept of "gene trees" is central to this issue; thus, sources of homoplasy in reconstructions of "phylogenies" of nucleic acid sequences will be reviewed and discussed. I will then discuss the relationship of homoplasy in single gene trees to homoplasy in multiple gene trees. The need for combining gene trees stems from the desire to reconstruct the phylogeny of something other than that of genes themselves, resulting in "species trees" or "organism trees." Problems arising from disconnection between gene and organismal levels are particularly acute with species, because it is there that the transition from reticulate to hierarchical patterns of relationship occurs. My discussion therefore focuses on species-level problems and deals with the issue of homoplasy in combined data sets involving closely related species. A major theme is character independence.

The "molecules vs morphology" debate in modern phylogenetic analysis often has emphasized the relative levels of homoplasy in two rather arbitrarily delineated classes of characters (e.g., Sanderson and Donoghue 1989). An underemphasized component of this comparison is that differences in homoplasy levels are expected when comparing single gene trees with trees summarizing information from multiple genes. This is true whether such trees combine information from several genes directly, as separate gene trees, or indirectly, as numerous independent morphological characters. Data sets from multiple genes may in some ways be more like morphological data sets than like individual gene sequence data, depending on how the data are utilized. A key element in comparing data sets that combine multiple genes or gene trees with those that consist of (putatively) independent morphological features is the molecular basis of phenotypic characters. Homoplasy in gene function is at best weakly correlated with homoplasy in the overall gene tree, and the further complexities of development responsible for screening off phenotype from genotype make any connection between gene tree homoplasy and phenotypic homoplasy extremely tenuous.

For the purposes of this paper, I will define homoplasy as character conflict on a cladogram. If the tree itself correctly represents the topology of hierarchically related terminal units ("taxa," whether genes or organisms), then homoplasy is observational error—"false homology." Whether compensation can be made for such error by weighting schemes, or the error "corrected" by *a posteriori* character analysis, are practical questions not treated here in any detail. The latter issue seems relevant in the context of whether homoplasy is "real" in some evolutionary sense. If, after detailed *a posteriori* analysis, no evidence of nonhomology can be discovered in a complex homoplastic character (e.g., a morphological or ecological attribute), one might conclude that either parallelism or reversal is a "real" feature of evolution in this instance. The systematist may then legitimately inquire as to its significance and underlying causes. Homoplasy thus viewed is an important and interesting phenomenon in its own right and is a source of opportunity for insights into the evolutionary process: phylogenetic "noise" is evolutionary "signal." However, the same "reality" could as well be attributed to a parallel base substitution at an unambiguously aligned nucleotide position in a gene. This is also "real" homoplasy, certainly, but of a kind so commonly encountered as to be, for the most part, quite unremarkable. In both cases, homoplasy results from the researcher's inability to distinguish homology from analogy and both are therefore observational errors.

Circumstances exist, however, in which relationships among some taxa are nonhierarchical. In such cases homoplasy will not be error in the above sense, but instead will represent conflict among potentially equally accurate observations, produced as a consequence of the artificial imposition of a single hierarchical structure. This "multiple signal" homoplasy is the basis for my discussion of data combination in phylogenetic analyses.

HOMOPLASY IN GENE TREES

In this first section, I briefly review homoplasy as it relates to reconstructing relationships among a set of DNA sequences. These are sampled from organisms, of course, but the sequences themselves are the taxa in the analysis; at this stage, no inferences are made about relationships of the organisms bearing these sequences. Sequences may be organelle genome haplotypes, alleles at a nuclear locus, or the paralogous and orthologous genes of a multigene family. The only type of homoplasy I will consider in this section is that involving changes of the primary sequence—base substitutions, insertions, deletions, etc. Homoplasy involving other attributes of genes (e.g., higher-level structure and gene function) are discussed in subsequent sections.

Nucleotide Substitutions

The simplest case for reconstructing gene phylogenies is a nonrecombining gene in which no length mutation occurs. In this case, relationships are strictly hierarchical and alignment is trivial. Genes from the chloroplast or animal mitochondrial genomes potentially fit these criteria, for example the chloroplast gene *rbcL* (excluding the length-variable extreme 3' end) in flowering plants, which has been exploited recently as a tool for phylogeny reconstruction (e.g., Chase et al. 1993). Under these conditions, homoplasy in the gene tree is due solely to parallelism and reversal involving apparently identical nucleotide substitutions in the genes of different taxa. As noted above, assuming that sequences have been produced and read accurately, there does not seem to be any way to perform *a posteriori* character analysis: an adenine is an adenine, whether it is homoplastic or uniquely synapomorphic. Homoplasy is therefore empirically irreducible and irremediable.

Alignment and "Structural" Characters

Most sequences evolve not only by nucleotide substitution, but also by structural changes. The most commonly observed structural changes are insertion and deletion, but larger transpositions and inversions may also occur. Insertion and deletion are particularly important because they create length variation among homologous sequences, producing ambiguity in sequence alignment. In the absence of length variation, positional homologies are maintained, and thus *characters* (nucleotide positions) must be homologous along the sequences being compared. Homoplasy is therefore confined to character *states,* as described above. Length variation, in contrast, obscures character homologies. For example (Fig. 1), a "GA" produced by slipped strand mispairing (Levinson and Gutman 1987) in one sequence, though perhaps iteratively homologous (to use Roth's [1988] term; being derived by duplication from an ancestral "GA"), is not homologous positionally to any other nucleotide in the set of sequences. Unless this is recognized and corrected by inserting a two-base-pair (bp) gap in all other sequences, mistakes in homology assessments will result for all downstream positions. The effect is identical to the accidental deletion of a character from one taxon in a morphological data matrix—all subsequent character states are dissociated

```
sequence 1  TTCAAGCTAGAGAGA--TGGCAAAT
sequence 2  TTCAAGCTAGAGAGAGATGGCAAAT
sequence 3  TTGAAACTGGAGAGA--TGCCAAAT
sequence 4  TTGAAGCTGGAGAGA--TGCCAAAT
```

FIGURE 1   Length variation caused by slipped-strand mispairing in a simple sequence (GA) repeat.

from their correct characters. The primacy of this initial stage of character homology assessment is apparent in that homoplasy is generated even in the absence of any character *state* homoplasy. For homoplasy produced by incorrect alignment, the obvious solution is to find a more appropriate alignment. Unfortunately, alignment has been one of the least understood areas of phylogenetic analysis (e.g., Felsenstein 1988), and considerable debate continues about methods for producing alignments (e.g., Mindell 1991; Thorne et al. 1992; Wheeler and Gladstein 1994) and about how to handle regions that are particularly difficult to align (Swofford and Olsen 1990; Gatesy et al. 1993; Wheeler et al. 1995). Despite this, the problem seems more tractable than that of parallelisms involving nucleotide character states that will persist even if the "true" alignment is found.

The gaps introduced by alignment routines are potentially useful phylogenetic characters. Lloyd and Calder (1991) have argued that multiresidue length variants constitute a particularly useful class of character, because their complexity, relative to individual substitutions, makes their homologies easier to hypothesize with confidence. Olmstead and Palmer (1994) have recently concluded that insertions and deletions observed in cpDNA restriction mapping studies are not, in general, any more prone to homoplasy than are restriction site gains or losses. Complex overlapping length variants, however, can be difficult to code in cladistic analyses (e.g., Freudenstein and Doyle 1994), as can the smaller insertions and deletions that are regularly observed in DNA sequence studies. Probably the most common "solution" here, as with restriction mapping data, has been effectively to ignore the information present in gapped regions, by coding gaps as strings of "?" and thereby allowing optimization algorithms to work with whatever information exists in other sequences in such regions (Nixon and Davis 1991; Platnick et al. 1991).

Allelic Recombination

Phylogenies are statements of hierarchical relationships, and to attempt to reconstruct phylogeny is to assume that hierarchy exists among the taxa being studied. Recombination violates this assumption by bringing together, in a single taxon, regions of sequences that possess different most recent common ancestors. Hudson's (1990) diagrams of coalescences under recombination illustrate the complexity of relationships that can exist along a single sequence. In the coalescence sense, segments with different histories are separate "genes," and I will call these "c-genes" (as in Doyle 1995) to distinguish them from the more common molecular and systematic usage ("a set of contiguous nucleotides having a particular function" [Rieger 1991]). Thus, an alcohol dehydrogenase (ADH) allele may consist of numerous c-genes; potentially, each nucleotide could be a separate c-gene. At the opposite extreme, the entire ca. 150-kb chloroplast genome, which possesses over 100 genes in the molecular sense, is generally considered to be effectively nonrecombining and therefore constitutes a single c-gene.

Recombination can produce homoplasy in a gene phylogeny, even when no other source of homoplasy is present (e.g., Nixon and Wheeler 1992). In Fig. 2, several alleles are shown for which no homoplasy exists among the characters that define them. Recombination among these alleles produces considerable homoplasy and concomitantly results in conflicting topologies, because different characters in the recombinant alleles are tracking different relationships than others in the same allele—each is a separate c-gene. The best (perhaps only) solution to problems introduced by recombination is to identify the individual c-genes and reconstruct their relationships separately; as noted by Hudson (1990), this can be very difficult when recombination frequency is high and c-genes are short and therefore possess few characters. Apart from nonphylogenetic statistical tests for detecting recombinant

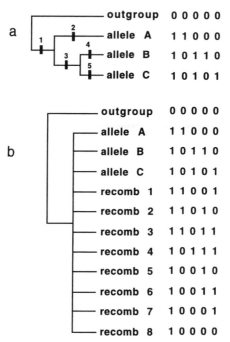

FIGURE 2   The effect of allelic recombination on homoplasy and resolution. (a) An outgroup and three ingroup alleles are scored for five variable sites; the single most parsimonious tree (ci = 1.0; ri = 1.0) shows their relationships. (b) The same four alleles, along with eight recombinant alleles formed from all possible combinations of the characters differentiating the three ingroup alleles; the strict consensus topology of the over 7000 most parsimonious trees (ci = 0.45; ri = 0.6) shows their relationships. Note that the only source of homoplasy is recombination.

regions (e.g., Sawyer 1989), reconstruction methods that attempt to deal with recombination have been proposed (e.g., Fitch and Goodman 1991; Hein 1993; Templeton and Sing 1993). Parsimony-based methods take advantage of patterns of homoplasy in such trees. DuBose et al. (1988), for example, note that character conflict can result from recombination among alleles at a locus, and that this conflict is indistinguishable from other sources of parallelism in its distribution on a parsimony tree. Dykhuizen and Green (1991) consider topological conflict among three genes in *E. coli* to be evidence of recombination.

Concerted Evolution

The above discussion of recombination dealt with genetic exchange among alleles at a locus. Concerted evolution (CE) is a similar process at the level of multiple related loci, in which repeated sequences become homogenized such that orthology/paralogy (OP) relationships (Fitch 1970) are obscured. In fully concerted gene families, paralogous genes within an individual are more similar to one another than to their presumed orthologues in individuals from other species, making them appear to be derived by recent, rather than ancient, duplication (Fig. 3). The primary mechanisms thought to be responsible for CE are unequal crossing over in tandem repeats, and gene conversion, which is not limited to contiguous sequences. The latter effectively is nonreciprocal recombination among paralogous sequences.

The influence of CE on homoplasy in gene trees depends on the degree to which it operates. When CE homogenizes sequences completely, and all copies of a given gene are identical in sequence within a species, then the entire gene family behaves as though it were a single sequence for the purposes of phylogeny reconstruction. In such cases homoplasy can still be present, due to parallelism and alignment mistakes. This can be seen in studies of genes

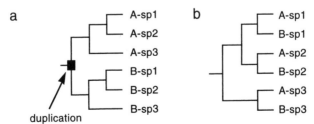

FIGURE 3   Concerted evolution. In (b), it appears that all sequences in each taxon share a most recent common ancestor with one another, when in fact they are derived from much earlier gene duplication events.

and spacer regions of the nuclear ribosomal rRNA family (e.g., Hamby and Zimmer 1992; Baldwin et al. 1995), which is the classic case of a concertedly evolving family. The probability of any given repeat type becoming fixed in a species should be identical whether or not the particular state is homoplastic when compared with the homologous character in other sets of homogenized arrays.

Intermediate levels of CE, however, can generate homoplasy (Sanderson and Doyle 1992), adding a new source to the homoplasy produced by more conventional sources such as parallel mutation. This is because, as is true for alleles at a locus, a separate phylogenetic history is contained in each paral-

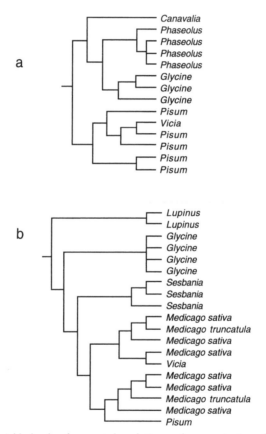

FIGURE 4 Variable levels of concerted evolutionary homogenization observed among legumes. (a) Vicilin-type seed storage protein genes evolve concertedly in *Glycine max* and *Phaseolus vulgaris*, whereas paralogy/orthology relationships can be observed in the gene family of *Pisum sativum* (Doyle 1992). (b) Concerted evolution has homogenized leghemoglobin proteins of some taxa (e.g., *Glycine max*), while paralogy/orthology relationships remain in *Medicago sativa* (Doyle 1994b).

ogous sequence. Concerted evolution mixes characters tracking different paralogous gene histories, just as allelic recombination creates hybrid alleles containing more than one historical signal. For phylogeny reconstruction, the result is the same, as can indeed be seen in methods such as those of Hein (1993) which can be used to detect either recombination or gene conversion. In the case of concerted evolution there exist two competing processes, either of which, alone, could produce an interpretable pattern. The first of these is simply divergence, which theoretically should produce a pattern in which hierarchical relationships within each set of orthologous genes should be interpretable phylogenetically, and relationships among paralogous groups should be interpretable as the historical sequence of gene duplication events. The second, CE, obscures these OP relationships by recombining the apomorphies of paralogs. Only when homogenization is so complete as to eliminate the first pattern is competing-signal homoplasy eliminated.

Homoplasy from incomplete CE may be a major problem for low-copy nuclear loci, where CE levels range from essentially none (e.g., plant actins: Shah et al. 1983) to very high levels (e.g., *rbcS*: Meagher et al. 1989). Even in ribosomal repeats, paralogy relationships may sometimes be maintained within species (e.g., Suh et al. 1993), suggesting that intermediate CE could also occur there. Moreover, in some gene families, it seems clear that concerted evolution operates in some taxa to a much greater degree than in others (Fig. 4). Even in gene families where high degrees of CE appear to operate, the process of homogenization may not keep pace with cladogenesis, nor are all parts of genes equally affected (e.g., *rbcS*: Meagher et al. 1989; Jamet et al. 1991). For protein-coding multigene families, the combination of divergence, allelic recombination, and partial concerted evolution would be expected to produce extremely complex patterns of allelic relationships. Reconstruction of allele histories would thus include homoplasy from a number of different sources, including both "error" and "mixed signal" homoplasy.

BEYOND THE GENE TREE: SOURCES OF HOMOPLASY ABOVE THE HIERARCHICAL LEVEL OF GENES

Stating the Problems: Tokogeny, Phylogeny, and Incongruence

Systematists, including those using molecular data, generally are interested in reconstructing the histories of something other than genes. Moreover, any attempt to combine information from more than one gene presupposes some higher level of organization than the single gene. Indeed, as I have already discussed, for nuclear loci this assumption is made still earlier, when it is assumed that "ADH" can be treated as a single entity when, in fact, it may more likely be composed of several c-genes. In sexually reproducing organisms, the "genome," defined as the sum total of genetic information in an organism (Rieger 1991), is quite transient, existing as a historical entity only for the duration of the individual's life. Even in the absence of crossing-over, independent chromosome assortment makes it highly unlikely that the haploid genomes transmitted in that individual's gametes will be identical to either of the genomes it received from its parents. And because recombination is a nonhierarchical process, tracing a "phylogeny" of genomes seems fruitless. This is merely a molecular restatement of Hennig's (1966) recognition that relationships among sexually reproducing individuals—the bearers of genomes—are tokogenetic (reticulate) rather than phylogenetic (hierarchical). Suggestions that sexually reproducing individuals can serve as terminal units in "phylogenetic" analyses in any capacity other than as mere stand-ins for some larger grouping (e.g., Vrana and Wheeler 1992) are thus self-contradictory. If homoplasy is defined as false synapomorphy, and synapomorphy itself is synonymous with *phylogenetic* homology (e.g., Patterson 1982; Roth 1988), then the term "homoplasy" perhaps should be limited to the occurrence of character conflict among entities whose relationships are phylogenetic and not tokogenetic.

The process of conducting a molecular systematic study involves several steps: (1) sampling individuals from higher taxa; (2) sampling a locus or loci from each individual; (3) constructing a gene tree from each locus; (4) inferring relationships among organismal taxa. The last stage, as noted elsewhere (Doyle 1992, 1995), requires an assumption that organismal taxa can be substituted directly for the alleles they bear, so that the topology of alleles is automatically the topology of the organismal taxa from which they were sampled. It is this assumption that is at the heart of the gene tree/species tree problem. If individuals are the taxa of such an analysis, the trees produced in step 4 might be seen as fundamentally "individual trees." However, individual trees have little biological meaning in sexually reproducing individuals, not only because individual relationships are tokogenetic, but quite simply because such trees will often be a technical impossibility. Individual trees based on nuclear loci cannot be uniquely hypothesized if the individuals sampled are heterozygous, because the two alleles that are both equally good place-holders for individuals will then occur in more than one place on the tree (Doyle 1995). This issue does not seem to have been addressed in the course of the current debate over methods of combining data. Yet the problem must be resolved for any single locus with heterozygous individuals, whether the approach to dealing with data from multiple loci is "total evidence" (e.g., Kluge 1989), data partitioning (e.g., Bull et al. 1993), or consensus (e.g., deQueiroz 1993). I will not deal further with this problem, because its direct connection to homoplasy is not obvious, except to suggest that some possible solutions, such as polymorphism coding or compartmentalization (e.g., Donoghue 1994), are likely to introduce ambiguity (Nixon and Davis 1991) and thus perhaps homoplasy.

It is the "species" that is generally accepted as the taxon that occupies the boundary between reticulate and hierarchical patterns of relationship: within species, relationships are tokogenetic; among them, relationships are phylogenetic. There is no necessary expectation that the topologies even of perfectly "correct" hierarchical gene trees (those that track the relationships of the alleles/haplotypes accurately) will have any meaning as regards the histories of the individuals from which they are sampled (e.g., Davis and Nixon 1992). For example, a mitochondrial gene tree of a human population will suggest that any given individual is more closely related to her great grandmother than to her own father, due simply to the maternal pattern of mitochondrial genome transmission.

The contrast between hierarchical gene trees and nonhierarchical relationships among individuals within species has recently led to some suggestions as to how molecular data might be used to delimit species (e.g., Dykhuizen and Green 1991). Baum and Shaw (1995) propose a method in which incongruence among gene trees from multiple alleles is used to define "genealogical" species—descendants of a single individual. Their idea relies on the very fact that relationships within species are tokogenetic, and therefore, as noted above, there is no single history of individuals to track. If multiple loci are sampled from a set of individuals, therefore, different topologies are expected for each locus. Baum and Shaw (1995) suggest that a strict consensus of these topologies will reveal a point of collapse that corresponds to the genealogical species. This topological conflict would also be reflected in homoplasy, were the different data sets combined directly. The relevant point here is not to discuss the merit of either the goals or the implementation of their method. Both, in my opinion, suffer from serious flaws (e.g., the problem of "individual trees" as well as general difficulties in using topological approaches for species delimitation; Doyle 1995). Rather, I simply wish to acknowledge the recognition by these and other authors (e.g., Dykhuizen and Green 1991) that character conflict is produced by discordance among gene trees in reticulating population systems.

Once the boundary between tokogeny and phylogeny is crossed, there should exist among terminal taxa—species—a historical sequence of cladogenic events that we can hope to reconstruct. However, the individual attributes that characterize species may not track this pattern. Allele topologies ("gene trees") are not expected to track species phylogenies ("species trees") at some loci because ancient allele polymorphisms are maintained across species, often, presumably by natural selection; examples include histocompatibility and incompatibility loci (e.g., Clark et al. 1991; Dwyer et al. 1991; Ioerger et al. 1991; Gaur et al. 1992). Such loci are unsuitable for organismal phylogenetic inference. Even for "suitable" loci, however, in which species possess monophyletic groups of alleles (Doyle 1995), discordance between allele histories and cladogenic histories may occur. It has long been recognized that rare gene flow produces such discordance; introgression is a well known cause of incongruence between character distributions and species boundaries (e.g., Anderson 1949; example in Doyle 1992). More recently, stochastic sorting of ancestral polymorphisms (lineage sorting, often discussed in the context of selective neutrality) has been noted as an additional problem that can produce patterns of incongruence virtually identical to those produced by introgression. Lineage sorting has been discussed not only for molecular data (e.g., Neigel and Avise 1986; Nei 1987), but also for morphological characters (Roth 1991), where, in my view, it is likely to be one source of so-called "homoplastic tendencies" (Sanderson 1991) among closely related species (Fig. 5). Sorting of polymorphisms from a common ancestor into its daughter species is inevitable during cladogenesis. The critical phylogenetic question is whether this process will produce patterns in which the allele or haplotype tree will suggest a sequence of cladogenic events different than that which actually occurred. This is a sampling problem, essentially, involving not only the systematist's sampling, but also sampling of lineages by drift leading to fixation of some alleles and extinction of others. For selectively neutral characters, the probability of discordance can be estimated, the critical parameters being effective population size and the amount of time between key cladogenic events (e.g., Pamilo and Nei 1988), which together determine the probability that polymorphic lineages will be maintained long enough to be sorted. Of course, in practice systematists are

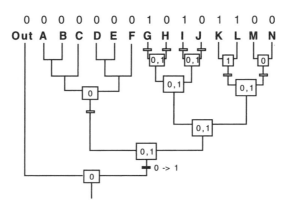

FIGURE 5   Lineage sorting as a source of "apomorphic tendency." An apomorphic character state arises (black bar) in an ancestor, making it polymorphic (0, 1); this is followed by several rounds of cladogenesis. Extinctions of lineages bearing either the apomorphic or plesiomorphic state occur (open bars) in the course of cladogenesis. Results of sorting of polymorphisms into modern species suggest parallel evolution of the apomorphic condition, clumped in one part of the tree, when in fact this state has arisen only once, and in a distant ancestor.

TABLE 1 Homoplasy Is Added by Combining Data Sets

Data set	Consistency index	Retention index	Steps	Extra steps
rbcL	0.503	0.444	279	—
ndhF	0.591	0.497	625	—
Combined	0.541	0.443	914	40

Note. Phylogenetic analyses of two chloroplast genes from genera of the plant family Solanaceae from Olmstead and Sweere (1994), Table 1. These genes necessarily have the same phylogenetic history, yet extra steps are added when data are combined.

not likely to know either of these parameters for the taxa under consideration. Moreover, the proportion of loci that actually behave neutrally remains an issue of considerable debate; selectively maintained polymorphisms will persist much longer, and will provide much more opportunity for discordance through lineage sorting than will neutral loci. Thus, it seems reasonable to assume that in some unknown percentage of systematic studies, particularly at the species level, lineage sorting problems will occur.

Typically, practical remedies suggested to the working systematist for avoiding or alleviating problems caused by incongruence between gene and species trees include: (1) additional sampling of individuals within taxa, in the hope that polymorphisms will be identified and rare introgressant individuals will not be mistaken for "typical" representatives of a species; (2) additional sampling of loci in individuals, in the hope that even if one locus is incongruent with the species tree, it will be seen as such by comparison with other, congruent loci. Both of these prescriptions are likely to add homoplasy even in the absence of actual incongruence. In any analysis, adding taxa increases the likelihood of observing homoplasy (Sanderson and Donoghue 1989). Homoplasy from the sources already described for gene trees will be added to each gene tree; no new sources are involved, as long as only one locus at a time is being considered. This source of added homoplasy applies, therefore, to any study, regardless of the method of handling multiple data sets. Moreover, when two data sets are combined, the resulting tree will require extra steps above those required by either data set alone, unless the two are topologically identical (Mickevich and Farris 1981). That homoplasy can result even when data from necessarily congruent genes are combined can be seen from analyses of multiple regions of a single c-gene, the effectively nonrecombining chloroplast genome (Olmstead and Sweere 1994), in which extra steps are attributable to data combination, causing a reduced retention index for the combined data set (Table 1). This source of homoplasy exists in "total evidence" analyses, where all characters from all loci are combined in a single data set and analyzed as a whole to identify the globally most parsimonious topology. As noted above, data combination is currently controversial and total evidence (character congruence) is only one of several methods that have been suggested (reviewed by Miyamoto and Fitch 1995). However, even some who disagree with the strict application of the total evidence approach (e.g., Bull et al. 1993) reject it only when topologies of different data sets are statistically incongruent. Thus homoplasy caused by data combination is as much an issue for them as for proponents of the strict total evidence approach. Combining fundamentally incongruent data sets will require many more extra steps than will congruent data sets, forming the basis for the Mickevich and Farris (1981; see also Kluge 1989) incongruence measures. Because topologies of independent loci in conspecific sexual individuals are expected often to be mutually incongruent due to tokogenetic relationships among such individuals (Hennig 1966; Davis and Nixon 1992; Baum and Shaw 1995; see above), increasing the number of individuals sampled per species is likely to introduce considerably more homoplasy than would increasing the number of species sampled.

Character Independence and Patterns of Homoplasy

Obtaining more data is often viewed as a solution for problems in reconstructing phylogeny, because, in general, stochastic noise can be overcome by amplifying the signal (but see Felsenstein 1978 and Steel et al. 1993). This should be true for a single gene, and for the single c-genes represented by the chloroplast or animal mitochondrial genomes. Even for nuclear genomes, if only a small fraction of all possible gene trees are fundamentally incongruent with the species tree, as might be true in some cases of introgression, combining data from different loci may yield a "correct" answer. We have no way of knowing how many loci are likely to have histories that do not track cladogenesis, in part because we have such poor definitions of species, but mainly because we do not know the true tree for any naturally occurring group of species, however defined. It is thus impossible to say whether this approach should reveal the truth. A total evidence approach may be preferred not because it is more likely to find the unknowable "true" tree, however, but because it seeks, in a single hypothesis, to account for all of the relevant data within a parsimony framework (e.g., Kluge 1989). In this context, consideration of the nature of the characters to be explained by "total evidence" could potentially lead to alternative ways of dealing with data. The characters in any phylogenetic analysis should be independent of one another, in that the state of one character should not automatically predict the state of another, as is the case in pleiotropy, for example. Lineage sorting or introgression can violate the requirement of character independence for the individual nucleotides of a gene (Doyle 1992). Nucleotides are generally assumed to evolve independently of one another, and as such are perfectly acceptable characters for gene tree reconstruction. However, what are sorted into different daughter species by lineage sorting are whole alleles, which are sets of linked nucleotides. All of the synapomorphies determining the place of the allele in the gene tree move as a set: the state of one automatically predicts the state of all others. They are not independent in the species tree. Ignoring, for simplicity's sake, the fact that the relevant units of sorting are c-genes and not genes in the standard molecular or systematic sense, it is alleles or haplotypes that are the units of synapomorphy or homoplasy. Whole alleles either track the species history or do not do so. Whole alleles at one locus are either congruent with alleles at another locus in their distributions or are not. Each allele is composed of many nucleotides, which together carry information concerning the placement of that allele in the gene tree for a given locus, but these move as a unit. Thus, when incongruence occurs, each nucleotide no longer provides independent evidence concerning the relationships of the species from which the gene is sampled. Thus, the molecular characters whose state distributions among taxa must be explained by total evidence are not individual nucleotides, but rather whole alleles.

A method has been proposed expressly to minimize the damage to parsimony analyses that can be caused by nonindependent gene tree characters (Doyle 1992). In this method, the character is defined as the genetic locus, with character states being the alleles or haplotypes of this locus. The gene tree relating these alleles gives the transformation series of the character states of this multistate character. The cladogram character can be recoded and combined with other such characters in phylogenetic analyses. Technically similar proposals for recoding of gene trees have been made, but based on different theoretical considerations (Baum 1992; Ragan 1992). I view recoding as entirely consistent with the spirit and objectives of the total evidence philosophy and not as an alternative to total evidence. Indeed it *is* total evidence, differing only in the definition of characters; it is certainly not consensus, nor does it advocate *not* combining data. Objections to this approach (e.g., de Queiroz 1993) have been based on the fact that (1) character information supporting or contradicting the topology at a given locus is lost when the gene tree is recoded and (2) most aspects of a given gene tree may be correct, even when some are fundamentally incongruent with the species tree. The elimination of character information, however, is not a by-product of the method, but rather is deliberate: neither synapomorphy nor homoplasy involving nucleotides in the gene tree is relevant to the species tree if there is fundamental incongruence between the two. The second objection would be more telling if there were some way to predict, *a priori, which* parts of the gene tree are misleading, the difficulty being that the gene tree itself may be perfectly correct as a gene tree while being an incorrect species tree. All of this is symptomatic of the disconnection of these hierarchical levels.

My own discomfort with the approach is not that it goes too far, as de Queiroz (1993) suggests, but that it does not go far enough. It must be recognized that *any* approach that combines data will generate homoplasy if there is incongruence among gene histories. Nonindependence of nucleotide characters exacerbates the problem, effectively weighting homoplasy and spreading it among different nodes of the species tree. The practical problem is how to minimize the potential damage caused by such homoplasy and still recover a topology that tracks species histories. Recoding as an ordered multistate character does not eliminate the problem of nonindependence, but only minimizes it by reducing the number of nonindependent characters. Where a total evidence approach might include 100 nonindependent characters, recoding reduces that number to the number of nodes on the tree. However, nonindependence remains, and weighted homoplasy will be generated when fundamentally incongruent recoded trees are combined as separate characters and the resulting data set is analyzed. Recoding, if done from considerations of nonindependence, views nucleotide substitutions as evidence for gene relationships, but as potentially disconnected from species relationships. It preserves only the topology of the gene tree, producing a single hypothesis, and eliminates all other sources of homoplasy in the gene tree. This may be a drawback, because to the extent that improved resolution *is* possible by combining independent characters, this benefit will be lost when recoded trees are combined. However, if recoding is done, it is done because it is felt that greater damage will be done by homoplasy stemming from mixed phylogenetic signals than from more conventional sources.

The following situation obtains: (1) if a locus is affected, its gene tree will yield misleading inferences of species relationships; (2) we cannot know *a priori* whether any given locus is affected by lineage sorting; (3) the unit of synapomorphy/homoplasy at affected loci is the allele or haplotype. To the extent that these considerations are relevant, a logical solution is to ignore topologies altogether and to treat each locus as an *unordered* multistate character in a total evidence analysis. The character states are, thus, alleles, and alleles are synapomorphic when shared by two or more species, regardless of their relationship to one another in the gene tree of that locus. Elsewhere (Doyle 1995) I have discussed, in the context of species delimitation, minimum requirements for loci suitable for phylogeny reconstruction and noted some significant difficulties with defining alleles. This problem is also relevant here. "Allele" is defined in genetics as "one of two or more alternate forms of a gene occupying the same locus on a particular chromosome or linkage structure and differing from other alleles of that locus at one or more mutational sites" (Rieger 1991). To be truly "the same" allele, there must be no "mutational site" differences (base substitutions) distinguishing sequences; in the strictest sense, only when alleles with *identical* sequences are found in two species is there a synapomorphy. With extremely fine resolution (e.g., long sequences including noncoding regions), it seems likely that species will not share alleles at all. A seemingly logical solution is to code as "identical" alleles that are closely related; an obvious way to do so would be by reference to the gene tree, with monophyletic groups of alleles being considered "the same." This, unfortunately, results in a degree of arbitrariness, because there is no cladistically objective way to decide *which* monophyletic groups of alleles in a nested hierarchy should be subsumed into a single operational allele; the lower the node, the more inclusive the definition of allele. This is presumably no different than the situation that obtains with morphological character state definition, which is indeed more subjective in that it is not done with reference to an objectively generated topology. It is apparent, however, that the level of inclusiveness in the definition of alleles (as with any character state definition) will have a profound effect on the level of homoplasy in the resulting tree (Fig. 6). At one extreme, if each allele sampled differs by at least one nucleotide substitution from every other, and if each is thus treated as a unique character state, then there could be no homoplasy, because each allele would be autapomorphic for the species from which it was sampled (leaving aside problems with polymorphism if multiple

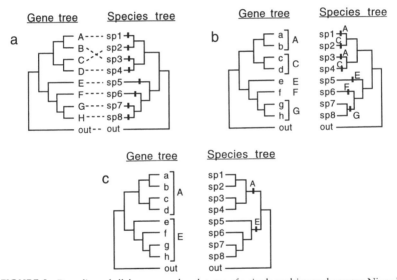

FIGURE 6 Recoding of alleles as unordered states of a single multistate character. Nine alleles at a locus are related to one another by the gene tree shown. These were sampled from eight different ingroup and one outgroup species, whose relationships are given by the species tree. The distribution of alleles in species are shown by the dashed lines in the top pair of trees. The species tree that would be inferred from the gene tree is identical to the true species tree except that the positions of species 2 and 3 would be reversed. (a) Coding each allele as an unordered state of the locus results in eight ingroup autapomorphies; there is no homoplasy, but there is also no resolution provided by the character. (b) A more inclusive coding, using the gene tree, groups alleles A and B as one state, A, C and D as a second state, C, and G and H as a third state, G. The distribution of recoded alleles A and C is homoplastic in species 1–4, while recoded allele G provides a synapomorphy for species 7 and 8. (c) A still more inclusive grouping of alleles provides synapomorphies, but removes homoplasy because recoded allele A is found in species 1–4.

individuals were sampled from each species). The potential for homoplasy is greatest when sister pairs of alleles are coded as a single state, but decreases with progressively inclusive groupings of alleles, until, at the other extreme, no homoplasy could occur if all ingroup alleles were coded as a single state.

To conclude this section, gene tree homoplasy is only a relevant indicator of homoplasy in the species tree if there is not fundamental incongruence between the two types of tree. In the absence of such fundamental incongruence, each nucleotide behaves as an independent character for the species tree, and both synapomorphies and homoplasies involving nucleotide transformations have their usual cladistic meaning. What this meaning actually is, however, is an issue of some debate; Nixon and Wheeler (1992), for example, note that the appearance of synapomorphies is more likely due to the extinction of lineages bearing plesiomorphies than to the actual appearance of the character state that is ultimately synapomorphic (see also Fig. 5). This appears to be a particularly useful observation, given that a major source of species tree character conflict is lineage sorting, a process that is based largely on stochastic elimination of polymorphisms. Such concerns complicate any attempt to optimize character state transformations on a cladogram in such a way that the result will be meaningful evolutionarily when the cladogram is viewed as a phylogeny. If these problems exist even for individual nucleotides when there is no fundamental incongruence, the problem is even more severe when the relevant unit of synapomorphy/homoplasy is the entire allele. Treating whole alleles as character states can reduce nonindependence problems, but optimization problems are likely to remain.

Where there is no fundamental incongruence between gene and species trees, the individual nucleotide is the relevant character, and homoplasy in the gene tree is precisely translated into homoplasy in the species tree. However, when there is fundamental incongruence, there is a disconnection: the amount of homoplasy in the gene tree does not accurately measure that in the species tree, and the amount of homoplasy in the species tree may not reflect that in the allele tree. If alleles are treated as the relevant character states of an ordered character (the locus), then homoplasy in the gene tree is of importance in its effect on reducing confidence in the topology of the gene tree. If homoplastic change in the gene tree results in incorrect hypotheses of allele relationships, then there will be a direct effect on homoplasy in the species tree. If, however, the locus is treated as an unordered character, then homoplasy in the gene tree may have no effect whatsoever on homoplasy levels in the species tree, unless it results in two alleles being considered identical. It is also possible to have a species tree in which there is no homoplasy, but individual gene trees in which there is, as for example in Fig. 6, where the gene tree would presumably include homoplastic substitutions, but there is no homoplasy in the species tree for this character (locus) in codings "a" or "c."

The possible solutions to independence problems discussed above, particularly ignoring gene tree topologies altogether, are responses to extreme situations. Perhaps most genes evolve neutrally, and the biology of most species is such that bottlenecks have occurred so as to reduce polymorphism. Perhaps patterns of molecular variation and of speciation usually have strong geographic components, such that variation will be apportioned into daughter species in a manner likely to minimize lineage sorting problems. Perhaps gene flow is rare between species. If these conditions obtain, then optimism may be warranted, and gene and species trees will generally be congruent. Our knowledge of any of the relevant parameters is so limited, however, that it seems worth at least exploring alternatives to the conventional approach should pessimism be more sensible.

GENE TREES AND PHENOTYPIC HOMOPLASY

The preceding section emphasized the disconnection in homoplasy that can occur when hierarchical boundaries are crossed. Much the same phenomenon occurs when moving from one organizational level to another. In this case, the disconnection is between gene tree and gene function, and the disconnection is relevant to the issue of molecular vs morphological characters in phylogenetic inference.

Molecules and Phenotype

Many types of data are considered "molecular," from electrophoretic assays of isozymic protein variation to a wide variety of DNA-based methods, including randomly amplified polymorphic DNA (RAPD), microsatellites, denaturing gradient gel or single-strand conformation polymorphism assays of allelic variation at known loci, restriction mapping, and nucleotide sequencing. The distinction between these and "morphological" approaches is well known to be more apparent than real. A more relevant distinction is perhaps "genotypic" vs "phenotypic," and anything that can be observed is phenotypic, from the size of a leaf to the mobility of a band on a gel. Of course, a genotypic character, the presence of a particular nucleotide at a certain position in a given gene, can be inferred directly from the phenotype of the sequencing gel band whereas the same cannot be said for leaf size. Other "molecular" techniques permit only much less precise determinations of genotype, however. Those that detect DNA variation using gene-specific primer pairs or probes narrow the source of variation to specific components of the genotype. RAPDs, used alone, do not afford this direct a connection to the genotype, because the method provides few criteria for homology assessment. RNA sequences, though tied directly to a genetic locus, may undergo editing, and so may not show a direct correspondence to the genotype. Similarly, protein electrophoretic mobility differences may be localized to changes at orthologous (allozymes) or paralogous (isozymes) loci, but are not generally directly attributable to specific base substitutions in the genotype.

What are typically considered morphological characters are clearly "phenotypic" in nature, but equally clearly have some genotypic basis, though the extent to which genes "determine" (as opposed to participate in) morphology is controversial (e.g., Nijhout 1990). Morphological characters and some of the above "molecular" character types, being phenotypic in nature, share several attributes, among which is the fact that they do not involve, or even permit, the construction of gene trees. There is no agreed-upon objective way, for example, to produce a "gene tree" directly from electrophoretic mobility variants of the protein ADH, as could be done from the nucleotide or amino acid sequences of the same variants. Those using isozyme data to infer relationships do not start by constructing, for each locus studied, a tree relating the observed mobility variants. Isozyme data may certainly be used to reconstruct phylogenies, but the phylogenies hypothesized are those of species, not of genes. It is true that a transformation series for variants at a particular locus can be hypothesized from the resulting tree, but this is no different than what can be done for any morphological character. It might be said that "morphological" characters differ from typical "molecular" data sources not in being phenotypic, but rather in being the product of gene function rather than simply of gene structure. This, too, however, seems a largely artificial distinction.

Homoplasy and Gene Structure and Function

The functioning of a gene is intimately related to its structure, though the connection may not always be obvious and easily observable. Secondary structure is critical to rRNA function, for example, while proper functioning of a protein encoded by a particular gene involves not only the primary amino acid (hence, DNA) sequence, but secondary through quaternary structures. The dependence of many enzymes on overall three-dimensional structure at catalytic or binding sites is well known. Higher level RNA or protein structures are all strongly influenced by the primary sequence, but can accommodate some changes, often by compensatory change at interacting residues. Structural features represent an additional class of phylogenetic character that, while not totally independent from substitutional transformations in the primary sequence, are not totally dependent, either. Moreover, there is no necessary correlation between the level of homoplasy in a structural character and that in the overall gene tree. For example, the electrophoretic mobility of a protein on a nondenaturing gel is a structural characteristic determined primarily by a relatively small number of amino acid residues that together produce an overall charge. If there is no homoplasy in the subset of nucleotide replacement substitutions affecting overall charge of an encoded protein, the structural character, "protein electrophoretic mobility," will show no homoplasy, despite the occurrence of homoplasy in the overall gene tree (Fig. 7). The reverse, however, is also true, and homoplasy may occur for such characters in the absence of any homoplasy at the sequence level (Fig. 8). Antigenicity is another higher-level character of a protein and is largely determined by the shapes of antigenic determinants (epitopes) on target molecules. Monoclonal antibodies directed against a particular epitope may cross-react with analogous epitopes of nonhomologous molecules, as in the case of the unrelated intensely sweet proteins thaumatin and monellin (Antonenko and Zanetti 1994). Here, too, there need be

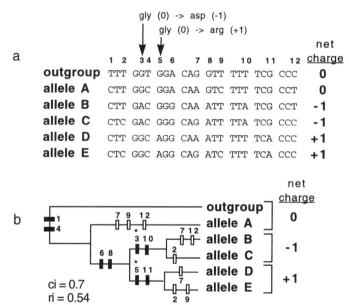

FIGURE 7 Homoplasy at the nucleotide sequence level does not necessarily translate into homoplasy in protein mobility. (a) Six allelic ingroup and one outgroup nucleotide sequences differ at 12 sites. Two of these changes (indicated by arrows) are nonsynonymous substitutions in which an uncharged amino acid, glycine, is replaced either by aspartate (negatively charged) or arginine (positively charged); these result in changes in the net charge of the encoded protein. (b) Allele tree topology, with character changes optimized to show both unique (filled boxes) and homoplastic (open boxes) character transformations; replacement substitutions are marked with asterisks. The higher-order character, "net charge of the encoded protein," (states shown to the right of the allele designations) is a prime determinant of electrophoretic mobility for a native protein and shows no homoplasy.

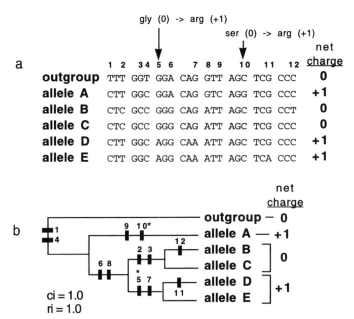

FIGURE 8   Homoplasy in a structural character can occur in the absence of gene tree homoplasy. (a) Six allelic ingroup and one outgroup nucleotide sequences differ at 12 sites. Two different nonsynonymous substitutions (indicated by arrows) replace an uncharged amino acid with a positively charged amino acid (arginine), changing the net charge of the encoded protein. (b) Allele tree topology, labeling as in Fig. 8; there is no homoplasy in this tree at the DNA sequence level. However, two unrelated groups of alleles have acquired a net positive charge in their encoded proteins.

no homoplasy in the gene tree, yet homoplasy can occur for the structural/functional character.

Gene function also requires sequences outside the coding region that are part of the contiguous nucleotide sequence that makes up the "gene" in the molecular sense; these include sequences both at the 3' end (e.g., polyadenylation signal) and, particularly, in the region upstream of the initiation codon. Many of the 5' sequences (e.g., binding sites for transcription factors and polymerase) are involved in the proper regulation and targeted expression of the gene. Physically more distant sequences may or may not be considered part of the gene, but some, including enhancers, can play important roles in gene expression. As with the more obviously structural characters of a gene, expression patterns need not show any direct correspondence to homoplasy in the gene tree. For example, orthology relationships in a gene family may be hypothesized by the standard procedure of constructing a gene tree from coding regions, but the expression patterns of these genes will appear to be homoplastic if recombination between paralogous loci switches upstream regulatory sequences between these paralogs (Fig. 9). In general, then, gene structure and function are closely related aspects of the gene and are simply a class of character that may show patterns of apparent homology that are incongruent with other aspects of that gene's structure (e.g., intron position) or coding region nucleotide sequence. Functional homoplasy seems sufficiently disconnected from gene tree homoplasy as to preclude any prediction of one from the other.

Morphology and Gene Trees

Morphological characters, being influenced by gene function (rather than by sequence alone), should show this same disconnection with gene tree homoplasy. Assume, in the simplest possible case, that a morphological feature is directly determined by the action of a particular gene, and expression of the gene will result in character state "x," while absence of expression results in state "y." Any of several possible mutations can result in state "y": deletion of all or part of the gene, mutations in the 5' regulatory region, substitutions producing termination codons, or replacement substitutions at any of several critical residues. Assuming further that state "x" is plesiomorphic, then state "y" will be homoplastic in taxa acquiring it by different pathways. As with the structural characters already described, this morphological homoplasy can occur in the absence of any homoplasy in the gene tree. Alternatively, the morphological character transformation x → y may be free of homoplasy, despite the occurrence of considerable homoplasy in the gene tree. The greater the complexity of the relationship between morphological character and individual genes, the more disconnected homoplasy of that character will be from homoplasy in any one of the genes involved. If two genes are involved in expression of a morphological character, such that both must be expressed in a particular way to obtain the x → y transformation, then changes at either one can now cause the transformation. Even if there is a correspondence between morphological homoplasy and homoplasy in one gene tree, there is unlikely to be such a correspondence with homoplasy in both. This may seem a trivial point, but it serves to illustrate that, just as moving from the hierarchical level of gene to that of species may cause a disconnection in homoplasy, so may moving from the organizational level of genotype to that of phenotype.

The point here is not to suggest that morphological characters are somehow superior to gene sequence data, but only to refute one simple fallacy: that morphological characters are far *worse* than molecular characters because, often being influenced by more than one gene, they must necessarily sum the homoplasy found in each of several gene trees. This is simply not true, as can be seen from the above discussion. Since it is gene function that is involved, and function can be disconnected from gene tree homoplasy, phenotypic characters are buffered from the homoplasy in either gene. Mor-

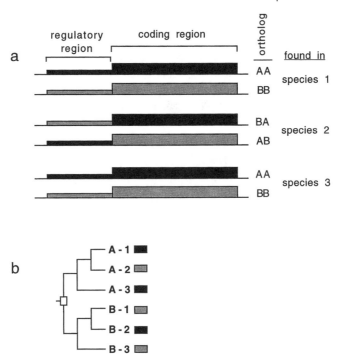

FIGURE 9 Homoplasy in expression patterns caused by recombination among paralogous sequences. (a) A two-member multigene family, found in three species, consists of "A" and "B" paralogous groups. In species 2, a reciprocal recombination has placed the regulatory region of the "B" paralog with the coding region of the "A" paralog, and vice versa. (b) The topology of the gene tree for the multigene family, as constructed from the coding regions of these six sequences, is due to an initial gene duplication (open box) followed by two speciation events (gene designations: A-1 = paralog "A" in species 1, etc.). The expression patterns of these genes, shown by shading of boxes to the right of the gene designations, are determined by the regulatory regions and so do not predict the orthology relationships of the recombinant genes in species 2.

phological characters have, instead, their own sources of homoplasy, some of which are likely to include the gene functional characters described above. For example, a phenotype requiring the expression of a particular target gene could be caused by mutations either in the target gene or in its regulator (Fig. 10); there need be no homoplasy associated with the actual mutations in either gene. The very complexity of developmental networks may introduce homoplasy, but it is not the same homoplasy as that found in the gene tree. This should be taken into account in any discussion of the relative utilities of DNA sequence data vs morphological characters, but, as yet, we still know too little about the molecular basis of morphological character transforma-

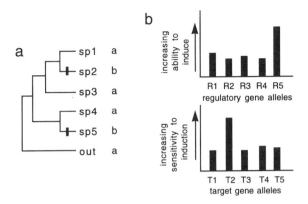

species	regulator allele	target gene allele	target gene expression	phenotype
sp1	R1	T1	-	a
sp2	R2	T2	+	b
sp3	R3	T3	-	a
sp4	R4	T4	-	a
sp5	R5	T5	+	b

FIGURE 10 Homoplasy in a phenotypic character dependent on expression of a regulated target gene. State "b" is produced if the target gene, T, is expressed; in the absence of T expression, state "a" occurs. For T to be expressed, it must be activated by the product of "regulatory" gene R. (a) Alleles of R and T are fixed in four ingroup and one outgroup species, whose relationship is given by the species tree; "a" is the plesiomorphic condition. (b) Products of R alleles differ from one another in their binding affinity for regulatory sites in T. Different alleles of R therefore differ in their ability to trigger expression of T; the product of allele R5 is a particularly strong inducer of T. Alleles of T differ in their sensitivity to R; allele T2 is particularly easily induced. (c) Expression of T therefore depends on the combination of alleles of R and T. A weak inducer (e.g., allele R1) combined with a less sensitive target (e.g., allele T1) results in no expression of T, and hence phenotype "a" is observed. Expression of T is observed where either a strong inducer (R5) or sensitive target (T2) allele is found. Thus, expression of T occurs only in species 2 and 5. These are not sister taxa, and state "b" thus appears as a parallelism in the species tree, but this homoplasy is due to independent mutations in different alleles at different loci. There is no connection between homoplasy in the phenotypic character and homoplasy in either gene R or gene T.

tions among species to generalize as to the sources of morphological homoplasy.

This information is significant because morphological characters could be an important source of data for phylogenetic analysis even in the face of the increasing ease of obtaining DNA sequence data. At the species level, as discussed above, numerous independent characters must be used if species trees are to be reconstructed accurately. Traditionally, morphology has provided the easiest access to such characters, but more recently, species trees increasingly have come to be inferred directly from gene trees. If, however, for the reasons discussed above, the relevant unit molecular character is not the individual nucleotide, but rather the entire locus, whose alleles should be coded as unordered states, then each locus surveyed becomes more directly analogous to a single morphological character. An extreme conclusion from these observations might be that the construction of gene trees is not cost-effective for species tree construction. Even with advances in sequencing technology it requires a major effort to sample many individuals in several different species and construct gene trees for them all at several loci. It may not be until much effort has been expended that particular loci will be found to be unsuitable, for example because polymorphisms transgress what are indisputable evolutionary boundaries. Perhaps morphological characters, for which variation patterns initially can be screened rapidly for many individuals, represent a better investment at the species level.

At the other end of the taxonomic spectrum, the problems with stochastically evolving molecular characters were noted some years ago by Lanyon (1988), who argued that adaptive morphological features may be more likely to be preserved and not be obscured by superimposed mutations, making them superior markers for ancient lineage divergences. The problem is that adaptively significant characters may evolve independently in more than one group. This, however, is presumably as much a problem for individual amino acid substitutions in particular genes as it is for morphological characters. The very complexity of morphological characters, relative to simple nucleotide substitutions, may be beneficial, in affording additional criteria for homology assessment (see McShea, 1996). There is nothing novel about the suggestion that increasingly precise character analysis may lead to discrimination between homoplasy and homology. However, as more is learned about the molecular underpinnings of morphological change, new criteria should be available to the systematist. For example, expression patterns and orthology relationships in the multigene families involved in nodulation (symbiotic fixation of atmospheric nitrogen) in the plant family Leguminosae may provide criteria for assessing the homologies of the syndrome, which appears most parsimoniously to have evolved several times in that family (Doyle 1994b). Knowledge about the molecular sources of homoplasy in morphological characters will be important in determining if or when molecular developmental genetic tools will be cost-effective in character analysis of morphological attributes (Doyle 1994a).

CONCLUSIONS

In one sense, homoplasy is simply homoplasy: character incongruence in a phylogenetic analysis. However, it seems important to seek to understand its many possible sources in such an analysis. Apart from irreducible sources of false homology, such as parallel nucleotide substitutions, there is also homoplasy due to mixing different historical signals. When a radio is poorly tuned, and is picking up two stations simultaneously, the solution is not simply to turn up the volume. Yet that may be analogous to trying to resolve a phylogeny by obtaining more sequence data and performing total evidence analyses when homoplasy is largely due to mixed phylogenetic signals. The other extreme is to assume that all data sets must be kept separate from one another. Between the Scylla and Charybdis lies the promise of using the patterns of homoplasy to diagnose the sources of incongruence and understand the biology involved. In this regard, it is important to recognize that crossing hierarchical or organizational levels can have profound effects on the meaning and amount of homoplasy. For both the gene tree/species tree and the molecules/morphology contrasts, it will often be very difficult to predict homoplasy at one level from that at another.

ACKNOWLEDGMENTS

The number of people who have (for better or worse) shaped the thoughts expressed here is too large for individual listing, but primary among them are my Hortorium colleagues Jerry Davis (whose encouragement upon reading an early draft is much appreciated), Kevin Nixon, Melissa Luckow, and my former students Anne Bruneau and John Freudenstein. I thank the editors for helpful comments in review and acknowledge NSF funding, most recently from Grants DEB 91-07480 and 94-20215.

REFERENCES

Anderson, E. 1949. Introgressive hybridization. Wiley, New York.
Antonenko, S., and M. Zanetti. 1994. Production and characterization of cross-reactive monoclonal antibodies to sweet proteins. Life Sciences 55:1187–1192.
Baldwin, B. G., M. J. Sanderson, J. M. Porter, M. F. Wojciechowski, C. S. Campbell, and M. J. Donoghue. 1995. The ITS regions of nuclear ribosomal DNA: A valuable source of evidence on angiosperm phylogeny. Annals of the Missouri Botanical Garden 82:247–277.
Baum, B. R. 1992. Combining trees as a way of combining data sets for phylogenetic inference and the desirability of combining gene trees. Taxon 41:3–10.
Baum, D. A., and K. L. Shaw. 1995. Genealogical perspectives on the species problem. Pp. 289 –303 in P. C. Hoch, A. G. Stevenson, and B. A. Schaal, eds. Experimental and molecular approaches to plant biosystematics. Missouri Botanical Garden, St. Louis, MO.
Bull, J. J., J. P. Huelsenbeck, C. W. Cunningham, D. L. Swofford, and P. J. Waddell. 1993. Partitioning and combining data in phylogenetic analysis. Systematic Biology 42:384–387.
Chase, M. W., D. E. Soltis, R. G. Olmstead, D. Morgan, D. H. Les, B. D. Mishler, M. R. Duvall, R. A. Price, H. G. Hills, Y.-L. Qiu, K. A. Kron, J. H. Rettig, E. Conti, J. D. Palmer, J. R. Manhart, K. J. Sytsma, H. J. Michaels, W. J. Kress, K. J. Karol, W. D. Clark, M. Hedrén, B. S. Gaut, R. K. Jansen, K.-J. Kim, C. F. Wimpee, J. F. Smith, G. R. Furnier, S. H. Strauss, Q.-Y. Xiang, G. M. Plunkett, P. S. Soltis, S. M. Swensen, S. E. Williams, P. A. Gadek, C. J.

Quinn, L. E. Eguiarte, E. Golenberg, G. H. Learn, Jr., S. C. H. Barrett, S. Dayanandan, and V. A. Albert. 1993. Phylogenetics of seed plants: An analysis of nucleotide sequences from the plastid gene *rbcL*. Annals of the Missouri Botanical Garden 40:528–580.

Clark, A. G., and T. H. Kao. 1991. Excess nonsynonymous substitution at shared polymorphic sites among self-incompatibility alleles of Solanaceae. Proceedings of the National Academy of Sciences USA 88:9823–9827.

Davis, J. I., and K. C. Nixon. 1992. Populations, genetic variation, and the delimitation of the phylogenetic species. Systematic Biology 41:421–435.

de Queiroz, A. 1993. For consensus (sometimes). Systematic Biology 42:368–372.

Donoghue, M. J. 1994. Progress and prospects in reconstructing plant phylogeny. Annals of the Missouri Botanical Garden 81:405–418.

Doyle, J. J. 1992. Gene trees and species trees: Molecular systematics as one-character taxonomy. Systematic Botany 17:144–163.

Doyle, J. J. 1994a. Evolution of a plant homeotic multigene family: Toward connecting molecular systematics and molecular developmental genetics. Systematic Biology 43:307–328.

Doyle, J. J. 1994b. Phylogeny of the legume family: An approach to understanding the origins of nodulation. Annual Review of Ecology and Systematics 25:325–349.

Doyle, J. J. 1995. The irrelevance of allele tree topologies for species delimitation, and a nontopological alternative. Systematic Botany 20:574–588.

DuBose, R. F., D. E. Dykhuizen, and D. L. Hartl. 1988. Genetic exchange among natural isolates of bacteria: Recombination within the *pho-A* gene of *Escherichia coli*. Proceedings of the National Academy of Sciences USA 85:7036–7040.

Dwyer, K. G., M. A. Balent, J. B. Nasrallah, and M. E. Nasrallah. 1991. DNA sequences of self-incompatibility genes from *Brassica campestris* and *B. oleracea*: Polymorphism predating speciation. Plant Molecular Biology 16:481–486.

Dykhuizen, D. E., and L. Green. 1991. Recombination in *Escherichia coli* and the definition of biological species. Journal of Bacteriology 173:7257–7268.

Felsenstein, J. 1978. Cases in which parsimony or compatibility methods will be positively misleading. Systematic Zoology 27:401–410.

Felsenstein, J. 1988. Phylogenies from molecular sequences: Inference and reliability. Annual Review of Genetics 22:521–566.

Fitch, D. H. A., and M. Goodman. 1991. Phylogenetic scanning: a computer-assisted algorithm for mapping gene conversions and other recombinational events. Computer Applications in the Biosciences 7: 207–216.

Fitch, W. M. 1970. Distinguishing homologous from analogous proteins. Systematic Zoology 19:99–113.

Freudenstein, J. V., and J. J. Doyle. 1994. Character transformation and evolution in *Corallorhiza* (Orchidaceae: Epidendroideae). I. Plastid DNA. American Journal of Botany 81: 1458–1467.

Gatesy, J., R. Desalle, and W. C. Wheeler. 1993. Alignment-ambiguous nucleotide sites and the exclusion of systematic data. Molecular Phylogenetics and Evolution 2:152–157.

Gaur, L. K., A. L. Hughes, E. R. Heise, and J. Gutknecht. 1992. Maintenance of DQB1 polymorphisms in primates. Molecular Biology and Evolution 9:599–609.

Hamby, R. K., and E. A. Zimmer. 1992. Ribosomal RNA as a phylogenetic tool in plant systematics. Pp. 50–91 *in* P. S. Soltis, D. E. Soltis, and J. J. Doyle, eds. Molecular systematics of plants. Chapman and Hall, New York.

Hein, J. 1993. A heuristic method to reconstruct the history of sequences subject to recombination. Journal of Molecular Evolution 36:396–405.

Hennig, W. 1966. Phylogenetic systematics. Univ. of Illinois Press, Urbana.

Hudson, R. R. 1990. Gene genealogies and the coalescent process. Oxford Surveys in Evolutionary Biology 7:1–44.

Ioerger, T. R., A. G. Clark, and T. H. Kao. 1990. Polymorphism at the self-incompatibility locus of Solanaceae predates speciation. Proceedings of the National Academy of Sciences USA 87:9732–9735.

Jamet, E., Y. Parmentier, A. Durr, and J. Fleck. 1991. Genes encoding the small subunit of rubisco belong to two highly conserved families in Nicotianeae. Journal of Molecular Evolution 33:226–236.

Kluge, A. G. 1989. A concern for evidence and a phylogenetic hypothesis of relationships among *Epicrates* (Boidae, Serpentes). Systematic Zoology 38:7–25.

Lanyon, S. M. 1988. The stochastic mode of molecular evolution: What consequences for systematic investigations? Auk 105:563–573.

Levinson, G., and G. A. Gutman. 1987. Slipped-strand mispairing: A major mechanism for DNA sequence evolution. Molecular Biology and Evolution 4:203–221.

Lloyd, D. G., and V. L. Calder. 1991. Multi-residue gaps, a class of molecular characters with exceptional reliability for phylogenetic analyses. Journal of Evolutionary Biology 4:9–21.

Meagher, R. B., S. Berry-Lowe, and K. Rice. 1989. Molecular evolution of the small subunit of ribulose bisphosphate carboxylase: Nucleotide substitution and gene conversion. Genetics 123:845–863.

McShea, D. W. 1996. Complexity and homoplasy. Pp. 207–225 in M. J. Sanderson and L. Hufford, eds. Homoplasy: The recurrence of similarity in evolution. Academic Press, San Diego.

Mickevich, M. F., and J. S. Farris. 1981. The implications of congruence in *Menidia*. Systematic Zoology 30:351–370.

Mindell, D. P. 1991. Aligning DNA sequences: Homology and phylogenetic weighting. Pp. 73–89 in M. M. Miyamoto and J. Cracraft, eds. Phylogenetic analysis of DNA sequences. Oxford Univ. Press, New York.

Miyamoto, M. M., and W. M. Fitch. 1995. Testing species phylogenies and phylogenetic methods with congruence. Systematic Biology 44:64–76.

Nei, M. 1987. Molecular evolutionary genetics. Columbia Univ. Press, New York.

Neigel, J. E., and J. C. Avise. 1986. Phylogenetic relationships of mitochondrial DNA under various demographic models of speciation. Pp. 515–534 in S. Karlin and E. Nevo, eds. Evolutionary processes and theory. Academic Press, New York.

Nijhout, H. F. 1990. Metaphors and the role of genes in development. BioEssays 12:441–446.

Nixon, K. C., and J. I. Davis. 1991. Polymorphic taxa, missing values, and cladistic analysis. Cladistics 7:233–242.

Nixon, K. C., and Q. D. Wheeler. 1992. Extinction and the origin of species. Pp. 119–143 in M. J. Novacek and Q. D. Wheeler, eds. Extinction and phylogeny. Columbia Univ. Press, New York.

Olmstead, R. G., and J. D. Palmer. 1994. Chloroplast DNA systematics: A review of methods and data analysis. American Journal of Botany 81:1205–1224.

Olmstead, R. G., and J. A. Sweere. 1994. Combining data in phylogenetic systematics: An empirical approach using three molecular data sets in the Solanaceae. Systematic Biology 43:467–481.

Pamilo, P., and M. Nei. 1988. Relationships between gene trees and species trees. Molecular Biology and Evolution 5:568–583.

Patterson, C. 1982. Morphological characters and homology. Pp. 21–74 in K. A. Joysey and A. E. Friday, eds. Problems of phylogenetic reconstruction. Academic Press, London.

Platnick, N. I., C. E. Griswold, and J. A. Coddington. 1991. On missing entries in cladistic analysis. Cladistics 7:337–344.

Ragan, M. A. 1992. Phylogenetic inference based on matrix representation of trees. Molecular Phylogenetics and Evolution 1:53–58.

Rieger, R. 1991. Glossary of genetics: Classical and molecular. Springer-Verlag, New York.

Roth, V. L. 1988. The biological basis of homology. Pp. 1–26 in C. J. Humphries, ed. Ontogeny and systematics. Columbia Univ. Press, New York.

Roth, V. L. 1991. Homology and hierarchies: Problems solved and unresolved. Journal of Evolutionary Biology 4:167–194.

Sanderson, M. J. 1991. In search of homoplastic tendencies: Statistical inference of topological patterns in homoplasy. Evolution 45:351–358.

Sanderson, M. J., and M. J. Donoghue. 1989. Patterns of variation in levels of homoplasy. Evolution 43:1781–1795.

Sanderson, M. J., and J. J. Doyle. 1992. Reconstruction of organismal phylogenies from multigene families: Paralogy, concerted evolution, and homoplasy. Systematic Biology 41: 4–17.

Sawyer, S. 1989. Statistical tests for detecting gene conversion. Molecular Biology and Evolution 6:526–538.

Shah, D. M., R. C. Hightower, and R. B. Meagher. 1983. Genes encoding actins in higher plants: Intron positions are highly conserved but the coding regions are not. Journal of Molecular and Applied Genetics 2:111–126.

Steel, M. A., P. J. Lockhart, and D. Penny. 1993. Confidence in evolutionary trees from biological sequence data. Nature 364:440–442.

Suh, Y., L. B. Thien, H. E. Reeve, and E. A. Zimmer. 1993. Molecular evolution and phylogenetic implications of internal transcribed spacer sequences of ribosomal DNA in Winteraceae. American Journal of Botany 80:1042–1055.

Swofford, D. L., and G. J. Olsen. 1990. Phylogeny reconstruction. Pages 411–501 *in* D. M. Hillis and C. Moritz, eds. Molecular systematics. Sinauer, Sunderland, MA.

Templeton, A. R., and C. F. Sing. 1993. A cladistic analysis of phenotypic associations with haplotypes inferred from restriction endonuclease mapping. IV. Nested analyses with cladogram uncertainty and recombination. Genetics 134:659–669.

Thorne, J. L., H. Kishino, and J. Felsenstein. 1992. Inching toward reality: An improved likelihood model of sequence evolution. Journal of Molecular Evolution 34:3–16.

Vrana, P., and W. Wheeler. 1992. Individual organisms as terminal entities: Laying the species problem to rest. Cladistics 8:67–72.

Wheeler, W. C., and D. S. Gladstein. 1994. MALIGN: A multiple sequence alignment program. Journal of Heredity 85:417–418.

Wheeler, W. C., J. Gatesy, and R. DeSalle. 1995. Elision: A method for accommodating multiple molecular sequence alignments with alignment-ambiguous sites. Molecular Phylogenetics and Evolution 4:1–9.

THE RELATIONSHIP BETWEEN HOMOPLASY AND CONFIDENCE IN A PHYLOGENETIC TREE

MICHAEL J. SANDERSON* AND MICHAEL J. DONOGHUE†

*University of California
Davis, California

†Harvard University
Cambridge, Massachusetts

INTRODUCTION

When the term "homoplasy" was coined by Lankester (1870), the evolutionary process was central:

> When identical or nearly similar forces, or environments, act on . . . parts in two organisms, which parts are exactly or nearly alike and sometimes homogenetic [homologous], the resulting correspondences called forth in the several parts in the two organisms will be nearly or exactly alike . . . I propose to call this form of agreement *homoplasis* or *homoplasy*. [p. 39]

However, the phylogenetic implications of this process were not lost on him:

> Zoology has for some time been embarrassed with the reference of all segmented Invertebrata to a common type, and the supposed homology of their segmented structures. This difficulty may, it is suggested, be possibly solved by the admission of true zooid-segmentation as being frequently due to homoplasy, and not by any means necessarily an indication of genetic affinity. [p. 43]

HOMOPLASY: The Recurrence of Similarity in Evolution, M. J. Sanderson and L. Hufford, eds.

67

To many phylogeneticists homoplasy is now viewed only as "error in homology assessment." Although this restricted definition tends to ignore the profound impact that homoplasy has on the generation of organismal diversity, the emphasis is appropriate when accurate phylogeny reconstruction is the central concern. Any source of error must be characterized, and, if possible, ameliorated. Surprisingly, little attention has been paid to a subtle but basic aspect of this "error": even if homoplasy is a mistaken hypothesis of homology for an individual character, does it generally impede efforts to reconstruct phylogenies using sets of many independent characters?

Because it is easy to imagine simple cases, such as a single character, in which homoplasy leads to a mistaken idea of relationships, it is tempting to generalize and equate *overall* levels of homoplasy with confidence in a tree. Such generalizations have been common. For many years, levels of homoplasy, as encapsulated in the consistency index (Kluge and Farris 1969), were routinely reported in phylogenetic studies, often accompanied by ad hoc explanations for low values (reviewed in Sanderson and Donoghue 1989; Donoghue and Sanderson 1992). More recently, surveys of consistency indices have served as the basis for suggestions that high CI's support the superiority of one kind of data over another. For example, Givnish and Sytsma (1992) and Jansen (1995) have both suggested that chloroplast restriction site data are better than sequence data for plant phylogenetics based at least in part on relative levels of homoplasy (Jansen also considered several other variables in addition to homoplasy). Clearly, many investigators view the level of homoplasy as a valuable indicator of the quality of the phylogenetic conclusions derived based on a data set.

Another reason that some workers have assumed a close relationship between homoplasy and confidence is the close relationship between rates of evolution and confidence. Analytical (Felsenstein 1978) and simulation studies (see Archie 1996 for a review) have generally agreed that it is more difficult to estimate trees accurately when rates of evolution are high, and rates of evolution ought generally to be correlated with levels of homoplasy. However, few studies have considered the distribution of homoplasy as a factor independent of rate (although see Landrum 1993; Archie 1996), and it is possible that rates, homoplasy, and confidence are intercorrelated in a way that masks the underlying relationships among these variables.

Critical analyses of the consistency index (Sanderson and Donoghue 1989; Archie 1989; Klassen et al. 1991; Meier et al. 1991) have prompted some reevaluation of the difference between homoplasy and "informativeness" and perhaps "accuracy" (Goloboff 1991a,b). In this paper we examine such issues further by considering the relationship, if any, between levels of homoplasy in data sets and confidence in the phylogenetic tree(s) estimated from those data. We begin where our previous study of levels of homoplasy (Sanderson and Donoghue, 1989) left off—with an examination of homoplasy in relation to an independent measure of the robustness of a tree.

Some authors have examined robustness in terms of the accuracy or "nearness" of the estimated tree to the true tree in simulation studies in which the true tree is known. Under those conditions the correspondence between the estimated tree and the true tree, as measured by a consensus index, for example, is obviously a reasonable indication of accuracy. However, we propose to examine a large set of real (not simulated) phylogenetic data sets in which the true tree is unknown. We hope that the disadvantages of an unknown underlying tree are offset by the generality entailed by the diversity of evolutionary processes likely to be uncovered in any large sample of real data sets.

Because we do not know the true tree in any real data set, we focus on "confidence," rather than accuracy. The bootstrap procedure (Felsenstein 1985; Sanderson 1989, 1995; Hillis and Bull 1993) is the most widely used statistical assessment of confidence in a phylogenetic tree. It aims to provide an indication of the size of the neighborhood of trees that are similar to the true tree. Specifically, the bootstrap attempts to estimate the effects of sampling error on phylogenetic inference. It aims to predict the probability that further sampling of characters will support clades discovered in the initial analysis. It is therefore highly dependent on the number of synapomorphies of a clade, for example—the idea being that a clade in which many synapomorphies have already been discovered is probably a clade in which there are also more undiscovered synapomorphies. The bootstrap has been criticized for many reasons (reviewed in Sanderson 1995)—some philosophical, some statistical, and some empirical (e.g., Cummings et al. 1995)—but its utility, or at least the perception of its utility, remains high. Some nonstatistical measures, such as the decay index (Bremer 1988; Donoghue et al. 1992) are of no utility for comparative purposes, because their magnitudes are data set dependent.

In principle, the bootstrap is decoupled from homoplasy. A data set with just one informative character (but several uninformative ones) will automatically have no apparent homoplasy (CI = 1.0) but its bootstrap value will be very low because the probability of future sampling overturning the relationship suggested by any single character is high. Conversely, it is not difficult to construct data matrices that have fairly low CIs but have high bootstrap values, if the homoplasy is dispersed randomly around a tree and is not correlated across characters. See Sanderson and Donoghue (1989: 1789) for an example.

We use the consistency index (CI: Kluge and Farris 1969) as a measure of homoplasy. It and the retention index (RI: Farris 1989) (see Archie 1996) are the most commonly cited statistics reported in phylogenetic studies. Both measure aspects of the goodness of fit of a data set to a hierarchical tree structure. However, CI is a better measure of overall homoplasy, because RI factors in the distribution of apomorphies among taxa. RI therefore reflects something in addition to homoplasy which is related to the probability that the data could actually be expected to exhibit a certain level of homoplasy given the distribution of character states among taxa. The difference can be seen in a simple example. If 2 of 10 taxa have the apomorphic (derived) state for a binary character, CI = 0.50 if the states are parallelisms, but RI = 0.00. If 4 of 10 taxa have the derived state, then CI = 0.25 but still RI = 0.00 when all four originations are independent. Clearly these two cases represent very different levels of homoplasy, but their retention indices are the same. Goloboff (1991b) suggests that CI provides more of an indication about homoplasy than it does about "informativeness" about relationships, implying that the two issues are logically separable.

As in our previous analysis (Sanderson and Donoghue 1989), we consider a set of phylogenetic analyses of real taxa taken from the primary literature. We use observational statistics to make inferences about the relationships between levels of homoplasy and bootstrap estimates of confidence. Each "observation" therefore represents an individual study sampled from a "population" of studies. This approach is not quite meta-analysis (Arnqvist and Wooster 1995). Meta-analysis, which is widely used to extract patterns from multiple studies, tests a hypothesis across a set of studies that is identical to a hypothesis tested (possibly by different methods) in each study. In the present analysis there is no "hypothesis" common to all studies. Instead we seek information about the relationships of variables in the population of studies. For example, is data set size correlated with robustness of the resulting trees?

However, several problems in true meta-analyses could be problematic here. Sometimes meta-analysis is applied to studies so different from each other that the only thing they share in common is a significance level (the "apples and oranges problem"). Phylogenetic analyses, however, share a fundamentally similar methodology. They use a data matrix of taxa by characters. In our study the characters are all discrete. Continuous characters and distance date are excluded. Phylogenetic studies generally employ some optimization strategy to estimate a phylogenetic tree. In our study maximum parsimony is the only algorithm used. Moreover, because the same algorithms and software are used, precisely the same set of statistics are gathered on each study.

Another issue in meta-analysis is the "file-drawer" problem, a bias introduced because studies that fail to find significance are rarely reported in the literature (they are relegated to the file drawer). However, perhaps because of the systematics community's collective uncertainty about how to assign significance to phylogenetic trees, systematists seem unlikely to fail to publish because of lack of significance. Only one journal that publishes many phylogenetic studies, *Molecular Biology and Evolution*, encourages as part of its editorial policy the reporting of significance levels (although reviewers of other journals certainly may encourage it), and we suspect that a significant number of systematists regards the reporting of bootstrap values as positively misguided (e.g., Wendel and Albert 1992).

MATERIALS AND METHODS

A set of 101 phylogenetic studies (Table 1) was extracted from the TreeBASE data base (WWW URL http: //phylogeny.harvard.edu/treebase). A summary of the characteristics of these data sets is provided in Table 2. They included 50 morphological, 29 RFLP, and 22 DNA sequence data sets, ranging in size from 5 to 68 taxa and from 10 to 2226 characters. All studies were on green plants.

Each data set was subjected to parsimony analysis using PAUP 3.1 (Swofford 1993). Heuristic search options consisted of simple addition sequence ("hold" set to 5) and "TBR" branch swapping with "maxtrees" set to 250. The following tree statistics were recorded for each run: consistency index (CI), retention index (RI), number of most parsimonious trees (Ntrees), and degree of resolution of one of the most parsimonious trees (arbitrarily, the first one, if more than one was found). Degree of resolution was measured by the normalized consensus fork index, which reports the fraction of possible clades that actually occur on a tree of given size (Swofford 1991). All runs were performed on the same computer.

Each data set was then bootstrapped (100 replicates) using the same set of heuristic search options. The list of clades appearing among the bootstrap replicates was saved and later examined by a UNIX shell script to summarize findings. For each complete bootstrap analysis, two statistics were extracted: (1) the maximum bootstrap level observed in the tree (B_{max}) and (2) the fraction of clades supported at the level of greater than 50% (B_{50}). Note that B_{50} is really the consensus fork index of the majority rule consensus tree (Swofford 1991). These two statistics convey different aspects of overall support for a tree. Neither of them is ideal, and other measures could be used, such as the average bootstrap value computed over all clades in the strict consensus tree (Jansen and Wee, personal communication), or Bremer's (1994) total support index. Of the two measures reported here, the B_{50} value should give a better overall summary of support but is influenced by the average degree of resolution possible with the given set of characters (i.e., low resolution in the most parsimonious trees probably will also lead to lower values of B_{50}. The B_{max} statistic, on the other hand, should be less sensitive to resolution and the distribution of character support among clades, but cannot discriminate among overall support when it is fairly high. There are many data sets for which B_{max} is 1.0.

Statistical analyses were performed using JMP version 3 (SAS Institute, Inc.). A matrix of partial correlation coefficients among all the variables is

TABLE 1   Raw Data from an Analysis of 101 Phylogenetic Data Sets[a]

Ref.	Data	No. characters	No. taxa	CI	RI	No. trees	Resolution (CFI)[b]	B_{50}	B_{max}
1	Morph	32	40	0.415	0.721	36	0.919	0.205	1.000
2	Morph	28	14	0.783	0.811	5	1.000	0.308	1.000
3	Morph	46	44	0.459	0.776	250	0.927	0.279	1.000
4	Morph	23	14	0.644	0.724	1	0.909	0.462	0.818
5	Morph	29	23	0.433	0.600	4	0.950	0.091	0.673
6	Morph	10	7	0.833	0.818	6	0.750	0.333	0.943
7	Morph	40	15	0.742	0.734	250	0.833	0.071	0.948
8	Morph	22	13	0.931	0.875	1	0.800	0.250	0.993
9	Morph	22	15	0.750	0.843	18	0.917	0.500	0.815
10	Morph	23	11	0.711	0.780	7	1.000	0.500	0.980
11	Morph	24	21	0.491	0.746	155	0.889	0.200	0.700
12	Morph	45	30	0.741	0.851	117	0.963	0.586	1.000
13	Morph	20	22	0.839	0.896	6	0.632	0.286	0.972
14	Morph	15	6	0.789	0.750	2	0.667	0.400	0.910
15	Morph	28	14	0.481	0.604	18	0.909	0.308	0.674
16	Morph	34	20	0.600	0.730	8	1.000	0.316	0.718
17	Morph	17	5	0.909	0.714	1	1.000	0.500	0.830
18	Morph	29	20	0.688	0.826	2	0.941	0.632	0.928
19	Morph	36	11	0.873	0.850	2	1.000	0.700	0.958
20	Morph	13	18	0.542	0.780	24	0.533	0.176	0.890
21	Morph	49	18	0.545	0.658	93	1.000	0.294	0.982
22	Morph	63	13	0.585	0.700	18	1.000	0.583	0.990
23	Morph	15	11	1.000	1.000	15	0.750	0.300	0.688
24	Morph	21	9	0.846	0.875	1	1.000	0.750	0.936
25	Morph	46	23	0.465	0.663	55	1.000	0.227	0.720
26	Morph	25	11	0.617	0.647	1	1.000	0.500	0.989
27	Morph	48	19	0.552	0.566	1	1.000	0.333	0.980
28	Morph	17	19	0.774	0.868	35	0.688	0.500	0.960
29	Morph	25	9	0.750	0.714	4	1.000	0.500	0.970
30	Morph	70	14	0.507	0.538	1	1.000	0.692	0.957
31	Morph	42	17	0.769	0.829	66	0.714	0.375	0.966
32	Morph	88	56	0.398	0.663	250	0.906	0.182	0.870
33	Morph	29	12	0.938	0.944	26	0.778	0.545	0.990
34	Morph	28	11	0.630	0.595	11	1.000	0.200	0.888
35	Morph	39	8	0.565	0.565	2	1.000	0.571	0.901
36	Morph	56	40	0.679	0.892	18	0.919	0.512	0.960
37	Morph	27	14	0.658	0.729	24	0.818	0.462	0.789
38	Morph	27	11	0.644	0.754	2	1.000	0.500	0.801
39	Morph	19	5	0.909	0.833	1	1.000	0.500	1.000
40	Morph	36	6	0.944	0.905	2	1.000	0.400	1.000
41	Morph	36	23	0.529	0.721	99	1.000	0.364	0.989
42	Morph	28	16	0.595	0.630	6	1.000	0.400	0.702
43	Morph	15	6	0.864	0.625	2	1.000	0.400	0.704
44	Morph	92	9	0.579	0.451	1	1.000	0.375	0.880
45	Morph	55	19	0.625	0.814	2	0.875	0.500	0.990
46	Morph	16	9	0.889	0.923	1	1.000	0.750	0.980

(continues)

TABLE 1—*Continued*

Ref.	Data	No. characters	No. taxa	CI	RI	No. trees	Resolution (CFI)[b]	B_{50}	B_{max}
47	Morph	23	16	0.487	0.557	4	1.000	0.133	0.560
48	Morph	35	14	0.631	0.730	2	1.000	0.846	0.960
49	Morph	60	24	0.301	0.553	12	1.000	0.174	0.995
50	Morph	24	15	0.617	0.727	7	0.917	0.429	0.941
51	Seq	118	43	0.674	0.851	8	0.650	0.452	1.000
52	Seq	671	24	0.798	0.901	21	0.857	0.609	1.000
53	Seq	82	6	0.689	0.367	1	1.000	0.400	0.663
54	Seq	17	8	0.758	0.692	1	1.000	0.571	0.959
55	Seq	547	23	0.623	0.458	1	1.000	0.318	1.000
56	Seq	86	12	0.625	0.674	1	1.000	0.727	1.000
57	Seq	358	12	0.854	0.806	3	1.000	0.727	0.980
58	Seq	88	12	0.515	0.470	4	1.000	0.545	1.000
59	Seq	174	18	0.580	0.663	2	1.000	0.411	1.000
60	Seq	631	22	0.768	0.776	250	0.947	0.524	1.000
61	Seq	138	31	0.615	0.757	48	0.821	0.433	1.000
62	Seq	718	19	0.622	0.714	6	0.875	0.722	1.000
63	Seq	1428	13	0.788	0.716	4	1.000	0.750	1.000
64	Seq	151	28	0.751	0.843	8	1.000	0.741	1.000
65	Seq	154	19	0.570	0.635	27	0.938	0.333	1.000
66	Seq	637	22	0.737	0.755	2	1.000	0.619	1.000
67	Seq	2167	33	0.559	0.434	2	1.000	0.656	1.000
68	Seq	720	41	0.550	0.729	3	1.000	0.800	1.000
69	Seq	486	9	0.923	0.942	1	1.000	0.750	1.000
70	Seq	2226	24	0.648	0.652	24	1.000	0.609	1.000
71	Seq	1428	21	0.615	0.558	31	1.000	0.400	1.000
72	Seq	270	28	0.636	0.749	3	0.960	0.111	1.000
73	Seq	652	35	0.752	0.654	250	0.906	0.529	1.000
74	RFLP	161	33	0.459	0.781	6	1.000	0.688	1.000
75	RFLP	82	68	0.562	0.892	250	0.708	0.433	1.000
76	RFLP	58	24	0.784	0.957	5	0.667	0.435	1.000
77	RFLP	65	11	0.755	0.835	4	0.875	0.300	1.000
78	RFLP	311	66	0.475	0.709	250	0.840	0.308	1.000
79	RFLP	43	12	0.872	0.935	1	1.000	0.818	1.000
80	RFLP	21	42	0.957	0.994	48	0.282	0.244	1.000
81	RFLP	104	6	0.889	0.917	1	1.000	0.600	1.000
82	RFLP	245	23	0.573	0.729	8	1.000	0.636	1.000
83	RFLP	110	27	0.894	0.845	6	0.625	0.538	1.000
84	RFLP	56	20	0.862	0.962	8	0.765	0.526	1.000
85	RFLP	45	32	0.381	0.746	6	0.724	0.290	0.970
86	RFLP	74	10	0.935	0.889	3	0.857	0.555	1.000
87	RFLP	115	16	0.846	0.868	19	0.615	0.400	1.000
88	RFLP	38	27	0.844	0.951	250	0.500	0.346	0.990
89	RFLP	29	11	0.967	0.957	1	0.500	0.400	1.000
90	RFLP	72	14	0.673	0.670	1	1.000	0.846	0.970
91	RFLP	44	33	0.740	0.949	1	0.533	0.437	1.000
92	RFLP	53	14	0.855	0.938	2	0.818	0.692	1.000

(continues)

TABLE 1—*Continued*

Ref.	Data	No. characters	No. taxa	CI	RI	No. trees	Resolution (CFI)[b]	B_{50}	B_{max}
93	RFLP	128	46	0.888	0.941	250	0.465	0.378	0.990
94	RFLP	91	45	0.561	0.834	250	0.810	0.432	0.990
95	RFLP	55	10	0.902	0.926	1	1.000	0.777	1.000
96	RFLP	194	9	0.975	0.970	1	0.667	0.500	1.000
97	RFLP	26	9	1.000	1.000	1	0.333	0.250	1.000
98	RFLP	72	38	0.459	0.824	64	0.829	0.541	1.000
99	RFLP	10	23	1.000	1.000	1	0.250	0.227	0.980
100	RFLP	30	10	0.762	0.792	1	0.857	0.333	0.900
101	RFLP	25	40	0.962	0.985	2	0.270	0.256	0.940

[a] See Appendix for references to studies.
[b] Normalized consensus fork index

provided in Table 3. These indicate the association among pairs of variables when all other variables are held constant. Significance tests for the matrix of partial correlations were not reported owing to multiple test issues (Sokal and Rohlf 1981). Instead, multiple regression was used. Principal component analysis was used to summarize variation patterns in the eight-dimensional space of the bootstrap indices plus the data set parameters (Table 4 and Fig. 1). Least squares multiple regression (Tables 5 and 6) of B_{50} and B_{max} against all the other variables simultaneously was used to examine the effect of the other variables on these measures of confidence. Finally, multiple regression was also used to repeat the analysis of Sanderson and Donoghue (1989), which examined the effect of these variables on the consistency index (Table 7). Unlike in Sanderson and Donoghue (1989) no attempt was made to factor out the effect that autapomorphies tend to inflate CI. The RI provides an index that is corrected for this (see Archie 1996), and a strong case can be made that an index of homoplasy per se, such as the CI, should in fact include autapomorphies (Goloboff 1991b).

RESULTS

The summary statistics reveal several interesting differences among the three different kinds of data. RFLP data sets have the highest average CIs and RIs, the largest number of most parsimonious trees, and the lowest average level of resolution in any given minimal tree. Sequence data sets have the highest bootstrap support levels for the B_{50} statistic, and they are about tied with RFLP studies for the B_{max} statistic. They also have the greatest resolution. Morphological data sets have the lowest bootstrap support, but their reten-

TABLE 2   Summary Statistics Classified by Type of Data

Statistic	Morphological	RFLP	DNA sequence
N	50	29	22
No. of characters			
Mean	33.8	86.8	606.4
Minimum	10	10	17
Maximum	92	311	2226
No. of taxa			
Mean	16.8	25.9	21.8
Minimum	5	6	6
Maximum	56	68	43
Consistency index (CI)			
Mean	0.67	0.77	0.68
Minimum	0.30	0.38	0.51
Maximum	1.00	1.00	0.92
Retention index (RI)			
Mean	0.74	0.88	0.69
Minimum	0.45	0.67	0.36
Maximum	1.0	1.0	0.94
No. of minimal trees			
Mean	33.5	49.9	30.5
Minimum	1	1	1
Maximum	250[a]	250[a]	250[a]
Resolution (CFI)[b]			
Mean	0.92	0.72	0.95
Minimum	0.53	0.25	0.65
Maximum	1.00	1.00	1.00
B_{50}			
Mean	0.41	0.48	0.55
Minimum	0.07	0.22	0.11
Maximum	0.85	0.84	0.80
B_{max}			
Mean	0.90	0.99	0.98
Minimum	0.56	0.90	0.66
Maximum	1.00	1.00	1.00

[a] Maximum set by MAXTREES option during PAUP runs.
[b] Normalized consensus fork index (see Swofford 1991).

tion indices are higher than those for sequence data sets. These findings hint at complex and subtle differences among different classes of data, which sample size may not have allowed us to detect in earlier work (Sanderson

TABLE 3 *Partial Correlation Coefficients for Variables Surveyed*

Variable	CI	No. characters	No. taxa	No. trees	B_{max}	B_{50}	RI	Resolution (CFI)
CI	0.308	−0.679	0.187	−0.007	0.170	0.619	−0.367
NChars		0.325	−0.082	0.195	0.197	−0.463	0.046
Ntaxa			0.546	0.116	0.077	0.361	−0.394
Ntrees				0.033	−0.234	−0.039	0.206
B_{max}					0.303	0.202	−0.076
B_{50}						0.202	0.529
RI							−0.238
Resolution							

and Donoghue 1989). Significance of these differences is tested in regression analysis (see below).

Partial correlation analysis reveals the association of pairs of these variables. The highest partial correlations are found between CI and number of taxa (−0.67), between CI and RI (0.61), between number of taxa and number of minimal trees (0.55), and between degree of resolution and B_{50} (0.53). A more synthetic view of the relationships among the variables is provided in a principal components analysis (Table 4 and Fig. 1). Plotted in the space of the first three principal components are the data points and vectors representing the directions in this space of the the original variables, measured as deviations from the mean. For example, the axis for CI points from the population mean to a point one standard deviation away in the positive

TABLE 4 *Component Loadings for First Five Principal Components of Variables Surveyed*

	Component 1	Component 2	Component 3	Component 4	Component 5
Eigen value	2.3226	2.0250	1.6027	0.7282	0.5634
Percentage	29.0327	25.3130	20.0342	9.1026	7.0422
CumPercent	29.0327	54.3457	74.3799	83.4825	90.5247
Eigenvectors					
CI	0.5193	−0.3057	−0.0439	−0.2011	0.3998
NChars	−0.1771	−0.0519	0.5507	−0.7408	0.2084
Ntaxa	0.0330	0.5864	0.3025	0.0634	−0.1324
Ntrees	0.0322	0.5502	0.1682	0.2401	0.6369
B_{max}	0.2805	−0.0132	0.5663	0.1742	−0.5196
B_{50}	0.0306	−0.4299	0.4804	0.3742	0.1955
RI	0.6066	−0.0077	0.0114	0.1619	0.1113
Resolution	−0.4990	−0.2685	0.1514	0.3904	0.2296

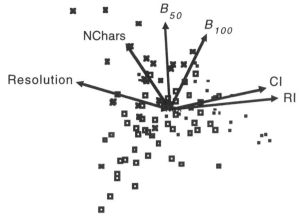

FIGURE 1 Principal component analysis of eight variables listed in Tables 3 and 4. Open squares are morphological data sets; crosses are DNA sequence data sets; closed squares are RFLP data sets. Plot was originally a three-dimensional plot of the first three principal components (Table 4), standardized so that each axis has the same standard deviation. The arrows represent vectors corresponding to a deviation of one standard deviation above the population mean for that variable, all others held constant. The graph is a projection of the original three-dimensional graph into two dimensions. Six of the eight variables have strong components in this plane; the other two, number of taxa and number of trees, are more or less perpendicular to the plane of the paper. Correlation among variables corresponds roughly to similarity in direction of vectors.

direction, all other variables being held constant. Variables with arrows in more or less the same direction are highly correlated. CI and RI are highly correlated, and this graph has the two measures of bootstrap support at roughly right angles to the measures of homoplasy, indicating poor association. Interestingly, degree of resolution is almost 180° away from the two measures of homoplasy, indicating that studies with high CIs or RIs tend to

TABLE 5   Multiple Regression Analysis^a of B_{50}

Source	No. params.	dF	Sum of squares	F ratio	Prob>F
Data type	2	2	0.2057	5.3576	0.0063
CI	1	1	0.0293	1.5288	0.2194
NChars	1	1	0.0478	2.4931	0.1178
Ntaxa	1	1	0.0024	0.1265	0.7229
Ntrees	1	1	0.0765	3.9857	0.0488
RI	1	1	0.1385	7.2172	0.0086
Resolution	1	1	0.8396	43.7233	0.0000

^{a}For 101 observations: $r^2 = 0.4900$. Whole model ANOVA F ratio = 11.0495 ($P < 0.0001$).

TABLE 6   Multiple Regression Analysis[a] of B_{max}

Source	No. params.	dF	Sum of squares	F ratio	Prob>F
Data type	2	2	784.9901	5.3164	0.0065
CI	1	1	2.9939	0.0406	0.8408
NChars	1	1	118.0663	1.5992	0.2092
Ntaxa	1	1	24.7098	0.3347	0.5643
Ntrees	1	1	0.3578	0.0048	0.9446
RI	1	1	554.8130	7.5151	0.0074
Resolution	1	1	121.9801	1.6522	0.2019

[a] For 101 observations: $r^2 = 0.32366$. Whole model ANOVA F ratio = 5.5034 ($P < 0.0001$).

have low resolution. The figure clearly shows that this effect is most pronounced for RFLP studies, and that those same studies with high CIs tend to have just average bootstrap support. The differences in CI and bootstrap support noted earlier between types of data can also be observed in this plot.

Multiple regression analysis (Tables 5 and 6) provides estimates of the significance of components in these associations. Both B_{max} and B_{50} depend strongly on the type of data and the retention index. Neither of them depends on CI, the number of characters, or the number of taxa. In addition, B_{50} depends on both the number of most parsimonious trees and the degree of resolution, whereas B_{max} does not. Although overall significance of both models is high, the amount of the variation explained is only 50% in the case of B_{50} and considerably less than that in the case of B_{max}. Clearly, other factors in these data sets must explain the remaining variation in bootstrap support.

TABLE 7   Parameter Estimates in Multiple Regression Analysis of CI on Variables in Survey[a]

Term	Estimate	Standard error	t ratio	Prob>|t|
Intercept	0.4223	0.1065	3.97	0.0001
Data[RFLP–seq]	−0.0123	0.0176	−0.70	0.4849
Data[morph–seq]	−0.0263	0.0159	−1.65	0.1017
NChars[a]	0.00007	0.00008	0.95	0.3460
Ntaxa	−0.0081	0.00093	−8.80	0.0000
Ntrees	0.00027	0.00015	1.80	0.0757
RI	0.7980	0.08351	9.56	0.0000
Resolution	−0.2087	0.06524	−3.20	0.0019

[a] For 101 observations: $r^2 = 0.7581$.

[b] Four sequence records excluded: these had over 1000 characters in their matrices, and when included lead to a marginally significant effect due to characters.

Multiple regression analysis of CI on various factors other than bootstrap support (Table 7) reconfirms the main findings of Sanderson and Donoghue (1989). CI depends markedly on the number of taxa ($P < 0.001$), but not on the number of characters (but see footnote to Table 7), the number of most parsimonious trees, or the type of data. It also depends on RI and on the degree of resolution; neither factor was included in our previous analysis. Oddly enough, the direction of the dependence of CI on degree of resolution is negative, meaning that homoplasy is correlated with level of resolution. Although one might predict that highly resolved cladograms would show low homoplasy, in fact, the reverse is true; the pattern is swayed by many studies that are very poorly resolved and therefore have high CIs. We believe that this is partly explained by the fact that these studies have fewer "effective" taxa than it appears, owing to large polytomies The polytomies arise because many of the taxa are actually indistinguishable on the basis of the characters in the matrix. The retention index is less sensitive to this phenomenon (see Table 3).

DISCUSSION

If it is agreed that the consistency index is the best measure of homoplasy per se; then these results suggest that homoplasy and confidence are decoupled. CI has no influence on either measure of bootstrap support for a tree. Table 1 includes numerous studies in which bootstrap support is high but CI is low and vice versa, which contradicts the conventional notion that the level of homoplasy has a substantial bearing on confidence in the associated phylogenetic hypothesis. This suggests that regardless of the amount of homoplasy observed in a data set, it has often proven possible to reconstruct trees with a requisite level of robustness. That this is true independent of the number of taxa in a data set should be some encouragement to investigators studying large and highly homoplastic clades.

This result apparently contradicts other findings that the level of homoplasy correlates directly with accuracy (not confidence) in the tree. For example, Archie (1996) reports results from simulation studies that show that accuracy of the tree estimate does indeed improve with lower levels of homoplasy. Although condidence and accuracy need not be the same thing, they should not be completely uncoupled. We suspect that the discrepancy arises from the different contexts in which homoplasy is being examined in a simulation versus a "meta-analysis." In a simulation, all results are conditional on a given model of evolution. Conditional on that model, the accuracy of tree estimation probably depends on the rate of evolution and thereby the level of homoplasy. In a survey such as ours the results are conditional not on a given evolutionary model but on a given survey sampling scheme.

Nothing about our sample survey methodology can guarantee that there is anything approaching a common model of evolution underlying all 101 data sets. The factors that do impinge on the issue of confidence in this survey emerge despite the vagaries of sample design and bias. The effect of the level of homoplasy may simply be too weak in the face of ordinary systematic practice (as mimicked by our survey design) to exert much influence on bootstrap values. Recall that even with seven independent variables we can explain at best only half the variation in bootstrap levels.

One result that was not entirely unexpected was the finding that some variables were influenced significantly by the type of data. Our earlier study had found no effect of type of data on level of homoplasy (CI), but the sample size for the molecular studies was small. The present, much larger analysis, also finds no effect on CI but the type of data does evidently affect bootstrap confidence levels. Sequence data are the "best" and morphological data the "worst" according to the bootstrap tests. The point of the present paper is not to examine the relative reliability of molecular versus morphological data, but we feel compelled to point out that our results can be viewed from at least two perspectives. One perspective would view our finding that molecular data sets have higher bootstrap values as confirmation of the view that molecular data are better than morphological data. However, morphological data are "nearly" as reliable as molecular data, and are usually considerably less expensive to obtain. This leads inevitably into consideration of the cost/benefit ratio measured in something like dollars per bootstrap percentage point.

Moreover, it is also clear that the overall resolution of individual maximum parsimony trees is worst for one class of molecular data, RFLP studies. This is a curiosity that remains to be explained adequately. Perusal of the trees obtained from chloroplast RFLP studies suggests that they are qualitatively different from trees obtained in either morphological or DNA sequence studies. They regularly contain one or more large polytomies, sometimes entailing 20–50% of the taxa. Morphological and sequence studies often have unresolved regions but they are rarely so unresolved as many RFLP studies. Whether this reflects an intrinsic bias in character "choice" in RFLP studies relative to other data or real differences in underlying evolutionary process is unknown. Even though individual trees from RFLP studies are the least resolved of the three kinds of data sets, it is important to note that the average resolution of the bootstrap majority rule trees (i.e, as measured by B_{50}) is still higher than those for morphological data sets.

This apparent contradiction might be explained as follows. Lack of resolution in individual most parsimonious trees (the factor surveyed as "Resolution" in this study) is probably due to either too few synapomorphies across the board or a very uneven distribution of synapomorphies among clades. Either would cause some taxa to be scored identically. Low resolution of the bootstrap majority rule tree can be due to this factor, too, which is why level of resolution and B_{50} are highly correlated (Table 3), but bootstrap resolution also reflects conflicting signals in the data owing to real character conflict. Thus, the bootstrap support values for RFLP studies may stem from lack of resolution due to the distribution of synapomorphies, whereas the somewhat lower bootstrap support values for morphological studies may just reflect real noise obscuring the phylogenetic signal.

In sum, we conclude that the conventional view that confidence is directly related to the level of homoplasy in a data set is not supported by available data from phylogenetic studies. The consistency index is a measure of homoplasy, not robustness. The implication of this is that homoplasy and confidence are two separate issues—not completely independent perhaps, especially in carefully controlled simulation experiments—but effectively so in real data sets.

ACKNOWLEDGMENTS

Thanks to T. Givnish and K. Sytsma, R. Jansen for providing us information on their studies of patterns of homoplasy in surveys of real data sets, and the Harvard Phylogeny Discussion Group for comments.

APPENDIX: SOURCES FOR SURVEY OF PHYLOGENETIC DATA SETS

Morphological Data
1. Anderberg, A. A. 1991. Taxonomy and phylogeny of the tribe Inuleae (Asteraceae). *Plant Systematics and Evolution* **176**:75–123.
2. Anderberg, A. A. 1994. Phylogeny of the Empetraceae, with special emphasis on character evolution in the genus *Empetrum*. *Systematic Botany* **19**:35–46.
3. Anderberg, A. A., and K. Bremer. 1991. Parsimony analysis and cladistic reclassification of the *Relhania* generic group (Asteraceae–Gnaphalieae). *Annals of the Missouri Botanical Garden* **78**:1061–1072.
4. Anderberg, A. A., and P. Eldenas. 1991. A cladistic analysis of *Anigozanthos* and *Macropidia* (Haemodoraceae). *Australian Systematic Botany* **4**:655–664.
5. An-Ming, L. 1990. A preliminary cladistic study of the families of the superorder Lamiiflorae. *Biological Journal of the Linnean Society* **103**:39–57.
6. Ayers, T. J. 1990. Systematics of *Heterotoma* (Campanulaceae) and the evolution of nectar spurs in the New World Lobelioideae. *Systematic Botany* **15**:296–327.

7. Barfod, A. S. 1991. A monographic study of the subfamily Phytelephantoideae (Arecaceae). *Opera Botanica* **105**:1–73.
8. Berry, P. E. 1989. A systematic revision of *Fuchsia* sect. *Quelusia* (Onagraceae). *Annals of the Missouri Botanical Garden* **76**:532–584.
9. Boufford, D. E., J. V. Crisci, H. Tobe, and P. C. Hoch. 1990. A cladistic analysis of *Circaea* (Onagraceae). *Cladistics* **6**:171–182.
10. Bruyns, P., and H. P. Linder. 1991. A revision of *Microloma* R. Br. (Asclepiadaceae–Asclepiadeae). *Botanische Jahrbücher für Systematik, Pflanzengeschichte und Pflanzengeographie* **112**:453–527.
11. Campbell, C. S. 1986. Phylogenetic reconstructions and two new varieties in the *Andropogon virginicus* complex (Poaceae: Andropogoneae). *Systematic Botany* **11**:280–292.
12. Caputo, P., and S. Cozzolino. 1994. A cladistic analysis of Dipsacaceae (Dipsacales). *Plant Systematics and Evolution* **189**:41–61.
13. Carr, B. L., J. V. Crisci, and P. C. Hoch. 1990. A cladistic analysis of the genus *Gaura* (Onagraceae). *Systematic Botany* **15**:454–461.
14. Cox, P. A. 1990. Pollination and the evolution of breeding systems in Pandanaceae. *Annals of the Missouri Botanical Garden* **77**:816–840.
15. Cox, P. B., and L. E. Urbatsch. 1990. A phylogenetic analysis of the coneflower genera (Asteraceae: Heliantheae). *Systematic Botany* **15**:394–402.
16. Crane, P. R. 1985. Phylogenetic analysis of seed plants and the origin of angiosperms. *Annals of the Missouri Botanical Garden* **72**:716–793.
17. Crisci, J. V., and P. E. Berry. 1990. A phylogenetic reevaluation of the Old World species of *Fuchsia*. *Annals of the Missouri Botanical Garden* **77**:517–522.
18. Crisp, M. D. 1991. Contributions towards a revision of *Daviesia* Smith (Fabaceae: Mirbelieae). II. The *D. latifolia* group. *Australian Systematic Botany* **4**:229–298.
19. Cruden, R. W. 1991. A revision of *Isidrogalvia* (Liliaceae): Recognition for Ruiz and Pavon's genus. *Systematic Botany* **16**:270–282.
20. Curry, K. J., L. M. McDowell, W. S. Judd, and W. L. Stern. 1991. Osmophores, floral features, and systematics of Stanhopea (Orchidaceae). *American Journal of Botany* **78**:610–623.
21. Doyle, J. A., and M. J. Donoghue. 1992. Fossils and seed plant phylogeny reanalyzed. *Brittonia* **44**:89–106.
22. Eriksson, R. 1994. Phylogeny of Cyclanthaceae. *Plant Systematics and Evolution* **190**:31–47.
23. Goldblatt, P., and J. E. Henrich. 1991. *Calydorea* Herbert (Iridaceae–Tigridieae): Notes on this New World genus and reduction to synonymy of *Salpingostylis*, *Cardiostigma*, *Itysa*, and *Catila*. *Annals of the Missouri Botanical Garden* **78**:504–511.

24. Graham, L. E., C. F. Delwiche, and B. D. Mishler. 1991. Phylogenetic connections between the 'green algae' and the 'bryophytes'. *Advances in Bryology* **4**:213–244.
25. Gustafsson, M. H. G., and K. Bremer. 1995. Morphology and phylogenetic interrelationships of the Asteraceae, Calyceraceae, Campanulaceae, Goodeniaceae, and related families (Asterales). *American Journal of Botany* **82**:250–265.
26. Hammel, B. E., and G. J. Wilder. 1989. *Dianthoveus*: A new genus of Cyclanthaceae. *Annals of the Missouri Botanical Garden* **76**:112–123.
27. Hickey, L. J., and D. W. Taylor. 1991. The leaf architecture of *Ticodendron* and the application of foliar characters in discerning its relationships. *Annals of the Missouri Botanical Garden* **78**:105–130.
28. Hoch, P. C., J. V. Crisci, H. Tobe, and P. E. Berry. 1993. A cladistic analysis of the plant family Onagraceae. *Systematic Botany* **18**:31–47.
29. Huck, R. B., W. S. Judd, W. M. Whitten, J. D. Skean, Jr. , R. P. Wunderlin, and K. R. Delaney. 1989. A new *Dicerandra* (Labiatae) from the Lake Wales ridge of Florida, with a cladistic analysis and discussion of endemism. *Systematic Botany* **14**:197–213.
30. Hufford, L. D., and P. R. Crane. 1989. A preliminary phylogenetic analysis of the 'lower' Hamamelidae. *In* P. R. Crane and S. Blackmore (Eds.), *Evolution, systematics, and fossil history of the Hamamelidae*, pp. 175–192. Clarendon Press, Oxford.
31. Judd, W. S. 1989. Taxonomic studies in the Miconieae (Melastomataceae). III. Cladistic analysis of axillary-flowered taxa. *Annals of the Missouri Botanical Garden* **76**:476–495.
32. Karis, P. O. 1989. Systematics of the genus *Metalasia* (Asteraceae–Gnaphalieae). *Opera Botanica* **99**:1–150.
33. Kenrick, P., and P. R. Crane. 1991. Water-conducting cells in early fossil land plants: Implications for the early evolution of tracheophytes. *Botanical Gazette* **152**:335–356.
34. Knapp, S. 1991. A cladistic analysis of the *Solanum* sessile species group (Section *Geminata pro parte*: Solanaceae). *Biological Journal of the Linnean Society* **106**:73–89.
35. Kress, W. J. 1990. The phylogeny and classification of the Zingiberales. *Annals of the Missouri Botanical Garden* **77**:698–721.
36. Kurzweil, H., H. P. Linder, and P. Chesselet. 1991. The phylogeny and evolution of the *Pterygodium–Corycium* complex (Coryciinae, Orchidaceae). *Plant Systematics and Evolution* **175**:161–223.
37. Ladiges, P. Y., and C. J. Humphries. 1983. A cladistic study of *Arillastrum*, *Angophora*, and *Eucalyptus* (Myrtaceae). *Botanical Journal of the Linnean Society* **87**:105–134.
38. Lavin, M. 1990. The genus *Sphinctospermum* (Leguminosae): Taxonomy and tribal relationships as inferred from a cladistic analysis of traditional data. *Systematic Botany* **15**:544–559.

39. Luteyn, J. L. 1991. A synopsis of the genus *Didonica* (Ericaceae: Vaccinieae) with two new species. *Systematic Botany* **16**:587–597.

40. Manning, J. C., and P. Goldblatt. 1991. Systematic and phylogenetic significance of the seed coat in the shrubby African Iridaceae, *Nivenia*, *Klattia*, and *Witsenia*. *Botanical Journal of the Linnean Society* **107**: 387–404.

41. Mishler, B. D. 1990. Reproductive biology and species distinctions in the moss genus *Tortula*, as represented in Mexico. *Systematic Botany* **15**:86–97.

42. Olmstead, R. 1989. Phylogeny, phenotypic evolution, and biogeography of the *Scutellaria angustifolia* complex (Lamiaceae): Inference from morphological and molecular data. *Systematic Botany* **14**:320–338.

43. Schutte, A. L., and B.-E. Van Wyk. 1990. Taxonomic relationships in the genus *Dichilus* (Fabaceae–Crotalarieae). *South African Journal of Botany* **56**:244–256.

44. Schwarzwalder, R. N., and D. L. Dilcher. 1991. Systematic placement of the Platanaceae in the Hamamelidae. *Annals of the Missouri Botanical Garden* **78**:962–969.

45. Simpson, M. G. 1990. Phylogeny and classification of the Haemodoraceae. *Annals of the Missouri Botanical Garden* **77**:722–784.

46. Smith, G. F., and B.-E. Van Wyk. 1991. Generic relationships in the Alooideae (Asphodelaceae). *Taxon* **40**:557–581.

47. Taylor, D. W., and L. J. Hickey. 1992. Phylogenetic evidence for the herbaceous origin or angiosperms. *Plant Systematics and Evolution* **180**:137–156.

48. Tucker, S. C., and A. W. Douglas. 1993. Utility of ontogenetic and conventional characters in determining phylogenetic relationships of Saururaceae and Piperaceae (Piperales). *Systematic Botany* **18**:614–641.

49. Varadarajan, G. S., and A. J. Gilmartin. 1988. Phylogenetic relationships of groups of genera within the subfamily Pitcairnioideae (Bromeliaceae). *Systematic Botany* **13**:283–293.

50. Wilken, D., and R. L. Hartman. 1991. A revision of the *Ipomopsis spicata* complex (Polemoniaceae). *Systematic Botany* **16**:143–161.

Sequence Data

51. Baldwin, B. G., and R. H. Robichaux. 1995. Historical biogeography and ecology af the Hawaiian silversword alliance (Asteraceae): New molecular phylogenetic perspectives. *In* W. L. Wagner and V. A. Funk (Eds.), *Hawaiian biogeography. Evolution on a hot spot archipelago*, pp. 259–287. Smithsonian Institution Press, Washington, DC.

52. Baum, D. A., K. J. Sytsma, and P. C. Hoch 1994. A phylogenetic analysis of *Epilobium* (Onagraceae) based on nuclear ribosomal DNA sequences. *Systematic Botany* 19:363–388.
53. Bremer, K. 1988. The limits of amino acid sequence data in angiosperm phylogenetic reconstruction. *Evolution* 42:795–803.
54. Bult, E. A., and E. A. Zimmer. 1993. Nuclear ribosomal RNA sequences for inferring tribal relationships within Onagraceae. *Systematic Botany* 18:48–63.
55. Campbell, C. S., M. J. Donoghue, B. G. Baldwin, and M. F. Wojciechowski. 1995. Phylogenetic relationships in Maloideae (Rosaceae): Evidence from sequences of internal transcribed spacers of nuclear ribosomal DNA and its congruence with morphology. *American Journal of Botany* 82:903–918.
56. Conti, E., A. Fischbach, and K. J. Sytsma. 1993. Tribal relationships in Onagraceae: Implications from *rbc*L sequence data. *Annals of the Missouri Botanical Garden* 80:672–685.
57. Cullings, K. W., and T. D. Bruns. 1992. Phylogenetic origin of the Monotropoideae inferred from partial 28S ribosomal RNA gene sequences. *Canadian Journal of Botany* 70:1703–1708.
58. Donoghue, M. J., R. G. Olmstead, J. F. Smith, and J. D. Palmer. 1992. Phylogenetic relationships of Dipsacales based on *rbc*L sequences. *Annals of the Missouri Botanical Garden* 79:333–345.
59. Doyle, J. A., M. J. Donoghue, and E. A. Zimmer. 1994. Integration of morphological and ribosomal RNA data on the origin of angiosperms. *Annals of the Missouri Botanical Garden* 81:419–450. [ribosomal RNA matrix, Table 3]
60. Eriksson, T., and M. J. Donoghue. Phylogenetic analysis of *Sambucus* and *Adoxa* (Adoxaceae) based on nuclear ribosomal ITS sequences and morphological characters. Submitted for publication.
61. Johnson, L. A., and D. E. Soltis. 1994. *mat*K DNA sequences and phylogenetic reconstruction in Saxifragaceae *s. str. Systematic Botany* 19:143–156.
62. Kim, K.-J., and R. K. Jansen. 1994. Comparisons of phylogenetic hypotheses among different data sets in dwarf dandelions (*Krigia*, Asteraceae): Additional information from internal transcribed spacer sequences of nuclear ribosomal DNA. *Plant Systematics and Evolution* 190:157–185.
63. Les, D., D. K. Garvin, and C. F. Wimpee. 1991. Molecular evolutionary history of ancient aquatic angiosperms. *Proceedings of the National Academy of Science USA* 88:10119–10123.
64. Manen, J.-F., A. Natali, and F. Ehrendorfer. 1994. Phylogeny of Rubiaceae–Rubieae inferred from the sequence of a cpDNA intergene region. *Plant Systematics and Evolution* 190:195–211.

65. Martin, P. G., and J. M. Dowd. 1991. A comparison of 18S ribosomal RNA and rubisco large subunit sequences for studying angiosperm phylogeny. *Journal of Molecular Evolution* **33**:274–282.

66. Nickrent, D. L., K. P. Schuette, and E. M. Starr. 1994. A molecular phylogeny of *Arceuthobium* (Viscaceae) based on nuclear ribosomal DNA internal transcribed spacer sequences. *American Journal of Botany* **81**:1149–1160.

67. Olmstead, R. G., and P. A. Reeves. 1995. Evidence for the polyphyly of the Schrophulariaceae based on chloroplast *rbc*L and *ndh*F sequences. *Annals of the Missouri Botanical Garden* **82**:176–193.

68. Sanderson, M. J., and M. F. Wojciechowski. 1996. Diversification rates in a temperate legume clade: Are there "so many species" of *Astragalus* (Fabaceae)? *American Journal of Botany*, in press.

69. Sang, T., D. J. Crawford, T. F. Stuessy, and M. S. O. 1995. ITS sequences and the phylogeny of the genus *Robinsonia* (Asteraceae). *Systematic Botany* **20**:55–64.

70. Scotland, R. W., J. A. Sweere, P. A. Reeves, and R. G. Olmstead. 1995. Higher-level systematics of Acanthaceae determined by chloroplast DNA sequences. *American Journal of Botany* **82**:266–275. [24 *ndh*F sequences, Fig. 4]

71. Scotland, R. W., J. A. Sweere, P. A. Reeves, and R. G. Olmstead. 1995. Higher-level systematics of Acanthaceae determined by chloroplast DNA sequences. *American Journal of Botany* **82**:266–275. [21 *rbc*L sequences, Fig. 7]

72. Steele, K. P., and R. Vilgalys. 1994. Phylogenetic analyses of Polemoniaceae using nucleotide sequences of the plastid gene *mat*K. *Systematic Botany* **19**:126–142.

73. Wojciechowski, M. F., and M. J. Sanderson. In preparation.

RFLP Data

74. Bremer, B., and R. K. Jansen. 1991. Comparative restriction site mapping of chloroplast DNA implies new phylogenetic relationships within Rubiaceae. *American Journal of Botany* **78**:198–213.

75. Bruneau, A., and J. J. Doyle. 1993. Cladistic analysis of chloroplast DNA restriction site characters in *Erythrina* (Leguminosae: Phaseoleae). *Systematic Botany* **18**:229–247.

76. Conant, D. S., D. B. Stein, A. E. C. Valinski, and P. Sudarsanam. 1994. Phylogenetic implications of chloroplast DNA variation in the Cyatheaceae. I. *Systematic Botany* **19**:60–72.

77. Crisci, J. V., E. A. Zimmer, P. C. Hoch, G. B. Johnson, C. Mudd, and N. S. Pan. 1990. Phylogenetic implications of ribosomal DNA restriction site variation in the plant family Onagraceae. *Annals of the Missouri Botanical Garden* **77**:523–538.

78. Doyle, J. J., and J. L. Doyle. 1993. Chloroplast DNA phylogeny of the papilionoid legume tribe Phaseoleae. *Systematic Botany* **18**:309–327.
79. Freudenstein, J. V., and J. J. Doyle. 1994. Character transformation and relationships in *Corallorhiza* (Orchidaceae: Epidendroideae). I. Plastid DNA. *American Journal of Botany* **81**:1449–1457.
80. Freudenstein, J. V., and J. J. Doyle. 1994. Plastid DNA, morphological variation, and the phylogenetic species concept: The *Corallorhiza maculata* (Orchidaceae) complex. *Systematic Botany* **19**:273–290.
81. Johansson, J. T., and R. K. Jansen. 1991. Chloroplast DNA variation among five species of Ranunculaceae: Structure, sequence divergence, and phylogenetic relationships. *Plant Systematics and Evolution* **178**: 9–25.
82. Kadereit, J. W., and K. J. Sytsma. 1992. Disassembling *Papaver*: A restriction site analysis of chloroplast DNA. *Nordic Journal of Botany* **12**:205–217.
83. Larson, S. R., and J. Doebley. 1994. Restriction site variation in the chloroplast genome of *Tripsacum* (Poaceae): Phylogeny and rates of sequence evolution. *Systematic Botany* **19**:21–34.
84. Lavin, M., S. Mathews, and C. Hughes. 1991. Chloroplast DNA variation in *Gliricidia sepium* (Leguminosae): Intraspecific phylogeny and tokogeny. *American Journal of Botany* **78**:1576–1585.
85. Manos, P. S., K. C. Nixon, and J. J. Doyle. 1993. Cladistic analysis of restriction site variation within the chloroplast DNA inverted repeat region of selected Hamamelididae. *Systematic Botany* **18**:551–562.
86. Morgan, D. R. 1993. A molecular systematic study and taxonomic revision of *Psilactis* (Asteraceae: Astereae). *Systematic Botany* **18**: 290–308.
87. Mummenhoff, K., and M. Koch. 1994. Chloroplast DNA restriction site variation and phylogenetic relationships in the genus *Thlaspi sensu lato* (Brassicaceae). *Systematic Botany* **19**:73–88.
88. Pennington, R. T. 1995. Cladistic analysis of chloroplast DNA restriction site characters in *Andira* (Leguminosae: Dalbergieae). *American Journal of Botany* **82**:526–534.
89. Peterson, G., and J. F. Doebley. 1993. Chloroplast DNA variation in the genus *Secale* (Poaceae). *Plant Systematics and Evolution* **187**:115–125.
90. Philbrick, C. T., and R. K. Jansen. 1991. Phylogenetic studies of North American *Callitriche* (Callitrichaceae) using chloroplast DNA restriction fragment analysis. *Systematic Botany* **16**:478–491.
91. Pillay, M., and K. W. Hilu. 1995. Chloroplast-DNA restriction site analysis in the genus *Bromus* (Poaceae). *American Journal of Botany* **82**:239–249.

92. Potter, D., and J. J. Doyle. 1994. Phylogeny and systematics of *Sphenostylis* and *Nesphostylis* (Leguminosae: Phaseoleae) based on morphological and chloroplast DNA data. *Systematic Botany* 19:389–406.
93. Soreng, R. J. 1990. Chloroplast-DNA phylogenetics and biogeography in a reticulating group: Study in *Poa* (Poaceae). *American Journal of Botany* 77:1383–1400.
94. Soreng, R. J., J. I. Davis, and J. J. Doyle. 1990. A phylogenetic analysis of chloroplast DNA restriction site variation in Poaceae subfam. Pooideae. *Plant Systematics and Evolution* 172:83–97.
95. Sytsma, K. J., and J. F. Smith. 1988. DNA and morphology: Comparisons in the Onagraceae. *Annals of the Missouri Botanical Garden* 75: 1217–1237.
96. Systma, K. J., J. F. Smith, and P. E. Berry. 1991. The use of chloroplast DNA to assess biogeography and evolution of morphology, breeding systems, and flavonoids in *Fuschia* sect. *Skinnera* (Onagraceae). *Systematic Botany* 16:257–269.
97. Talbert, L. E., G. M. Magyar, M. Lavin, T. K. Blake, and S. L. Moylan. 1991. Molecular evidence for the origin of the S-derived genomes of polyploid *Triticum* species. *American Journal of Botany* 78:340–349.
98. Vaillancourt, R. E., and N. F. Weeden. 1993. Chloroplast DNA phylogeny of Old World *Vigna* (Leguminosae). *Systematic Botany* 18:642–651.
99. Whittemore, A. T., and B. A. Schaal. 1991. Interspecific gene flow in sympatric oaks. *Proceedings of the National Academy of Sciences USA* 88:2540–2544.
100. Wolfe, A. D., and W. J. Elisens. 1994. Nuclear ribosomal DNA restriction-site variation in *Penstemon* section *Peltanthera* (Scrophulariaceae): An evaluation of diploid hybrid speciation and evidence for introgression. *American Journal of Botany* 81:1627–1635.
101. Wolf, P. G., P. S. Soltis, and D. E. Soltis. 1993. Phylogenetic significance of chloroplast restriction site variation in the *Ipomopsis aggregata* complex and related species (Polemoniaceae). *Systematic Botany* 18: 652–662.

REFERENCES

Arnqvist, G., and D. Wooster. 1995. Meta-analysis: Synthesizing research findings in ecology and evolution. Trends in Ecology and Evolution 10:236–240.
Archie, J. W. 1989. Homoplasy excess ratios: New indices for measuring levels of homoplasy in phylogenetic systematics and a critique of the consistency index. Systematic Zoology 38: 235–269.
Archie, J. W. 1996. Measures of homoplasy. Pp. 153–188 *in* M. J. Sanderson and L. Hufford, eds. Homoplasy: The recurrence of similarity in evolution. Academic Press, San Diego.

Bremer, K. 1988. The limits of amino acid sequence data in angiosperm phylogenetic reconstruction. Evolution 42:795–803.

Bremer, K. 1994. Branch support and tree stability. Cladistics 10:295–304.

Cummings, M. P., S. P. Otto, and J. Wakeley. 1995. Sampling properties of DNA sequence data in phylogenetic analysis. Molecular Biology and Evolution 12:814–822.

Donoghue, M. J., R. G. Olmstead, J. F. Smith, and J. D. Palmer. 1992. Phylogenetic relationships of Dipsacales based on *rbc*L sequences. Annals of the Missouri Botanical Garden 79: 333–345.

Donoghue, M. J., and M. J. Sanderson. 1992. The suitability of molecular and morphological evidence in reconstructing plant phylogeny. Pp. 340–368 *in* P. S. Soltis, D. E. Soltis, and J. J. Doyle, eds. Molecular systematics in plants. Chapman and Hall, New York.

Farris, J. S. 1989. The retention index and rescaled consistency index. Cladistics 5:417–419.

Felsenstein, J. 1978. Cases in which parsimony or compatibility methods will be positively misleading. Systematic Zoology 27:401–410.

Felsenstein, J. 1985. Phylogenies and the comparative method. American Naturalist 125:1–15.

Givnish, T. J., and K. J. Sytsma. 1992. Chloroplast DNA restriction site data yield phylogenies with less homoplasy than analyses based on morphology or DNA sequences. American Journal of Botany 79:145 [abstract]

Goloboff, P. 1991a. Homoplasy and the choice among cladograms. Cladistics 7:215–232.

Goloboff, P. 1991b. Random data, homoplasy and information. Cladistics 7:395–406.

Hillis, D. M., and J. J. Bull. 1993. An empirical test of bootstrapping as a method for assessing confidence in phylogenetic analysis. Systematic Biology 42:182–192.

Jansen, R. K. 1995. DNA sequence data isn't really better than cpDNA restriction site data for studies of plant phylogeny at the intrafamilial level. American Journal of Botany 82:137 [abstract]

Klassen, G. J., R. D. Mooi, and A. Locke. 1991. Consistency indices and random data. Systematic Zoology 40:446–457.

Kluge, A., and J. S. Farris. 1969. Quantitative phyletics and the evolution of anurans. Systematic Zoology 18:1–32.

Landrum, L. 1993. Factors influencing th accuracy of the parsimony criterion and a method for estimating true tree length. Systematic Botany 18:516–524.

Lankester, E. R. 1870. On the use of the term homology in modern zoology. Annals and Magazine of Natural History 6:34–43.

Meier, R., P. Kores, and S. Darwin. 1991. Homoplasy slope ratio: A better measurement of observed homoplasy in cladistic analyses. Systematic Zoology 40:74–88.

Sanderson, M. J. 1989. Confidence limits on phylogenies: The bootstrap revisited. Cladistics 5: 113–129.

Sanderson, M. J. 1995. Objections to bootstrapping phylogenies: A critique. Systematic Biology 44:299–320.

Sanderson, M. J., and M. J. Donoghue. 1989. Patterns of variation in levels of homoplasy. Evolution 43:1781–1795.

Sokal, R. R., and F. J. Rohlf. 1981. Biometry, 2nd ed. W. H. Freeman, New York.

Swofford, D. 1993. PAUP: Phylogenetic analysis using parsimony. llinois Natural History Survey, Champaign, Illinois.

Swofford, D. L. 1991. When are phylogeny estimates from molecular and morphological data incongruent? Pp. 295–333 *in* M. M. Miyamoto and J. Cracraft, eds. Phylogenetic analysis of DNA sequences. Oxford Univ. Press, New York.

Wendel, J. F., and V. A. Albert. 1992. Phylogenetics of the cotton genus (*Gossypium*): Character-state weighted parsimony of chloroplast-DNA restriction site data and its systeamtic and biogeographic implications. Systematic Botany 17:115–143.

*"Inconsistencies cannot both be right, but
imputed to man they may both be true."*
Samuel Johnson, Rasselas

NONFLORAL HOMOPLASY AND EVOLUTIONARY SCENARIOS IN LIVING AND FOSSIL LAND PLANTS

RICHARD M. BATEMAN

*Royal Botanic Garden and Royal Museum of Scotland,
Edinburgh, Scotland, United Kingdom*

INTRODUCTION

General Precepts

Any phylogenetic study inevitably includes many precepts—assumptions that must be accepted *a priori* if the ideas of greatest interest are to be developed satisfactorily. The key precepts of this paper are as follows:

1. The distribution of morphological character-state transitions across the topology of a cladogram is biologically meaningful. In particular, it can be used to discriminate among contrasting evolutionary mechanisms possessing the potential to explain the inferred pattern of character-state transitions.

2. Such interpretations are most informative if the maximum possible number of species in the clade under scrutiny is included in the analysis. This conclusion generates a strong incentive to incorporate fossil as well as living taxa.

HOMOPLASY: The Recurrence of Similarity in Evolution, M. J. Sanderson and L. Hufford, eds.
Copyright © 1996 All rights of reproduction in any form reserved.

3. Homoplastic character-states can be as informative about evolutionary mechanisms as synapomorphies, despite being less valuable for determining relationships and thereby erecting monophyletic classifications.

Consequently, in this chapter homoplasy is viewed primarily as the most interpretationally important of several forms of ambiguity in the topological position (i.e., placement across the cladogram) of specific character-state transitions.

Homoplasy, Positional Ambiguity, and Optimization

Homoplasy is here defined as more than one transition between the same two states of a single coded character across a particular topology. These multiple transitions do not necessarily occur in the same direction. The simplest possible case of homoplasy is two origins (parallelism *sensu lato*: 0>1; 0>1) or an origin followed by a loss (reversal: 0>1; 1>0) in a single binary character (Fig. 1). In a cladistic context, homoplasy is traditionally regarded as reflecting (1) conflict among characters revealed by the application of the Wagner parsimony algorithm (Fig. 1A). However, there are two further causes of ambiguity in the placement of specific character-state transitions across a particular topology (here termed *positional ambiguity*): (2) Missing values (e.g., Nixon and Davis 1991). A primary cladistic data matrix consists of rows of coded taxa (termed "terminal taxa" by most cladists, despite the fact that they do not terminate anything at the time they are being coded) and columns of coded characters. The absence of a value in a particular cell of such a matrix need not reflect complete lack of knowledge of that organism—it may simply reflect inability to recognize homologous character states or, more importantly, the absence of the coded feature (e.g., Doyle and

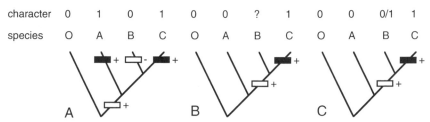

FIGURE 1   Three causes of ambiguity in the positions of character-state transitions on cladograms. (A) Character conflict. Because it is incongruent with other characters (not shown), this character is depicted as homoplastic. It undergoes two state changes: two origins under Deltran optimization (black boxes) and an origin and a loss under Acctran optimization (white boxes). (B) Missing value. Inability to score species B for this character results in arbitrary insertion of 0 under Deltran, yielding an early transition, or 1 under Acctran, yielding a later transition. (C) Polymorphism. A situation analogous to that of (B), but caused by the presence of both the primitive and the derived states in species B. Again, 0 is preferred under Deltran but 1 under Acctran (after Bateman and DiMichele 1994a, Fig. 1).

Donoghue 1986). (3) Polymorphic values. The converse of a missing value, a cladistic polymorphism involves the presence of two or more character states in a single cell of a primary matrix. This represents variation in the coded feature within the coded taxon that is perceived as too great to shoehorn into a single character state.

Both missing (Fig. 1B) and polymorphic (Fig. 1C) values also result in positional ambiguities in character-state transitions that must be solved arbitrarily using optimization algorithms, though polymorphism is perhaps better negated *a priori* by splitting coded taxa until all cells contain single fixed values (e.g., Nixon and Davis 1991). Although several optimization options exist (e.g., Swofford and Maddison 1987; Wiley et al. 1991), the full range of potential solutions to a positional ambiguity is spanned by the two logical extremes: Acctran offers the basalmost first transition of the character and thus favors reversals over parallelisms, whereas Deltran offers the apicalmost first transition of the ambiguous character and thus favors parallelisms over reversals.

The choice of Acctran or Deltran is irrelevant to cladists interested only in the sister-group relationships inherent in the topology but crucial to those cladists who wish to use the topology as a framework for interpreting the distribution and sequence of origin of character states (see also da Pinna 1991). The Deltran-generated parallelism illustrated by the black boxes in Fig. 1A is interpreted as refuting the initial hypothesis of homology of the derived character state in the two groups that possess it. Note, however, that this does not impugn its value as a synapomorphy *within* the two groups—it is simply automatically reinterpreted as *two* nonhomologous synapomorphies. Given sufficient enthusiasm, additional evidence can then be sought to test this new hypothesis of homology generated by the cladogram, constituting a classic example of reciprocal illumination.

However, the same character examined by Acctran results in a perceived origin followed by a reversal (Fig. 1A, white boxes), an explanation that does *not* refute the initial hypothesis of homology. Rather, the character is usually perceived as (though rarely explicitly stated to be) a nested pair of related synapomorphic states, with the presence of the feature being more general than its subsequent loss. In this case the net result is that (1) the primary absence of the character state is plesiomorphic, (2) the secondary absence (loss) delimits a monophyletic group (in the particular example of Fig. 1A it is autapomorphic), and (3) its earlier origin delimits an intermediate paraphyletic group of species that possess the feature. Note that the distinction between absence and presence is influenced by *a priori* homology decisions, whereas the distinction between primary absence and secondary absence is wholly a function of the topology of the cladogram and hence made *a posteriori*.

In summary, homoplasy is one form of positional ambiguity of character-state transitions. In this computerized age the ambiguity is almost invariably "solved" by *ad hoc* application of one of several optimization algorithms, most commonly Acctran or Deltran. Different optimization algorithms lead to different and fundamentally incompatible biological interpretations of the ambiguous character-state transitions and hence of the origins of the taxonomic groups that they delimit. This is one paradox of homoplasy as viewed in an evolutionary cladistic context.

Six Serendipitous Case Studies

Illustrating several important aspects of homoplasy in the context of scenario building, the following six brief case studies focus largely on vegetative transitions in nonflowering plants, in deference to the accompanying chapters on reproductive transitions in angiosperms (Armbruster 1996; Endress 1996; Hufford 1996). The relevance of the case studies may not be immediately apparent, as they are not presented as self-contained discussions. Rather, sufficient information is provided on each empirical study to provide a framework for the concluding general discussion.

Topics covered in the discussion include the evolutionary significance of long branches and their relationship to taxonomic sampling density, the importance of including fossil taxa to shorten long branches and thus more effectively determine the sequence of character-state acquisitions in the clade, the relative merits of morphological and molecular phylogenetic data for unraveling contrasting modes of evolution, and the breadth of applicability of the "total evidence" approach to phylogeny reconstruction.

TWO STEPS FORWARD AND ONE STEP BACK: ARCHITECTURAL TRANSITIONS AMONG THE RHIZOMORPHIC LYCOPSIDS

The series of papers investigating various implications of a phylogeny of the rhizomorphic lycopsids (e.g., Bateman et al. 1992; DiMichele and Bateman 1992; Bateman 1994; Bateman and DiMichele 1994a) remains perhaps the most comprehensive exploration of the topological ambiguities and interpretative difficulties that can be caused by repeated vegetative simplification events within a clade.

The rhizomorphic lycopsids display an early (Devonian) evolution of the secondarily thickened tree habit and centralized (bipolar) determinate growth from a small, nonwoody, rhizomatous lycopsid ancestor. This was achieved independently of similar acquisitions in the better known lignophyte clade (progymnosperms plus seed plants: e.g., Doyle and Donoghue 1992; Rothwell and Serbet 1994). The group retained some form of secondary thickening and centralized determinate growth in all of its many subsequent architectural manifestations (including its only extant member, the diminutive, ecologically relictual *Isoetes*), suggesting that these features constitute an unbreakable developmental constraint (Bateman 1994). However, phylogenetic analysis of the 10 best–known fossil genera (Fig. 2) revealed a minimum of two (more probably three) reversions approximating the much smaller ancestral body size, including two independent origins of the recumbent woody growth habit termed pseudoherbaceous by Bateman and DiMichele (1991). As a result of these architectural changes, homoplasy is widespread in the group.

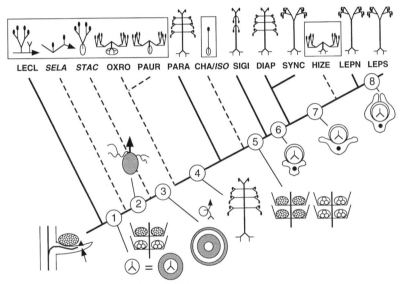

FIGURE 2 Morphological phylogeny of the ligulate lycopsids. The preferred most-parsimonious cladogram of 10 rhizomorphic lycopsid genera (clade 3) is appended to an intermediate portion of a cladogram from a broader analysis of the Lycopsida. Unlike trees, small-bodied genera are boxed; within these boxes, ellipses denote compact, "cormose" rootstocks. Taxa whose placement are doubtful due to repeated vegetative simplification are subtended by dashed branches; extant genera are italicized. Key synapomorphies following the evolution of the ligule in the coneless, homosporous *Leclercqia* are shown on the major axis: (1) heterosporous bisexual cone, (2) centralized growth, (3) true bipolar growth with secondary thickening, (4) tree habit, (5) segregation of megasporophylls and microsporophylls into separate unisexual cones, (6) reduction to a single functional megaspore, dispersal unit switched from megaspore only to megaspore plus megasporangium plus megasporophyll, (7) lateral expansion of sterile megasporophyll alations, (8) enclosure of megasporangium by alations. Genera as follows: LECL, *Leclercqia*; SELA, *Selaginella s.s.*; STAC, *Stachygynandrum* (heterophyllous Selaginellaceae); OXRO, *Oxroadia*; PAUR, *Paurodendron*; PARA, *Paralycopodites*; CHA, *Chaloneria*; ISO, *Isoetes* (extant); SIGI, *Sigillaria*; SYNC, *Synchysidendron*; DIAP, *Diaphorodendron*; HIZE, *Hizemodendron*; LEPN, *Lepidodendron*; LEPS, *Lepidophloios* (modified after Bateman 1994, Fig. 2A; Bateman and DiMichele 1994b, Fig. 10).

Each evolutionary simplification event involved several correlated losses of characters. Some were a direct result of the meristematic suppressions thought to have driven these architectural changes. (Plant architecture is largely determined by the relative positions of different types of localized generative tissue termed meristems. Suppression of one type of meristem can eliminate the organ(s) that it generated and modify the behavior of the remaining meristems (e.g., Bateman in press).) Subsequent selectively mediated losses affected characters that had become functionally redundant as a result of the much smaller body size (and consequent modified life history and habitat preference) of the lineage. These architectural and vegetative simplifications occurred repeatedly in the rhizomorphic lycopsids, leading to the rarely discussed scenario of parallel losses—in this case, of many features that define vegetative complexity.

This interpretation in turn led to the realization that the topology shown in Fig. 2 is probably inaccurate in detail. Repeated multiple losses of character states through architectural simplification yield a "two steps forward, one step back" pattern of architectural evolution (see Discussion) that tends to lead to the placement of each reduced shrub/pseudoherb below its most probable position in the phylogeny—in this case, the position indicated primarily by its relative degree of reproductive "sophistication" (admittedly, this is a subjective concept). Furthermore, *Oxroadia* was chosen as an outgroup to root the cladogram because, among all of the coded taxa, it had both the oldest first appearance in the fossil record and the greatest morphological simplicity. However, subsequent interpretations (e.g., Bateman 1994) suggest that this simplicity is far more consistent with secondary reduction than plesiomorphy, analogous to the secondarily reduced simplicity unequivocally shown by the more reproductively derived genus *Hizemodendron* (Bateman and DiMichele 1991). (This is by no means the only use of paedomorphs as outgroups; see for example the use of the heterosporous water ferns, discussed below, to root the *rbc*L tree for Dennstaedtioid ferns by Wolf et al. 1994.)

This example shows how extensive vegetative homoplasy, reflecting a comprehensible evolutionary cause, apparently undermined the detail (though not the broad sweep) of the topology of the cladogram. Inevitably this homoplasy adversely affects the accuracy of the sequence of character-state transitions dictated by that unreliable topology.

Nucleic acid sequence data are potentially the most effective test of morphologically based hypotheses of simplification. The morphological and molecular data must be gathered for the same range of coded taxa but then analyzed separately to test the hypothesis that the morphological phylogeny is broadly accurate but the putatively simplified taxa may be placed erroneously close to the base of the cladogram. Thus, we would expect a broadly similar molecular phylogeny but with the simplified taxa placed more apically on the topology. Unfortunately, the rhizomorphic lycopsids cannot be tested in this manner, as only one highly derived and highly simplified genus has survived to the present; we cannot compare its genes with those of its more stately and more ecologically significant arboreous forebears. Fortunately, as the pteridophytic "living fossils" case study demonstrates, other lineages of putatively simplified pteridophytes have left a wider range of extant descendants.

THE RAMPANT PARALLELISM OF HETEROSPORY

Heterospory encompasses a suite of several reproductive innovations that together reflect the increasing morphological and physiological distinctiveness of the male and female gametophytes and the transfer of control of gametophytic development and gender from the gametophyte to the sporophyte.

In a detailed review of heterosporic phenomena that integrated living and fossil evidence, Bateman and DiMichele (1994b) recognized a minimum of 11 origins of heterospory (defined as unambiguously bimodal spore size distributions, with the large and small spore morphs likely to engender female and male gametophytes respectively). Although this figure is herein revised to 10 (following the revelations regarding water fern phylogeny that are detailed in the next case study), heterospory remains arguably the most iterative (i.e., repeatedly evolving) potential key innovation in the evolutionary history of the plant kingdom. Moreover, 8 of the 10 inferred origins occurred during a relatively brief period of the Late Devonian and Carboniferous (ca. 380–310 Ma), early in the evolutionary history of vascular land plants.

To understand the underlying causes of this highly iterative evolutionary pattern, Bateman and DiMichele (1994b) dissected heterospory and the seed habit into 12 phenomena that were more discrete and definable, and then sought repetitive patterns among the putative phylogenetic lineages. The maximum number of heterosporic phenomena acquired by each lineage is summarized in Fig. 3. The potentially most valuable information would be the *sequence* in which these characters were acquired, but few morphological phylogenetic studies have been performed at appropriate taxonomic and phylogenetic levels. One exception is a composite cladogram of the ligulate lycopsids (Fig. 2) that indicates a wholly nonhomoplastic pattern of reproductive synapomorphies (this pattern would probably survive repositioning of the ambiguously placed, secondarily reduced genera discussed in the last section). Heterospory, dioicy, heterosporangy, and endospory first appear in *Stachygynandrum* (heterophyllous *Selaginella*), monosporangiate cones in *Sigillaria*, monomegaspory and endomegasporangy in *Diaphorodendron* plus *Synchysidendron*, and integumentation in *Lepidophloios*. This sequence

CLASS Order	Heterospory*	Dioicy	Heterosporangy*	Endospory	Monomegaspory*	Endomegasporangy	Integumentation*	Lagenostomy*	In situ pollination	In situ fertilisation	Pollen tube formed	Siphonogamy
ZOSTEROPHYLLOPSIDA[†] Barinophytales[†]	X	X?	O	O?	O	O	O	O	O	O	O	O
LYCOPSIDA (Clubmosses) Selaginellales	**X**	**X**	**X**	**X**	O	O	O	O	O	O	O	O
Rhizomorphales[1]	**X**	**X**	**X**	**X**	X	X?	X	O	O	O	O	O
SPHENOPSIDA (Horsetails) Equisetales	X	X?	X	X?	X	X?	O	O	O	O	O	O
Sphenophyllales[†]	X?	X?	O	O?	O	O	O	O	O	O	O	O
PTEROPSIDA (Ferns) Stauropteridales[†]	X	X?	X	X?	O[2]	X?	O	O	O	O	O	O
Hydropteridales (Salv.)	**X**	**X**	**X**	**X**	X[3]	X	O	O[4]	O	O	O	O
Hydropteridales (Mars.)	**X**	**X**	**X**	**X**	X	O	O	O	O	O	O	O
Filicales (*Platyzoma*)	**X**	**X**	**X**	O	O	O	O	O	O	O	O	O
PROGYMNOSPERMOPSIDA[†] Aneurophytales[†]	X	X?	O[5]	O?	O	O	O	O	O	O	O	O
Archaeopteridales[†]	X	X?	X	O?	O	O	O	O	O	O	O	O
Protopityales[†]	X?	X?	X?	O?	O	O	O	O	O	O	O	O
Noeggerathiales[†]	X	X?	X	X?	X	X?	O	O	O	O	O	O
Cecropsidales[†]	X	X?	X	X?	X	X?	O	O	O	O	O	O
GYMNOSPERMOPSIDA **(Seed-plants)**	**X**	**X**	**X**	**X**	**X**	**X >**	**X**	**X**	**X**	**X**	**X**	**X**

FIGURE 3   Maximum numbers of heterosporic characters acquired by specific taxonomic orders, listed in *approximate* sequence of acquisition. Origins of heterospory are phylogenetically independent except for the two families of hydropteridalean ferns and some progymnosperm orders, plus their descendants, the gymnosperms. Daggers indicate extinct higher taxa; asterisks indicate heterosporic characters most likely to be detected in fossils. Boldface entries indicate the maximum number of characters exhibited by extant members of the orders.

[1]Lepidodendrales plus Isoetales of most authors. [2]Strictly, reduction in *Stauropteris* is to two viable megaspores rather than one (e.g., Chaloner and Hemsley 1991). [3]*Salvinia* only. [4]*Salvinia* possesses a cellular perispore that superficially resembles a pteridospermalean nucellus. [5]Sporangia of *Chaleuria* contain spores that are dominantly but not exclusively of one gender (modified after Bateman and DiMichele 1994b, Fig. 13).

of character-state transitions illustrates several important concepts pertaining to homoplasy:

1. The apparent sequence is consistent with the pattern predicted from several similar, broadly adaptive evolutionary scenarios (cf. Chaloner and Pettitt 1987; Haig and Westoby 1988, 1989; DiMichele et al. 1989; Chaloner and Hemsley 1991; Hemsley 1993; Stewart and Rothwell 1993; Bateman and DiMichele 1994b).

2. The specified acquisition of four heterosporic phenomena in a single evolutionary step between the Devonian fossil *Leclercqia* and the extant *Stachygynandrum* (Fig. 2) could be interpreted as a saltational (and probably nonadaptive) rather than a gradual (and probably adaptive) event, but the occurrence at this internode of many vegetative innovations in addition to the reproductive acquisitions indicates an improbably long morphological branch and suggests rather that intermediate taxa once existed.

3. The terms used to describe heterosporic phenomena in the lycopsid clade have also been applied to comparable phenomena in other lineages, yet by definition these phenomena are nonhomologous among the many disparate lineages. Clearly, it would be impractical to coin different terms for each biologically definable phenomenon in each of these many remarkable examples of parallelism. Nonetheless, it is beneficial to state which clades are being discussed wherever nonspecific, grade-based heterosporic terms are employed.

4. Adaptive explanations advanced for the evolution of heterospory in the lycopsids gain credence if similar patterns and evolutionary–ecological consequences can be documented in other lineages (Fig. 3); some repetition would be expected as different lineages attempt to rise to similar environmental challenges.

THE ORIGIN OF THE SPOROCARP-BEARING WATER FERNS

Recent molecular studies have greatly advanced our knowledge of the phylogeny of the ferns, arguably the extant group of vascular plants most recalcitrant to morphological phylogenetic analysis due to extensive phenotypic homoplasy (e.g., Rothwell in press).

Until recently the classic example of fern homoplasy was the independent origins of two apparently distinct groups of water ferns, both highly adapted to riparian and fully aquatic habitats. The Marsileales includes three extant genera (*Marsilea, Regnellidium, Pilularia*) and the Salviniales two (*Salvinia, Azolla*; e.g., Bierhorst 1971; Sporne 1975). Although vegetative simplification is evident in the leaf morphology and stem anatomy of both groups (see below), their many reproductive specializations are even more striking. First, the Marsileales and Salviniales possess the most strongly developed and sophisticated heterospory of all extant ferns, probably a testament to the relative success of this life history strategy in aquatic habitats (Chaloner and Pettitt 1987; DiMichele et al. 1989; Bateman and DiMichele 1994b; DiMichele and Bateman 1996). Large megaspores well endowed with nutrients are equipped with various flotation devices. Much smaller microspores are produced *en masse* from separate microsporangia; in the Salviniales they are equipped with organic grappling hooks and/or adhesive surfaces to enable microspore capture by the megaspores prior to fertilization. The resulting gametophytes have evolved into aquatically dispersed disseminules.

As already noted, heterospory *sensu lato* is a highly iterative phenomenon. Comparative studies of morphology, anatomy, and development performed over the past century demonstrated to the satisfaction of most interested parties that heterospory had evolved independently in the Marsileales and Salviniales, presumably in response to similar selection pressures induced by the aquatic life habit. The most compelling evidence for the remarkable functional convergence of the two groups was the independent origin of the sporocarp—a tough protective structure surrounding the sporangia. The sporocarp of the Marsileales encloses both genders of sporangium (Fig. 4A) and is believed to be evolutionarily derived from a leaf segment (pinna) or an entire leaf, whereas that of the Salviniales encloses a single functional megasporangium or several functional microsporangia (Fig. 4B) and is thought to have evolved from an indusium—the specialized sterile cover that

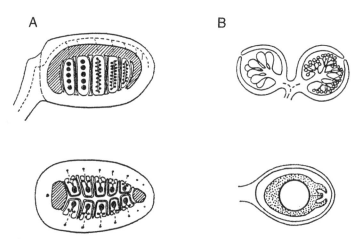

FIGURE 4   Sporocarps of (A) *Marsilea vestita* (Marsileaceae) and (B) *Salvinia natans* (Salviniaceae). These reproductive units were widely regarded as convergent due to structural and developmental dissimilarities, but recent fossil (Fig. 5) and molecular phylogenetic (Fig. 6) evidence shows that they are homologous (after Sporne 1975, Figs. 35, 37).

protects the sporangial clusters (sori) of many of the more conventional terrestrial filicalean ferns. Not surprisingly, this apparently clear-cut example of convergence has become a textbook classic (e.g., Bierhorst 1971; Sporne 1975; Bold et al. 1987; Gifford and Foster 1989).

However, this elegant and compelling hypothesis has recently been dealt two blows in quick succession that together are fatal. The first was a morphological cladistic analysis of the heterosporous water ferns. This included a range of suggested filicalean outgroups and, more significantly, a newly discovered Late Cretaceous fossil, *Hydropteris pinnata*, that uniquely combines vegetative and reproductive features of the Marsileales, Salviniales, and Filicales (Rothwell and Stockey 1994). *Hydropteris* was shown as basal to the Salviniales in the cladogram of 17 unordered morphological characters (Fig. 5), which revealed the Marsileales and Salviniales (reclassified as Marsileaceae and Salviniaceae plus Azollaceae respectively in Fig. 5) as sistergroups (admittedly, reanalysis of the data by the present author shows that merely ordering a single foliar character is sufficient to place *Hydropteris* as sister taxon to the Salviniales plus the Marsileales). Either topology reveals that the water ferns (now assigned to a single order, Hydropteridales) are

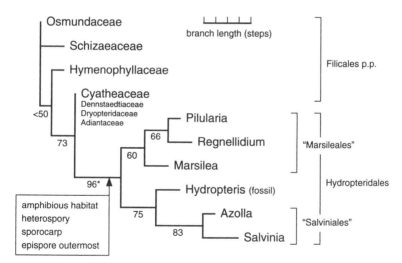

FIGURE 5 Morphological phylogeny of the water ferns and several potential sister-groups, showing a single origin of the heterosporous sporocarps illustrated in Fig. 4. Based on the 17-character matrix of Rothwell and Stockey (1994, Table 1). Characters were unordered and Osmundaceae the outgroup; the search was conducted using the branch-and-bound option of PAUP 3.1.1 (Swofford 1993). The figure illustrates one of three most-parsimonious trees of length 32 steps, consistency index 0.78, and retention index 0.79. Branch lengths reflect Deltran optimization and bear percentage recovery values for 1000 bootstrap replicates. Removal of the questionable extrinsic character "amphibious habitat" only slightly reduced the bootstrap value for the key node supporting the newly recognized hydropteridalean clade.

monophyletic and hence requires only one origin of the many heterosporic features of the water ferns, including the much discussed sporocarp. Any doubts surrounding this startling conclusion were rapidly dispelled by several molecular phylogenetic studies of the extant genera that were presented at the 1994 AIBS meeting in Knoxville, Tennessee, and are summarized as a nonnumerical hypothetical phylogeny in Fig. 6. Although individual trees tend to be poorly supported, the broader trends are credible. RFLP mapping of the chloroplast genome (Raubeson and Jansen 1992; Stein et al. 1992; Raubeson et al. 1994) revealed two independent gene inversions

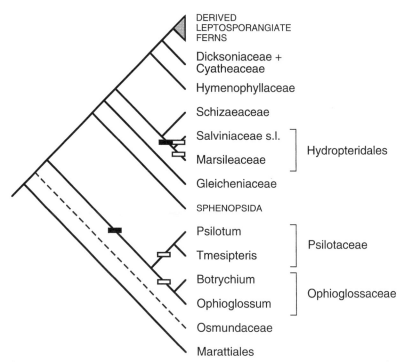

FIGURE 6   Author's approximation of a molecular phylogeny of the basal ferns, reflecting diverse data on plastid genome structure (Raubeson et al. 1994) and *rbc*L sequencing (Hasebe et al. 1994; Manhart 1994; Pryer et al. 1994). Note that the topology, for example the near-basal position of the Osmundaceae, is strongly discordant with the morphological phylogeny of Rothwell (1994, in press) and strongly dependent on the use of the Marattiales as outgroup. The fern clade *sensu lato* may also include the sphenopsids (horsetails) and lignophytes (progymnosperms plus seed plants). If accurate, this tree reduces previous hypotheses of four independent cases of vegetative reduction (white boxes) to two independent origins (black boxes). In the Hydropteridales, simplification is associated with the acquisition of heterospory and a shift to amphibious–aquatic habitats. Note that the phylogenetically controversial Psilotaceae and Ophioglossaceae are derived relative to the Marattiales and Osmundaceae, and therefore by definition are *bona fide* ferns.

followed by duplication of the *chl*L gene that together give considerable structure to preliminary molecular phylogenies of the basal ferns. Additional phylogenies based on *rbc*L sequencing showed a monophyletic Hydropteridales with a 100% bootstrap confidence value and an estimated divergence time of 140 Ma (Hasebe et al. 1994). A combined morphological and *rbc*L study (Pryer et al. 1994) and a *gap*dh study (J. Pahnke personal communication 1994) also confirmed a monophyletic Hydropteridales and suggested a sister-group relationship with the Schizaeaceae (see also Garbary et al. 1993; Mishler et al. 1994).

In summary, molecular data from several living genera and morphological data from their fossil progenitors together refute a popular *a priori* hypothesis of convergence; the perceived homoplasy has been dispelled in favor of *bona fide* synapomorphy. Morphological homologies within the group can now be carefully reassessed in the light of this knowledge.

THE ORIGIN OF SOME PTERIDOPHYTIC "LIVING FOSSILS"

The composite molecular pseudophylogeny summarized in Fig. 6 encompasses another intriguing hypothesis of monophyly that requires radical revision of previous scenarios of vegetative simplification.

The phylogenetic position of the Psilotaceae (extant genera *Psilotum* and *Tmesipteris*, whose sporophytes superficially appear strikingly different: Figs. 7A, 7B) has long challenged comparative morphologists (e.g., Bierhorst 1971; Kaplan 1977). Again, the salient problem underlying the historical controversies was distinguishing, in the absence of a phylogeny, between genuinely plesiomorphic characters and secondary losses with consequent reduction in overall phenotypic complexity, given only a small number of extant "long-branch" taxa (i.e., a taxon separated from its sister-group by many character-state transitions). The popular hypothesis that the simple sporophytes of the Psilotaceae are primitive among land plants led many textbook authors (Sporne 1975; Bold et al. 1987; Gifford and Foster 1989) to place them in the otherwise wholly fossil group of the earliest vascular land plants, the "rhyniophyte" plexus (a group that radiated in the Siluro-Devonian, ca. 400 Ma). Even when a derived position of the Psilotaceae relative to the rhyniophytes was acknowledged (Taylor and Taylor 1992; Stewart and Rothwell 1993), the family was still discussed in the inappropriate *context* of the rhyniophytes. The more recent hypothesis of secondary simplification to explain the simple shoot systems of the Psilotaceae would imply a much younger divergence from an ancestor that was a *bona fide* fern. This view has gained ground in recent years, following developmental

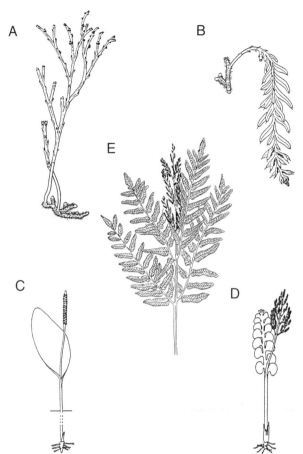

FIGURE 7   Sporophytes of (A) *Psilotum*, (B) *Tmesipteris* (both Psilotaceae), (C) *Ophioglossum*, (D) *Botrychium* (both Ophioglossaceae), and (E) *Osmunda* (Osmundaceae). (A)–(D) are probably the collective products of a single vegetative simplification event that affected a more complex marattialean or ?osmundacean ancestor at least 230 MY ago (A–D after Sporne 1975, Figs. 7, 29; E after Gifford and Foster 1989, Fig. 13.45E).

morphological investigations of *Psilotum* that reinterpreted supposed sporangiophores and sporophylls as reduced branches and leaves respectively (Bierhorst 1971; Kaplan 1977; Rouffa 1978). Nonetheless, old hypotheses of plesiomorphic simplicity rather than secondary loss die hard.

A second problem of long-branch taxa potentially showing radical simplification is presented by the Ophioglossaceae (extant genera *Ophioglossum*, *Botrychium*, and the monotypic *Helminthostachys*: Figs. 7C, 7D). In these pteridophytes a subterranean rhizome generates repeated aerial developmental units that consist of a single laminate megaphyllous leaf; the more vigorous units also develop a single fertile axis bearing many large sporangia.

In *Ophioglossum* this fertile axial system is reduced and the sporangia are set in two opposite rows. Once again, the large phenotypic distance separating these genera from all others obscures homologies and thus weakens their phylogenetic placement; this has led to an even wider array of hypotheses of relationship. If the ophioglossacean genera represent primary simplicity they could be descended directly from the wholly extinct trimerophyte plexus, which may have generated several fern lineages (e.g., Stewart and Rothwell 1993, chart 20.1). If their simplicity is secondary, reflecting phenotypic reduction, they could be descended from more complex ferns (e.g., Sporne 1975; Stewart and Rothwell 1993) or, more controversially, the progymnosperms (Bierhorst 1971; Kato 1988).

The recent molecular studies discussed above support previous hypotheses of monophyly of (1) Psilotaceae and (2) Ophioglossaceae (albeit on the basis of limited taxonomic sampling (cf. Hasebe et al. 1994; Manhart 1994).) What was wholly *un*expected was the well supported monophyly of the Psilotaceae *plus* the Ophioglossaceae (Fig. 6), based on the chloroplast gene *rbc*L, 18S RNA, and chloroplast DNA structural evidence (Manhart 1994; Mishler et al. 1994; Pryer et al. 1994). These two fern families differ radically in phenotype and had been hypothesized by many pteridologists to have originated from different taxonomic classes, yet the molecular data suggest that they have a shared origin in a single initial simplification event. Admittedly, several subsequent character-state losses must have been superimposed on subsets of the clade in order to generate their now very different phenotypes; loss of the *chl*L plastid gene from *Psilotum* has also been documented (Raubeson et al. 1994).

Current evidence suggests that the Marattiales (Fig. 7E) and (more controversially) the Osmundaceae may be the extant groups most closely related to this morphologically reduced clade (Stein et al. 1992; Hasebe et al. 1994; Raubeson et al. 1994; but see Rothwell 1994, in press). Scenarios involving reduction in size and vegetative complexity of these taxa can in theory generate the vegetative morphology of the Ophioglossaceae (Figs. 7C, 7D), though the pattern of derivation of the exclusively extant Psilotaceae (Figs. 7A, 7B) is more obscure. *Botrychium* extends back in time to at least ca. 60 Ma (Rothwell and Stockey 1989), the Ophioglossaceae perhaps to ca. 230 Ma (Millay and Taylor 1992), the Osmundaceae certainly to ca. 215 Ma (Harris 1961) and possibly to ca. 260 Ma (Gould 1970), and the Marattiales arguably to ca. 350 Ma (Stewart and Rothwell 1993). Thus, although these morphologically simple families may qualify as "living fossils," they are unlikely to have originated during the great land plant radiation of the Siluro-Devonian (ca. 400 Ma: Gensel and Andrews 1984; Bateman 1991; DiMichele et al. 1992; DiMichele and Bateman 1996); their phenotypic simplicity is secondary rather than primary.

As in the example of the heterosporous water ferns, the *a priori* hypothesis of secondary simplification is supported by molecular phylogenetic analyses, but the hypothesis of parallel origins by phenotypic reversals is refuted. Consequently, the observed level of homoplasy is less than was expected *a priori*. In this second example of fern simplification, the phylogenetically established morphological homologies between two supposedly independently derived groups—the Psilotaceae and Ophioglossaceae—may remain less intuitive even when the morphological data are carefully reexamined.

ARE THE BRYOPHYTES VASCULAR PLANT PRECURSORS OR SECONDARILY SIMPLIFIED RHYNIOPHYTES?

One of the most consistent features of reasonably detailed morphological phylogenies of the land plants (Mishler and Churchill 1985b; Bremer et al. 1987; Crane 1990; Kenrick and Crane 1991; Albert et al. 1994; Donoghue 1994; Kenrick 1994; Mishler et al. 1994) is the interpolation, between the basal paraphyletic group of charophycean green algae and apical monophyletic group of vascular plants, of a paraphyletic group of bryophytes (admittedly, there has been less consensus on key relationships *within* the bryophytes—that is, among the liverworts, hornworts, and mosses (e.g., van der Peer et al. 1990; Waters et al. 1992; Garbary et al. 1993)). This intermediate placement of the bryophytes on the main axis of the land plant phylogeny generates an inherently appealing "ladder of progression" in life history evolution: dominance by the haploid gametophytic phase in aquatic environments (charophytes) is transferred to terrestrial environments (bryophytes), where it then progressively gives way to increased dominance by the diploid sporophytic phase (eutracheophytes) (e.g., Graham et al. 1991; Kenrick and Crane 1991; Hemsley 1994; Kenrick 1994) (Fig. 8).

However, bryophytes are notoriously elusive in fossil assemblages of early land plants (e.g., Hueber 1961; Taylor 1988; Hemsley 1990; Taylor and Taylor 1992; Kenrick 1994). Consequently, the occasional heretic has suggested that one or more of the three main bryophytic groups may be secondarily reduced, and that gametophytic dominance in the terrestrial realm may therefore have arisen from ancestral life histories that gave a more prominent role to the sporophyte (e.g., Robinson 1985; Taylor 1988; Garbary et al. 1993). This possibility has been given greater credence by recent recognition of fossil tracheophytes of Devonian age (ca. 400 Ma) that clearly exhibit near-isomorphic alternation of generations (Remy and Haas 1991; Remy et al. 1993; Kenrick 1994). Although distinguishable, the gametophytic and sporophytic generations are of similar size and structural complexity (unfortunately, we cannot observe the third criterion relevant to dominance, their relative longevity). Of course, these near-isomorphic "rhyniophyte" genera

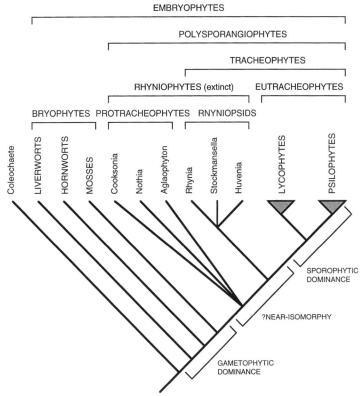

FIGURE 8 Morphological phylogeny of representative early land plants, demonstrating the pivotal phylogenetic position of the Siluro–Devonian rhyniophytes that possessed putatively isomorphic alternation of generations. The implication that the gametophyte-dominated bryophytes evolutionarily preceded the isomorphic rhyniophytes is challenged by the scenario summarized in Fig. 9. Note that molecular resolution of this fundamental problem is hampered by the exclusively fossil nature of the rhyniophytes (modified after Kenrick and Crane 1991, Fig. 26; Kenrick 1994, Fig. 3).

("protracheophytes" plus rhyniopsids *sensu,* Kenrick and Crane (1991) and Kenrick (1994)) can readily be shoe-horned into the preexisting phylogenies as an extra paraphyletic group between the bryophytes and eutracheophytes (Fig. 8), but the rejuvenated rhyniophytes also offer an alternative evolutionary scenario. Specifically, could the rhyniophytes be both direct descendants of the charophytes and the ancestral plexus of *independently derived* bryophytic and eutracheophytic clades?

A supracladistic evolutionary and ecological scenario that would favor this hypothesis is summarized in Fig. 9. The evolutionary relationship between sporophyte and gametophyte can be viewed as antagonistic, as each phase attempts to dominate the life history of the other (Bateman and

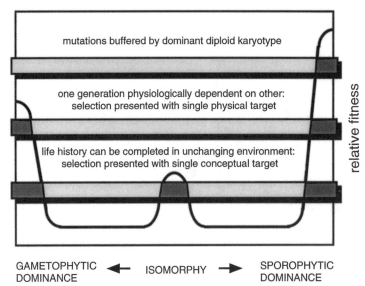

GAMETOPHYTIC ◀— ISOMORPHY —▶ SPOROPHYTIC
DOMINANCE   DOMINANCE

FIGURE 9   Hypothetical relative fitness of contrasting life histories, indicating peaks and valleys in overall fitness. The relative heights of the peaks are justified on the basis of three advantageous properties—similar habitat preferences of sporophyte and gametophyte, physical dependence of one life-history phase on the other, and buffering of mutations in the diploid phase—none of which is hypothesized to characterize the intervening fitness valleys of incomplete dominance by one generation over the other (see also Bateman and DiMichele 1994b).

DiMichele 1994b; DiMichele and Bateman 1996). During their evolution the pteridophytes and their derivatives show increasingly strong sporophytic dominance, with the result that selection becomes increasingly focused on the diploid phase. The derived gametophytes of seed plants are unisexual and highly reduced, and thus effectively parasitic on the sporophyte. This presents selection with a unified organic target, improving the probability that the organism can achieve optimal fitness in its preferred habitat.

This rationale also applies to the extreme gametophytic dominance of the bryophytes. Here, in contrast, the sporophyte is near-parasitic on the gametophyte, and hence it is the gametophyte rather than the sporophyte that determines habitat preference. As in sporophytic dominance, environmentally mediated selection is effectively presented with a single target. However, the long life of the haploid gametophyte relative to the diploid sporophyte means that it will accumulate mutations unbuffered by wild-type alleles (e.g., Knoll et al. 1986).

In contrast to sporophytic and gametophytic dominance, near-isomorphy presents two more-or-less equal targets to selection within a single habitat. This is not necessarily a serious handicap if the two generations possess similar adaptations to that habitat and hence have roughly equal overall fitnesses. They then continue to present a single conceptual target to selection, even though the two generations are physically separate and economically independent (Fig. 9).

Rather, first principles suggest that the most problematic life-history strategy would be incomplete dominance of one generation by the other; that is, where the haploid and diploid generations are physically separate and phenotypically distinct, thereby offering separate targets for selection, but one is significantly larger bodied, longer lived, more phenotypically complex, and/ or has greater control over meiosis and syngamy than the other. The habitat preferences of the two generations are then likely to diverge, threatening the long-term survival of the lineage (unless it can short-circuit the full sexual life history).

Perhaps the extant examples most similar to such life histories are among the Hymenophyllaceae (Fig. 6). These filmy ferns are characterized by small sporophytes possessing thin megaphyllous leaves that lack the protective waxy cuticle that is otherwise ubiquitous among eutracheophytic sporophytes. In this clade the gametophytes and sporophytes are strongly dimorphic, more similar in size than those of most extant pteridophytes, and economically independent. Recent startling evidence suggests that in at least some species the gametophytes are on average longer lived, more persistent, and more geographically widespread than the marginally more complex sporophytes (Farrar 1990; Sheffield 1994). Even the combined habitat preferences of sporophyte and gametophyte are unusually narrow, generally requiring constant atmospheric moisture and minimal frost exposure to compensate for the absence of a cuticle.

If the hypothetical scenario outlined (Fig. 9) is valid, it would predict greater overall evolutionary (and certainly ecological) success for extreme gametophyte or sporophyte dominance than for isomorphism, which in turn would be more successful than incomplete dominance of either life history phase over the other. This interpretation is supported by the relative abundance of species known to possess these contrasting life histories; the modern terrestrial flora consists almost exclusively of species with strong sporophytic or gametophytic dominance.

Thus, in theory the way is clear to derive the bryophytes from one or more isomorphic ancestors in the Siluro-Devonian by sporophytic reduction and ecological specialization. How strong is the evidence against such a theory? Two recent studies suggested unconventional placements for the bryophytes, but both are open to criticism. Garbary et al. (1993) placed the bryophytes as a monophyletic sister-group to the tracheophytes, but this study was restricted to data on spermatozoan ultrastructure, ignoring other characters and fossil taxa. Manhart's (1994) *rbc*L summary yielded highly improbable diphyly of both (1) the liverworts and (2) the tracheophytes, but again was obliged to omit fossil taxa; moreover, the gene appeared incapable of satisfactorily resolving divergences of such profound time depths.

It may be more relevant to examine the evidence in favor of the placement of a paraphyletic Bryophyta immediately below the polysporangiophyte clade (rhyniophytes plus eutracheophytes). When the rhyniophytes are omitted the several synapomorphies of the eutracheophytes look impressive (e.g., Bremer et al. 1987) and militate strongly against hypotheses of secondary reduction from these relatively complex plants. However, when the rhyniophytes are considered (Crane 1990; Kenrick and Crane 1991; Kenrick 1994) the only concrete synapomorphies of the tracheophytes *sensu lato* are physically independent sporophytes bearing multiple sporangia—the very characters whose polarity and apocryphal irreversibility are being questioned here.

Hence, the inclusion of the fossil rhyniophytes in recent phylogenetic analyses of land plant origins has rendered far more probable the reductive origin of one or more bryophyte lineages by requiring significantly fewer character losses (arguably only two) than any previous hypothesis of reduction, from the more phylogenetically distant eutracheophytes. Given the potential ecological constraints on incomplete dominance, the paucity of reliable early bryophytic fossils, and the clear predilection for morphological simplification within the main bryophyte lineages (cf. Robinson 1985; Mishler and Churchill 1985a), present consensus character coding merits reexamination in the context of reductive hypotheses. Choice of outgroups would be especially crucial in allowing the bryophytes the opportunity to be placed *among* the isomorphic taxa in a morphological cladistic analysis. Unfortunately, any attempts to test the possible origin(s) by simplification of the bryophytes using molecular data will be seriously handicapped by the lack of extant isomorphic rhyniophytes.

PROFOUND SHIFTS IN ECOLOGICAL ROLES: AQUATICS AND HETEROTROPHS

Many of the most radical evolutionary simplifications in land plant morphology are associated with profound shifts in ecological roles. The derived, ecologically specialized products of such shifts have attracted preferential attention from morphological and especially molecular phylogeneticists, perhaps in part because their vegetative modification (generally simplification, with the obvious exception of insectivores) has rendered their phylogenetic relationships especially ambiguous.

One such example has already been discussed in the context of the rhizomorphic lycopsids and hydropteridalean water ferns—extreme vegetative reduction and specialization for semiaquatic and aquatic habitats. Common consequences of such changes are reduction or loss of rooting systems (associated with transfer of absorptive functions to modified leaves in the Salviniaceae), decreased proportions of strengthening and/or vascular tissues, elongation and thinning of leaf-bearing axes, webbing of surficial leaves and dissection of subsurficial leaves, modification of stomata and their concentration in parts of the plant still frequently exposed to the atmosphere, and in fully aquatic species the evolution of flotation aids such as internal air spaces and ribs or large trichomes on leaves (e.g., Crawford 1987).

Apart from the aforementioned hydropteridalean ferns (Raubeson et al. 1994; Rothwell and Stockey 1994) and *Ceratopteris* (Page 1979), the best known examples of aquatic adaptations occur among the angiosperms. In most cases, highly reduced genera have proved to be phylogenetically nested within typically less specialized nonaquatic families. On fossil (G. W. Rothwell personal communication 1995), extant morphological (Mayo 1993) and chloroplast DNA (French et al. 1993) evidence, the tiny floating duckweeds of the genus *Lemna* (Fig. 10A) are nested within the Araceae; the cryptocorynid aroids (Fig. 10B) form the most likely sister-group to *Lemna*. Other examples of vegetatively reduced hydrophilic angiosperms include the submerged aquatic *Callitriche*, which is monophyletic and nested within the Scrophulariaceae (Reeves and Olmstead 1993; Philbrick and Les 1995), and the riparian annual *Navarettia*, nested within the Polemoniaceae (Spencer 1993).

Of greater phylogenetic significance are two groups of vegetatively simple aquatics that have each been implicated on both morphological and molecular evidence as potentially the most primitive extant angiosperms: the Nympheaceae (Hamby and Zimmer 1992) and *Ceratophyllum* (Les 1988; Crepet et al. 1993; Chase et al. 1993; Endress 1994; Crane et al. 1995). *Ceratophyllum*, in particular, is still strongly favored as a near-basal angiosperm, implying that its vegetative simplicity could be plesiomorphic rather than secondary within the angiosperms. Resolution of this question is necessary to infer whether the angiosperms could have originated in an aquatic habitat (a similar aquatic–terrestrial transition has been inferred for the apomictic heterosporous pteridophytes and earliest gymnosperms: Bateman and DiMichele 1994b; DiMichele and Bateman 1996). However, *Ceratophyllum* occupies an extremely long branch of the *rbc*L tree (e.g., Chase et al. 1993) and its perceived phylogenetic position may therefore be an erroneous case of long-branch attraction (e.g., Hendy and Penny 1989; Donoghue 1994; Penny et al. 1994). Whatever the true nature of the oldest angiosperm (Doyle 1994), transitions between terrestrial and aquatic habitat preferences—together with the associated radical vegetative and life history transitions—have occurred many times during the evolutionary history of the plant kingdom. Similar arguments can be made for iterative origins of tropical mangrove trees and of annual from perennial life histories among angiosperms.

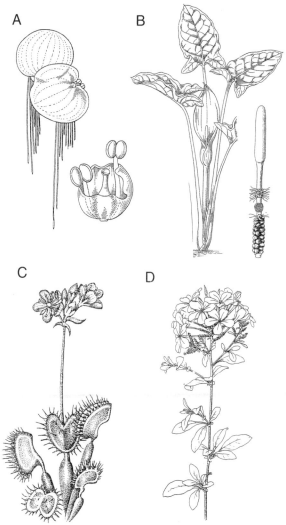

FIGURE 10   Positive correlations between ecological shifts and parallel radical changes in phenotype. Hydrophily (French et al. 1993; Mayo 1993); (**A**) the vegetatively reduced aquatic *Lemna polyrrhiza*, and (**B**) its nonaquatic aroid sister-group, here illustrated by *Arum italicum*. Heterotrophy (Williams et al. 1994); (**C**) the complex, ecologically specialized spring-trap insectivore *Dionaea muscipula* and (**D**) its noninsectivorous sister-group in the Plumbaginaceae, here illustrated by *Plumbago auriculata* (A and B after Ross-Craig 1973, Figs. XXXI plate 40B, XXX plate 4; C after Cheers 1992, p. 62; D after Luteyn 1990, Fig. 5).

Another suite of radical morphological changes associated with major shifts in ecological roles is the acquisition of at least partially heterotrophic nutrition in previously autotrophic plant lineages.

Saprophytism and parasitism in many phylogenetically disparate ground-dwelling genera (e.g., *Orobanche, Monotropa, Rafflesia*) generally lead to greatly reduced rooting structures and leaves (typically functionally replaced by various forms of haustoria), and substantial or even complete loss of chlorophyll. Aerial epiphytic parasites such as *Viscum* are more likely to retain chlorophyllous leaves, but in both cases flowers become the main morphological indicators of phylogenetic relationship. The hemiparasitic and parasitic Scrophulariceae and Orobanchaceae have recently been shown to be collectively monophyletic (DePamphilis and Young 1995; Wolfe and DePamphilis 1995). With this exception, these phenomena have not yet been subjected to systematic phylogenetic surveys; nonetheless, clearly they have evolved independently on many occasions.

An alternative approach to heterotrophic nutrition in plants, especially in nitrogen deficient soils, is insectivory. Unlike all of the case studies discussed above, this life-history transition tends to increase rather than decrease overall morphological complexity (Figs. 10C, 10D). The tremendous range of complex adaptations to insectivory—in particular the various spring, bladder, passive and active flypaper, and pitcher traps, and their underlying physiological specializations—have long fascinated comparative morphologists (e.g., Darwin 1875; Juniper et al. 1989). Recent *rbc*L phylogenies (Albert et al. 1992; Williams et al. 1994) have identified approximately seven independent evolutionary origins of carnivory; some encompass only one mode of insect capture, but three modes occur in different members of the Droseraceae.

Interestingly, where traditional morphological opinions on the phylogenetic positions of various types of heterotrophs have been tested using molecular data, the original morphological inferences have been largely upheld. Moreover, no clear reversals to autotrophy have been documented in any heterotrophic clade, suggesting that these nutritional adaptations rapidly escalate to inescapable physiological constraints—presumably their loss inevitably causes the death of the deficient individual. Thus, at the coarsest level of resolution of nutritional homologies, present evidence suggests that saprophytism–parasitism and insectivory are both highly iterative and reliably irreversible.

DISCUSSION

It would be difficult to find a set of botanical cladistic case studies more diverse than those outlined above. I have deliberately avoided extensive discussion within each case study, intending merely that they should provide an empirical background for the more conceptual discussions that follow.

Parallelisms, Reversals, and Parallel Reversals

Many of the above case studies focus on examples of the repeated acquisition of particular phenotypic features (heterospory in pteridophytes, insect–trapping devices in carnivorous plants) or their repeated loss (the tree habit in the rhizomorphic lycopsids, leaves in saprophytes and parasites, extensive root systems in aquatics). In other case studies, presumed examples of parallelism were disproved by cladistic analysis (the sporocarps of the water ferns, simplification of the pteridophytic "living fossils"); in yet another, reversals were suggested despite contrary cladistic evidence (bryophyte origins).

Together, these examples amply demonstrate the simplistic nature of the contrast between the supposed opposites of parallelism and reversal implicit in cladistic optimization procedures when applied to morphological data. Specifically, reversals in the cladistic sense of reversion to a relatively plesiomorphic character state do not necessarily equate with the physical loss of the specified feature; this depends on the polarity of the character, which in most cases is determined by outgroup comparison. Furthermore, reversals often occur in parallel, particularly where each coincides with the occupation of a radically different niche. This observation explains the repeated loss of the tree habit among the rhizomorphic lycopsids, of leaves in saprophytes–parasites, and of extensive rooting systems in aquatics.

As has long been recognized, morphological parallelisms (including parallel reversals) must be regarded as errors of *a priori* homology assessment (e.g., Patterson 1988). Thus, there may be 10 origins of heterospory, but further careful scrutiny will identify differences between modes of heterospory in different lineages. Indeed, such differences are already evident when the phenomenon is divided into more meaningful characters (Bateman and DiMichele 1994b), and the term "heterospory" becomes of purely functional rather than phylogenetic significance (Fig. 3). Other botanical features that are pragmatically (terminologically) unified but phylogenetically iterative include leaves, roots, bipolar growth, and secondary thickening (Bateman 1994), together with the carnivorous angiosperm case study already discussed (Albert et al. 1992).

Morphological losses of features are less amenable to homology reassessment, as it is difficult to find subtle differences among morphological features that do not exist in any of the species in question! Thus, any hope of identifying and rectifying such homoplasies must lie in studying the underlying genetic changes, particularly if it can be demonstrated that the gene(s) that coded for the secondarily absent feature in sister-groups that possessed it have been suppressed in different ways in the clades under comparison. Unraveling the causal factors underlying parallel gains or parallel losses of a particular feature or function is worth the effort if such patterns are indeed the best (albeit circumstantial) available evidence of adaptation.

Morphological versus Molecular Data: Testing Hypotheses of Phenotypic Simplification

Molecular data have other roles to play in the investigation of homoplasy. Phylogenies based on nucleic acid sequence data incur relatively high levels of homoplasy, an inevitable consequence of translating a language that uses an alphabet of only four letters. Moreover, if the lengths of all possible trees are computed and the number of topologies of each length is plotted, the most parsimonious end of the resulting, approximately bell-shaped distribution of tree lengths is usually very steep—that is, there are many most-parsimonious trees and very many almost parsimonious trees. Thus, molecular phylogenies are often presented as some form of consensus tree. This decision has two important consequences: the inevitable induction of polytomies reduces topological resolution, and it becomes impossible to accurately fit ambiguous character-state distributions to the topology because optimizations are likely to differ among the constituent most-parsimonious trees (e.g., Swofford 1991; Donoghue and Sanderson 1992; Patterson et al. 1993). In a purely molecular phylogenetic context this does not matter, given that (with the possible exception of first and second versus third positions in codons: e.g., Albert et al. 1993) all base pairs are equally significant. However, it is the morphological characters that more directly reflect, and more accurately pace, the rate of evolution (cf., Donoghue 1989; Harvey et al. 1994; Milligan 1994; Bateman in press). Also, in comparison with morphological phylogenies, individual sequenced genes such as the erstwhile plastid panacea rbcL are perhaps most appropriately perceived conceptually as single characters with many component states (e.g., Doyle 1992, 1994, this volume; Albert et al. 1994).

For these reasons, I regard rigorous morphological phylogenies (i.e., those based on well founded phenotypic data for all organs and all life history phases) as likely to yield on average topologies that are more accurate than single-gene molecular phylogenies based on the same range of coded taxa. The notable exception is profound reduction, where under some conditions parsimony is liable to misrepresent secondary losses as plesiomorphies and hence place simplified taxa inappropriately close to the base of the tree. First principles suggest that several factors decrease the likelihood of a strongly secondarily reduced coded taxon being placed accurately in a phylogeny.

Some mutually reinforcing factors can generate nonparsimonious evolutionary trees, which by definition cannot be recovered by parsimony analysis; they place each secondarily reduced lineage one or more nodes below the correct position (pseudoplesiomorphy). This is most likely to occur when synapomorphies recently acquired by the lineage are lost before they have been inherited by more than very few coded taxa (1). The probability that such losses will undermine the topology is increased if many characters are lost on the same branch of the cladogram, and the probability of such simultaneous losses is in turn increased if the affected characters prove to be developmentally correlated. Losses of similar sets of characters in several lineages of the clade (i.e., parallel reversals, again often developmentally correlated) can also lead to pseudoplesiomorphy, particularly when the affected branches are clustered on the cladogram so that it is more parsimonious to postulate multiple acquisitions of the actual synapomorphies than multiple losses of the actual homoplasies (2). Proximity of the reversal-bearing branches to the (by definition plesiomorphic) base of the cladogram also promotes pseudoplesiomorphy (3). Perhaps most serious is the erroneous identification of a secondarily reduced coded taxon as plesiomorphic and its consequent use as an outgroup to root the phylogeny and polarize the characters (4). (Secondary simplification is, of course, equally damaging to attempts to polarize characters via ontogenetic transformations.)

Inaccurate topologies (though without the basal bias induced by factors (1)–(4)) are further promoted by generally high levels of homoplasy, relative to its size, in the primary matrix (5). These increase the potential number of parsimonious distributions of character states competing for nonhomoplastic status. Large numbers of missing values in the matrix, reflecting the inability to code the lost feature(s), tend to engender ambiguities in both the topologies and the position of character-state transitions (6).

Reviewing the six case studies reveals that factors (1)–(4) are suspected of influencing the rhizomorphic lycopsid study (each dashed branch in Fig. 2 could reasonably be attached to the branch immediately above it in the near-pectinate cladogram). Factors (1) and (4) together may call into question the current near-consensus on bryophyte origins, (1) and perhaps (2) and (4) have prevented successful morphological resolution of the sister group relationships of the water ferns and several pteridophytic "living fossils," (1) and (2) have conspired to obscure the origins of aquatic and saprophytic–parasitic species, and (2) has helped to retard the production of phylogenies suitable for interpreting the many origins of heterospory. Admittedly, the heterospory problem in particular will not be adequately solved until we have a well-founded and detailed phylogeny of all pteridophytes.

Thus, repeated simplification is arguably the class of evolutionary changes most likely to undermine the accuracy of a morphological cladogram. There is in effect a threshold of "two steps forward, one step back" character change beyond which the evolutionary pattern is no longer parsimonious. In this situation, process-based reasoning strongly suggests that (1) molecular and morphological data are likely to generate conflicting most-parsimonious trees and (2) molecular trees are on average more likely to be reliable, at least for segments of the cladogram that contain nodes supported by multiple reversals. This is essentially the argument that has been advanced here in the case studies of water fern and "living fossil" pteridophytes. Moreover, this dual analytical approach of morphology and molecules will increasingly resolve the iterative evolution of heterospory and of aquatic and heterotrophic life histories. However, the rhizomorphic lycopsid phylogeny cannot be tested in this way as only one atypical lineage has survived to the present day and thus remains molecularly testable. And although bryophytes are well represented in the modern flora, it is their relationship to the wholly extinct (and therefore molecularly impervious) rhyniophyte plexus that is crucial to understanding their evolutionary origins.

Homoplasy, Long Branches, and Evolutionary Scenarios

Long branches of phylogenies are much discussed but rarely defined. Here, for the sake of argument, they are defined as individual branches within a phylogeny that exceed the upper bound of one standard deviation from the mean number of character-state transitions per branch, including autapomorphies, within that phylogeny (thus, corresponding short branches fail to reach the lower bound of one standard deviation from the mean). When considering seed-plant relationships, good examples of morphological long branches subtend the anthophytes, Gnetales, and angiosperms (Doyle and Donoghue 1986; Stein and Beck 1987; Nixon et al. 1994; Rothwell and Serbet 1994; Crane et al. 1995), and of molecular (*rbc*L) long branches the Gnetales, angiosperms, and the putative basal angiosperm *Ceratophyllum* (Chase et al. 1993) (also the lycopsid *Selaginella*: Manhart 1994).

Discussion of long branches has generally focused on their tendency to artificially attract each other during tree construction (e.g., Hendy and Penny 1989). Falsely joining two long branches both misleadingly shortens them and yields an incorrect topology. Less often discussed is the fact that these two consequences together inevitably lead to erroneous patterns of character-state transitions—the all-important arbiters of underlying evolutionary mechanisms.

There has been much recent debate concerning the distinction between "hard" and "soft" polytomies in phylogenies (e.g., Maddison and Maddison 1992; Archibald 1994; Hoelzer and Melnick 1994). A hard polytomy is viewed as an accurate representation of one or more speciation events that did not involve the acquisition of synapomorphies, whereas a soft polytomy simply reflects the temporary absence of data capable of resolving the relationships of the three or more daughter lineages under scrutiny. Here, I will argue that this terminological distinction between genuine irresolution and (hopefully transient) ignorance can be extended to long branches.

A "hard" long-branch represents a saltational speciation event—a profound morphological transition involving the immediate evolutionary modification of several characters (Bateman and DiMichele 1994a). Perhaps the best and most frequent examples of such evolutionary steps are radical phenotypic changes reflecting genome suppression events, which simultaneously deprive a lineage of many characters; several examples consistent with such changes are highlighted in the above case studies.

In contrast, a "soft" long-branch reflects a series of speciation events that cannot be distinguished because some of the daughter species have been omitted from the phylogenetic analysis. Such omissions are accidental if the intermediate species, living or fossil, have not yet been discovered or are not known to the analyst, but they can also be deliberate. For example, the analyst is likely to be obliged to omit fossil species from an exclusively molecular phylogeny (see below), and phylogenetic analyses at all but the lowest taxonomic levels are obliged to use "place-holding" species to represent much larger, putatively monophyletic groups (such as the single species that routinely represent taxonomic classes or orders in the many high-level cladistic analyses spanning the land plants or seed plants). Selective sampling from among the known species of a clade is an operational necessity, given that computational constraints generally prevent a global parsimony analysis of more than 25–30 coded taxa (e.g., Pankhurst 1995). To minimize soft long branches, it is desirable that only the more primitive coded taxa should graduate to subsequent cladistic analyses conducted at progressively higher taxonomic levels.

Including the most complete range of relevant coded taxa is unimportant for determining sister-group relationships and thereby classifying taxa on the basis of monophyly, but it is essential for the interpretation of underlying evolutionary mechanisms. On average, maximizing the number of coded taxa included in an analysis minimizes the number of character-state transitions on each branch of the most-parsimonious cladogram(s) (although each new taxon added to a matrix can in theory also add new characters or character states, there is in practice a lower probability that this will occur with each additional taxon). This in turn maximizes the probability of determining the *sequence* of acquisition of character states within a lineage by distributing them on successive branches of the tree.

Here, the greatest relevance of the distinction between hard and soft long branches is that morphological long branches are the best phylogenetic indication of profound simplification events (and the assumed underlying saltational mechanisms). Indeed, Bateman and DiMichele (1994a) argued that saltation should be the null hypothesis for explaining a morphological long branch, to be tested by the addition of further coded taxa in an attempt to dissociate the character-state transitions supporting the long branch (Fig. 11). Dissociation falsifies the hypothesis that all of the co-occurring character-state transitions occurred in a single speciation event. It also increases our knowledge of the sequence of character-state acquisitions through time. Interpretationally, dissociation of multiple character-state transitions shifts the emphasis away from a saltational scenario (of the kind advocated here for rhizomorphic lycopsid vegetative architecture) and toward a more gradual adaptive scenario (a more likely explanation for the origins of heterospory and carnivory). This places the focus firmly on the function of the few remaining synapomorphies supporting the internode (for an unduly critical review of such approaches see Frumhoff and Reeve 1994).

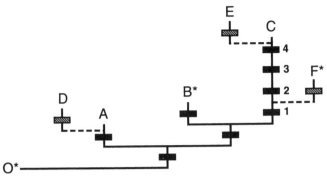

FIGURE 11 Cladograms as tests of saltational evolutionary hypotheses. The initial hypothetical analysis of outgroup species O and ingroup species A–C yields a single fully resolved topology. The relatively long branch subtending species C is supported by transitions in four potentially developmentally correlated characters, for example: 1, suppression of dichotomy in root apical meristems; 2, suppression of dichotomy in shoot apical meristems; 3, switch from multiple to single reproductive cones; 4, switch from polycarpy to monocarpy. Simultaneous change in these four characters by saltation is assumed as the null hypothesis, and then tested by reanalyzing the matrix with three additional ingroup species (labeled D–F). Species D is attached to an irrelevant branch. Although species E is connected to the relevant branch, it fails to dissociate the four correlated character-states. However, addition of species F separates character 1 from characters 2–4. This implies that the state transition occurred earlier in character 1 and thereby falsifies that part of the saltational hypothesis. Note that the fossil status of three of the species (asterisked) has no bearing on their analytical performance (after Bateman and DiMichele 1994a, Fig. 3).

Last, there are good reasons for believing that hard (saltational) long branches are more common among higher plants with their modular construction than among the more developmentally integrated higher animals (for more extensive discussions see van Steenis 1976; Bateman and DiMichele 1994a). In particular, plants have far greater ability to undergo mosaic evolution—radically different evolutionary rates and trajectories in different organs. The semi-independence of their component organs reflects meristematic growth. This in turn is a consequence of the sessile nature of each plant, which limits opportunities for direct competition among plants.

Homoplasy and the Role of Fossils in Phylogeny Reconstruction

The optimal role of fossils in phylogeny reconstruction has long been debated (e.g., Patterson 1981; Donoghue et al. 1989; Forey 1992; Fisher 1994; Smith and Littlewood 1994). Fossil coded taxa played a major role in the first five case studies of homoplasy presented here and have important, if not yet adequately explored, implications for the sixth (DiMichele and Bateman 1996).

Without fossil data, the only living genus of rhizomorphic lycopsids (*Isoetes*) would appear extremely morphologically divergent from its putative closest living relatives of the Selaginellaceae (Fig. 2). Indeed, the branch of the morphological phylogeny separating the Isoetaceae and Selaginellaceae would be so long that few homologous characters could be identified with confidence. It is the fossil evidence that clearly demonstrates that this is a "soft" long-branch, reflecting major phenotypic gaps in the modern flora. In the case of the water ferns, the fossil *Hydropteris* constituted a key link in the refutation of a long-standing hypothesis of morphological and functional convergence in the Salviniaceae and Marsileaceae (Fig. 5), again providing valuable information on the sequence of character acquisition as well as clarifying evolutionary relationships. In the analysis of heterospory, fossils not only yield intermediates between extant heterosporous groups but they also reveal several wholly extinct heterosporous lineages (Fig. 3). Thus, fossils are vital for assessing the minimum number of origins of heterospory in the plant kingdom, and for determining approximate dates for those origins.

In contrast, for the pteridophytic "living fossils" of the Psilotaceae and Ophioglossaceae, superficial morphological comparison proved extremely misleading (extant Psilotaceae with the long-extinct rhyniophytes; extant Ophioglossaceae with the long-extinct progymnosperms) or minimally informative (extant and extinct species of the ophioglossacean *Botrychium*); the key breakthrough came from molecular phylogenetic analysis (Fig. 6). Although the scenario of a possible reductive origin of the bryophytes outlined above was prompted largely by the existence of the wholly fossil rhyniophytes and by the negative evidence of the paucity of convincing bryophytes in early terrestrial plant assemblages, fossils have played negligible roles in most recent cladistic studies addressing this question. And convincing tracheophytic heterotrophs have not yet been found in the fossil record, though hydrophilic plants are preferentially preserved and hence may yet provide information on the ancestry, timing, and habitat of origin of the many documented aquatic lineages.

Overall, fossil plants are important primarily for increasing the number of related species that can be included in an analysis of a clade with a known fossil record. Some of these extinct species provide unique character states (and, more importantly, unique *combinations* of character states) not found among extant species, a consequence of including a greater proportion of primitive species relative to the extant flora. However, the plesiomorphy of plant fossils can be overemphasized; many of the examples in the case studies are specialized, formerly ecologically codominant clades that have left few if any extant descendants. Fortunately, this historical selectivity of the modern flora does not prevent fossils from providing some of the most graphic illustrations of various manifestations of homoplasy. Moreover, the inclusion of all related taxa, living and fossil, is essential in any attempt to assess the number of origins (or, equally importantly, losses) of a particular feature or its presumed function.

Reciprocal Illumination, Evolutionary Mechanisms, Total Evidence, and Causal Characters

The above arguments supporting the inclusion of fossils in phylogeny reconstruction rest on the strong desire for completeness of taxonomic coverage in any cladistic analysis. The second axis of completeness in a cladistic matrix is the number of coded characters and the range of properties of the plant that they represent. Many published cladograms are based on very limited spectra of evidence (recently published examples include leaf trichomes, apical meristem geometry, spermatozoid ultrastructure, and especially the rapidly increasing numbers of single-gene nucleic acid sequence trees). Such studies may or may not yield accurate hypotheses of relationship—that is not the primary issue here. Rather, assessing the degree of phenotypic divergence in a speciation event, or the genetic changes that underlie that event, evidently requires a far greater amount of both morphological and genetic information. To restate the argument, cladistic analyses based on limited data yield the various homoplasy indices (e.g., Swofford and Maddison 1987) as readily as more inclusive matrices, but in terms of interpreting the underlying *causes* of homoplasy they are of little value.

Also relevant is the need to maximize internal completeness in a cladistic matrix—that is, to minimize the proportion of missing values, which tend to introduce ambiguity into both phylogenetic relationships and the optimization of character-state transitions. Thus, two parallel approaches to phylogeny reconstruction are desirable: (1) matrices consisting exclusively of the widest possible range of morphological data but including extinct as well as extant species (assuming that they are available) and (2) matrices combining morphological and molecular data but restricted to extant species. Note that matrices consisting only of gene sequence data without corresponding phenotypic data provide no information on evolutionary mechanisms, as the sequenced gene is extremely unlikely to have played any direct role in the evolutionary transitions that generated the range of taxa under scrutiny.

The products of type (2) matrices are commonly dubbed "total evidence" trees (e.g., Kluge 1989; Swofford 1991; Patterson et al. 1993; Williams 1994) but this term is inappropriate; they are by no means "total" in terms of the range of possible species under comparison. More importantly, total evidence cladograms may (as is often argued) yield on average a more accurate hypotheses of relationship than any of the constituent submatrices of data, but by definition they fail to provide an opportunity for comparing contrasting hypotheses of relationship based on qualitatively different submatrices (cf. Bull et al. 1993; de Queiroz 1993; Chippindale and Wiens 1994; Olmstead and Sweere 1994). More specifically, molecular data cannot be used to test morphologically based hypotheses of secondary loss if the two suites of data are routinely amalgamated *a priori*.

It is more valuable to be able to predict the conditions under which one type of phylogenetic data is more likely to be reliable than any other. For example, nucleic acid sequence data are likely to be more accurate in clades believed to have experienced repeated morphological simplification, whereas deep, closely spaced evolutionary radiations are likely to be tracked more accurately by the irregular beat of the punctuationist phenotype than by the more constant, clocklike base pair substitutions. Yet other types of data are better mapped onto cladograms rather than included in the primary matrix (e.g., continuously variable characters, behavioral characters, dates of first accepted appearances in the fossil record).

Last, all of the phylogenies discussed in this paper are probably inaccurate in topology and/or patterns of character-state transitions. Specifically, I have challenged the rhizomorphic lycopsid phylogeny on the grounds of repeated simplification, and recent phylogenies that include bryophytes on the grounds that the resulting most-parsimonious trees may reflect subjective character coding, and may make less evolutionary sense than alternative hypotheses that are slightly less parsimonious by present coding. This conclusion reflects a philosophy that parsimony is the best starting point for investigating patterns of evolutionary relationship, but should not be regarded as the end point. Near-most parsimonious cladograms also merit investigation, particularly where there is strong evidence outside the narrow confines of the cladogram in support of the subparsimonious phylogeny (as in the case of possible bryophyte origins from rhyniophytes by simplification). In other words, cladograms are useful as *tests* of externally generated scenarios, as well as *sources* of such scenarios.

The phylogenies of hydropteridalean ferns (Fig. 5) and land plants (Fig. 8) were selected from among a range of available topologies for those groups, and that of the basal ferns (Fig. 6) summarizes much topological controversy. Thus, if perceived homoplasy is to be viewed both positively, as an indicator of evolutionary mechanisms, and negatively, as erroneous assertions of homology in need of reassessment, cladograms should be interpreted in the context of competing potential underlying mechanisms.

Ideally, characters in future cladistic matrices will increasingly reflect reciprocal illumination between observed evolutionary changes in phenotype and the underlying genetic mutations. In this context, developmental genetic data directly reflecting the *cause(s)* of specific evolutionary transitions (developmental parsimony *sensu,* Bateman 1994, in press) may ultimately provide more reliable character states for phylogenetic analysis than either pseudoparticulate morphological characters or the blurred phylogenetic fingerprints provided by single-gene sequence data. Such data should minimize homoplasy. Specifically, they will be more likely to avoid the erroneous identification as homologous of nonhomologous character states that is inevitable in both static descriptions of morphology and coding of the depauperate four-letter alphabet inherent in nucleic acid sequences. However, as noted by Williams et al. (1990) and Doyle (1994), causal genomic parsimony will be more easily said than done.

SUMMARY

1. Six case studies have been selected to illustrate various forms of potential homoplasy in plants: architectural transitions among the rhizomorphic lycopsids, the iteration of heterospory, the origins of the sporocarp-bearing water ferns, of some pteridophytic "living fossils," and of the bryophytes, and repeated ecological shifts to aquatic and heterotrophic life habits. Together, these studies emphasize vegetative rather than reproductive characters and character losses rather than gains; they also focus on nonflowering plants with a significant fossil record. Homoplasy is considered not as a class of abstract philosophical concepts but as a suite of factors influencing in various ways the evolutionary interpretation of cladograms; thus, character-state distributions and optimization procedures are given greater emphasis than topological representations of relationship.

2. Exclusively morphological and combined morphological–molecular matrices can be used to infer underlying evolutionary mechanisms, whereas exclusively molecular matrices cannot. On the one hand, the relative positions on a cladogram of form and related putative function have been used to falsify adaptational hypothesis. On the other hand, it can be argued that the evolutionary null hypothesis in a cladogram should be that all characters occupying a single internode changed in the same evolutionary event unless proved otherwise; the falsification of this hypothesis requires the addition of taxa in an attempt to dissociate the apparently correlated character-state transitions. In effect, long branches previously perceived as "hard" (real) are thereby demonstrated to be "soft" (artifacts of inadequate sampling). This approach requires maximal sampling of taxa and hence the inclusion of extinct species; these are also required to estimate the number of origins of particular features and functions. Ecological factors, notably their sessile life habits, suggest that hard (saltational) long branches are more common among higher plants than higher animals.

3. The distinction made by optimization algorithms between parallelisms and reversals is wholly arbitrary but leads to very different evolutionary interpretations; single reversals are *bona fide* evolutionary events, whereas parallelisms and multiple reversals (in effect parallel reversals) refute the *a priori* hypothesis of homology. Reversals are perceived in relation to relatively plesiomorphic states of the character and do not necessarily reflect loss of a morphological feature. Morphological homology reassessment is easier for multiple acquisitions than for multiple losses, given the difficulty of studying in detail a feature that does not exist!

4. However, independent losses of similar features may be demonstrable by distinguishing different modes of developmental gene suppression at the genetic level. Molecular data also offer the best independent test of hy-

potheses of simplification developed from morphological data, where the topology of the cladogram can be undermined (in broadly predictable circumstances that lead to "pseudoplesiomorphy") by repeated morphological reduction. This observation militates against "total evidence" as the only approach to phylogeny reconstruction. The alternative approach of consensus trees is also inappropriate, as character-state transitions cannot be satisfactorily mapped across the incompletely resolved consensus topology. Rather, the most effective evolutionary interpretations result from identifying and interpreting *differences* among cladograms generated from different types of data for the same range of coded taxa.

5. The justifiable desires for maximum completeness of data, maximum resolution of the resulting tree, and minimum homoplasy levels together dictate a dual approach to phylogeny reconstruction. The first suite of matrices should include the greatest possible range of morphological data and encompass both extant and extinct species. The second suite of matrices should be confined to extant species and include both morphological and molecular data. The best longer term method of minimizing homoplasy will be a more dynamic approach to character coding that focuses on the developmental genes that underlie specific phenotypic transitions, supported by reciprocal illumination between evolutionary theory and phylogenetic practice.

NOTE ADDED IN PROOF

The pteridophyte molecular tree presented in Fig. 6 is significantly incongruent with the thorough and detailed phylogenetic analysis of ferns recently published by Pryer et al. Rooted using Lycopodiaceae, their *rbc*L strict consensus tree (Fig. 10) shows the topology [(((Marattiales (Ophioglossaceae, Psilotaceae) (Sphenopsida, Hymenophyllaceae)) (Osmundaceae (Gleicheniaceae s.l. (Schizaeaceae s.l. (Dicksoniaceae + Cyatheaceae (Hydropteridales, "Higher Ferns")))]. The high degree of homoplasy, paucity of nontrivial nodes with bootstrap values >50%, and contrasts in topology with both the morphological tree (Fig. 9) and the "total evidence" tree (Fig. 11) are noteworthy. Nonetheless, both of the morphologically reduced clades discussed in this paper remain monophyletic: the Hydropteridales (bootstrap 91%) is placed significantly above the Schizaeaceae and the Psilotaceae + Ophioglossaceae (bootstrap <50%) as sister-group to the improbable clade Hymenophyllaceae + Sphenopsida.

ACKNOWLEDGMENTS

I thank Larry Hufford, Toby Pennington, Nick Rowe, Mike Sanderson, and an anonymous reviewer for comments on the manuscript, Gar Rothwell for access to his unpublished morphological phylogeny of extant and extinct ferns, and Bill DiMichele for suggesting the definitions of long and short branches.

REFERENCES

Albert, V. A., S. E. Williams, and M. W. Chase. 1992. Carnivorous plants: Phylogeny and structural evolution. Science 257:1491–1495.

Albert, V. A., M. W. Chase, and B. D. Mishler. 1993. Character-state weighting for cladistic analysis of protein-coding DNA sequences. Annals of the Missouri Botanical Garden 80: 752–766.

Albert, V. A., A. Backlund, and K. Bremer. 1994. DNA characters and cladistics: The optimization of functional history. Pp. 249–272 *in* R. W. Scotland, D. M. Williams, and D. J. Siebert, eds. Models in phylogeny reconstruction. Oxford University Press, Oxford.

Archibald, J. D. 1994. Metataxon concepts and assessing possible ancestry using phylogenetic systematics. Systematic Biology 43:27–40.

Armbruster, W. S. 1996. Biochemical homoplasies in plant reproductive biology. Pp. 227–243 *in* M. J. Sanderson and L. Hufford, eds. Homoplasy: The recurrence of similarity in evolution. Academic Press, New York.

Bateman, R. M. 1991. Palaeoecology. Pp. 34–116 in C. J. Cleal, ed. Plant fossils in geological investigation: The Palaeozoic. Horwood, Chichester.

Bateman, R. M. 1994. Evolutionary–developmental change in the growth architecture of fossil rhizomorphic lycopsids: Scenarios constructed on cladistic foundations. Biological Reviews 69:527–597.

Bateman, R. M. In press. Architectural radiations cannot be optimally interpreted without morphological and molecular phylogenies. Pp. 000–000 *in* M. H. Kurmann and A. R. Hemsley, eds. The evolution of plant architecture. Academic Press, London.

Bateman, R. M., and W. A. DiMichele. 1991. *Hizemodendron*, gen. nov., a pseudoherbaceous segregate of *Lepidodendron* (Pennsylvanian): Phylogenetic context for evolutionary changes in lycopsid growth architecture. Systematic Botany 16:195–205.

Bateman, R. M., and W. A. DiMichele. 1994a. Saltational evolution of form in vascular plants: A neoGoldschmidtian synthesis. Pp. 63–102 *in* D. S. Ingram and A. Hudson, eds. Shape and form in plants and fungi. Academic Press, London.

Bateman, R. M., and W. A. DiMichele. 1994b. Heterospory: The most iterative key innovation in the evolutionary history of the plant kingdom. Biological Reviews 69:345–417.

Bateman, R. M., W. A. DiMichele, and D. A. Willard. 1992. Experimental cladistic analyses of anatomically preserved arborescent lycopsids from the Carboniferous of Euramerica: An essay in paleobotanical phylogenetics. Annals of the Missouri Botanical Garden 79: 500–559.

Bierhorst, D. W. 1971. Morphology of vascular plants. Macmillan, New York.

Bold, H. C., C. J. Alexopoulos, and T. Delevoryas. 1987. Morphology of plants and fungi, 5th ed. Harper and Row, New York.

Bremer, K., C. J. Humphries, B. D. Mishler, and S. P. Churchill. 1987. On cladistic relationships in green plants. Taxon 36:339–349.

Bull, J. J., J. P. Huelsenbeck, C. W. Cunningham, D. L. Swofford, and P. J. Waddell. 1993. Partitioning and combining data in phylogenetic analysis. Systematic Biology 42:384–397.

Chaloner, W. G., and A. R. Hemsley. 1991. Heterospory: Cul-de-sac or pathway to the seed? Pp. 151–167 *in* S. Blackmore and S. H. Barnes, eds. Pollen and spores: Patterns of diversification. Oxford University Press, Oxford.

Chaloner, W. G., and J. M. Pettitt. 1987. The inevitable seed. Bulletin de la Societé Botanique de la France 134:39–49.

Chase, M. W., D. E. Soltis, R. G. Olmstead, D. Morgan, D. H. Les, B. D. Mishler, M. R. Duvall, R. A. Price, H. G. Hills, Y.-L. Qiu, K. A. Kron, J. H. Rettig, E. Conti, J. D. Palmer, J. R. Manhart, K. J. Sytsma, H. J. Michaels, W. J. Kress, K. J. Karol, W. D. Clark, M. Hedrén, B. S. Gaut, R. K. Jansen, K.-J. Kim, C. F. Wimpee, J. F. Smith, G. R. Furnier, S. H. Strauss, Q.-Y. Xiang, G. M. Plunkett, P. S. Soltis, S. M. Swensen, S. E. Williams, P. A. Gadek, C. J. Quinn, L. E. Eguiarte, E. Golenberg, G. H. Learn, Jr., S. C. H. Barrett, S. Dayanandan, and V. A. Albert. 1993. Phylogenetics of seed plants: An analysis of nucleotide sequences from the plastid gene *rbc*L. Annals of the Missouri Botanical Garden 40:528–580.

Cheers, G. 1992. A guide to the carnivorous plants of the world. Angus and Robertson, Prymble, Australia.

Chippindale, P. T., and J. J. Wiens. 1994. Weighting, partitioning, and combining characters in phylogenetic analysis. Systematic Biology 43:278–287.

Crane, P. R. 1990. The phylogenetic context of microsporogenesis. Pp. 11–41 *in* S. Blackmore and S. B. Knox, eds. Microspores: Evolution and ontogeny. Academic Press, London.

Crane, P. R., E. M. Friis, and K. R. Pederson. 1995. The origin and early diversification of angiosperms. Nature 374:27–33.

Crawford, R. M. M., ed. 1987. Plant life in aquatic and amphibious habitats. British Ecological Society Special Publication 5. Blackwell, Oxford.

Crepet, W. L., K. C. Nixon, D. W. Stevenson, and E. M. Friis. 1993. The relationships of seed plants in reference to angiosperm outgroups. American Journal of Botany 80:123. [abstracts]

da Pinna, M. C. C. 1991. Concepts and tests of homology in the cladistic paradigm. Cladistics 7:367–394.

Darwin, C. 1875. Insectivorous plants. Murray, London.

DePamphilis, C. W., and N. D. Young. 1995. Evolution of parasitic Scrophulariaceae/Orobanchaceae: Evidence from sequences of chloroplast ribosomal gene *rps*2 and a comparison with traditional classification schemes. American Journal of Botany 82:126. [abstracts]

de Queiroz, A. 1993. For consensus (sometimes). Systematic Biology 42:368–372.

DiMichele, W. A., and R. M. Bateman. 1992. Diaphorodendraceae, fam. nov. (Lycopsida, Carboniferous): Systematics and evolutionary relationships of *Diaphorodendron* and *Synchysidendron*, gen. nov. American Journal of Botany 79:605–617.

DiMichele, W. A., and R. M. Bateman. 1996. Plant paleoecology and evolutionary inference: Two examples from the Paleozoic. Review of Palaeobotany and Palynology 90:223–247.

DiMichele W. A., J. I. Davis, and R. G. Olmstead. 1989. Origins of heterospory and the seed habit. Taxon 38:1–11.

DiMichele, W. A., R. W. Hook, R. Beerbower, J. Boy, R. A. Gastaldo, R. A., N. Hotton III, T. L. Phillips, S. E. Scheckler, W. A. Shear, and H.-D. Sues. 1992. Paleozoic terrestrial ecosystems: Review and overview. Pp. 204–325 *in* A. K. Behrensmeyer, J. D. Damuth, W. A. DiMichele, R. Potts, H.-D. Sues, and S. L. Wing, eds. Terrestrial ecosystems through time. Chicago University Press, Chicago.

Donoghue, M. J. 1989. Phylogenies and the analysis of evolutionary sequences, with examples from seed plants. Evolution 43:1137–1156.

Donoghue, M. J. 1994. Progress and prospects in reconstructing plant phylogeny. Annals of the Missouri Botanical Garden 81:405–418.

Donoghue, M. J., J. A. Doyle, J. Gauthier, A. G. Kluge, and T. M. Rowe. 1989. The importance of fossils in phylogeny reconstruction. Annual Review of Ecology and Systematics 20:431–460.

Donoghue, M. J., and M. J. Sanderson. 1992. The suitability of molecular and morphological evidence in reconstructing plant phylogeny. Pp. 340–368 *in* P. S. Soltis, D. E. Soltis, and J. J. Doyle, eds. Molecular systematics of plants. Chapman and Hall, London.

Doyle, J. A. 1994. Origin of the angiosperm flower: A phylogenetic perspective. Plant Systematics and Evolution (supplement) 8:7–29.

Doyle, J. 1996. Homoplasy connections and disconnections. Pp. 37–66 *in* M. J. Sanderson and L. Hufford, eds. Homoplasy: The recurrence of similarity in evolution. Academic Press, San Diego, CA.

Doyle, J. A., and M. J. Donoghue. 1986. Seed plant phylogeny and the origin of angiosperms: An experimental cladistic approach. Botanical Review 52:321–431.

Doyle, J. A., and M. J. Donoghue. 1992. Fossils and seed plant phylogeny reanalyzed. Brittonia 44:89–106.

Doyle, J. J. 1992. Gene trees and species trees: Molecular systematics as one-character taxonomy. Systematic Botany 17:144–163.

Doyle, J. J. 1994. Evolution of a plant homeotic multigene family: Toward connecting molecular systematics and molecular developmental genetics. Systematic Biology 43:307–328.

Endress, P. 1994. Evolutionary aspects of the floral structure in *Ceratophyllum*. Plant Systematics and Evolution (supplement) 8:175–183.

Endress, P. 1996. Homoplasy in the angiosperm flower neontological record. Pp. 303–325 *in* M. J. Sanderson and L. Hufford, eds. Homoplasy: The recurrence of similarity in evolution. Academic Press, San Diego, CA.

Farrar, D. R. 1990. Species and evolution in asexually reproducing independent fern gametophytes. Systematic Botany 15:98–111.

Fisher, D. C. 1994. Stratocladistics: Morphological and temporal patterns and their relation to phylogenetic process. Pp. 133–171 *in* L. Grande and O. Rieppel, eds. Interpreting the hierarchy of nature. Academic Press, London.

Forey, P. L. 1992 Fossils and cladistic analysis. Pp. 124–136 *in* P. L. Forey, C. J. Humphries, I. J. Kitching, R. W. Scotland, D. J. Siebert, and D. W. Williams, eds. Cladistics: A practical course in systematics. Oxford University Press, Oxford.

French, J. C., Y. Hur, and M. Chung. 1993. cpDNA phylogeny of the Ariflorae. P. 5 *in* P. Wilkins, S. Mayo, and P. Rudall, eds. Monocotyledons: An international symposium abstracts. Royal Botanic Gardens, Kew.

Frumhoff, P. C., and H. K. Reeve. 1994. Using phylogenies to test hypotheses of adaptation: A critique of some current proposals. Evolution 48:172–180.

Garbary, D. J., K. S. Renzaglia, and J. G. Duckett. 1993. The phylogeny of land plants: A cladistic analysis based on male gametogenesis. Plant Systematics and Evolution 188: 237–269.

Gensel, P. G., and H. N. Andrews. 1984. Plant life in the Devonian. Praeger, New York.

Gifford, E. M., and A. S. Foster. 1989. Morphology and evolution of vascular plants, 3rd ed. W. H. Freeman, New York.

Gould, R. E. 1970. *Palaeosmunda*, a new genus of siphonostelic osmundaceous trunks from the Upper Permian of Queensland. *Palaeontology* 13:10–28.

Graham, L. E., C. F. Delwiche, and B. D. Mishler. 1991. Phylogenetic connections between the "Green Algae" and the "Bryophytes". Advances in Bryology 4:213–244.

Haig, D., and M. Westoby. 1988. A model for the origin of heterospory. Journal of Theoretical Biology 134:257–272.

Haig, D., and M. Westoby. 1989. Selective forces in the emergence of the seed habit. Biological Journal of the Linnean Society 38:215–238.

Hamby, R. K., and E. A. Zimmer. 1992. Ribosomal RNA as a phylogenetic tool in plant systematics. Pp. 50–91 *in* P. S. Soltis, D. E. Soltis, and J. J. Doyle, eds. Molecular systematics of plants. Chapman and Hall, London.

Harris, T. M. 1961. The Yorkshire Jurassic flora, I. Thallophytes–Pteridophytes. British Museum (Natural History), London.

Harvey, P. H., E. C. Holmes, A. Ø. Mooers, and S. Nee. 1994. Inferring evolutionary processes from molecular phylogenies. Pp. 313–333 *in* R. W. Scotland, D. M. Williams, and D. J. Siebert, eds. Models in phylogeny reconstruction. Oxford University Press, Oxford.

Hasebe, M., P. G. Wolf, W. Hauk, J. R. Manhart, C. H. Haufler, R. Sano, G. J. Gastony, N. Crane, K. M. Pryer, N. Murakami, J. Yokoyama, and M. Ito. 1994. A global analysis of fern phylogeny based on *rbc*L nucleotide sequences. American Journal of Botany 81: 120–121. [abstracts]

Hemsley, A. R. 1990. *Parka decipiens* and land plant spore evolution. Historical Biology 4:39–50.

Hemsley, A. R. 1993. A review of Palaeozoic seed-megaspores. Palaeontographica B 229:135–166 + 5 pls.

Hemsley, A. R. 1994. The origin of the land plant sporophyte: An interpolational scenario. Biological Reviews 69:263–273.

Hendy, M. D., and D. Penny. 1989. A framework for the quantitative study of evolutionary trees. Systematic Zoology 38:297–309.

Hoelzer, G. A., and D. J. Melnick. 1994. Patterns of speciation and limits to phylogenetic resolution. Trends in Ecology and Evolution 9:104–107.

Hueber, F. M. 1961. *Hepaticites devonicus*, a new fossil liverwort from the Devonian of New York. Annals of the Missouri Botanical Garden 48:125–132.

Hufford, L. 1996. Ontogenetic evolution underlies floral homoplasy. Pp. 271–301 *in* M. J. Sanderson, and L. Hufford, eds. Homoplasy and the evolutionary process. Academic Press, London.

Juniper, B. E., R. J. Robins, and D. M. Joel. 1989. The carnivorous plants. Academic Press, London.

Kaplan, D. R. 1977. Morphological status of the shoot systems of Psilotaceae. Brittonia 29: 30–53.

Kato, M. 1988. The phylogenetic relationship of Ophioglossaceae. Taxon 37:381–386.

Kenrick, P. 1994. Alternation of generations in land plants: New phylogenetic and palaeobotanical evidence. Biological Reviews 69:293–330.

Kenrick, P., and P. R. Crane. 1991. Water-conducting cells in early land plants: Implications for the early evolution of tracheophytes. Botanical Gazette 152:335–356.

Kluge, A. G. 1989. A concern for evidence and a phylogenetic hypothesis of relationships among Epicrates (Boidae, Serpentes). Systematic Zoology 38:7–25.

Knoll, A. H., S. W. F. Grant, and J. W. Tsao. 1986. The early evolution of land plants. Pp. 45–63 *in* R. A. Gastaldo, and T. W. Broadhead, eds. Land plants: Notes for a short course. University of Tennessee Department of Geological Sciences Studies in Geology 15.

Les, D. H. 1988. The origin and affinities of the Ceratophyllaceae. Taxon 37:326–345.

Luteyn, J. L. 1990. The Plumbaginaceae in the flora of the southwestern United States. Sida 14: 167–178.

Maddison, W. P., and Maddison, D. R. 1992. MacClade, version 3.0: Analysis of phylogeny and character evolution. Sinauer, Sunderland, Mass.

Manhart, J. R. 1994. Phylogenetic analysis of green plant *rbc*L sequences. Molecular Phylogenetics and Evolution 3:114–127.

Mayo, S. J. 1993. What is the Ariflorae? P. 11 *in* P. Wilkins, S. Mayo, and P. Rudall, eds. Monocotyledons: An international symposium abstracts. Royal Botanic Gardens, Kew.

Millay, M. A., and T. N. Taylor. 1992. New fern stems from the Triassic of Antarctica. Review of Palaeobotany and Palynology 62:41–64.

Milligan, B. G. 1994. Estimating evolutionary rates for discrete characters. Pp. 299–311 *in* R. W. Scotland, D. M. Williams, and D. J. Siebert, eds. Models in phylogeny reconstruction. Oxford University Press, Oxford.

Mishler, B. D., and S. P. Churchill. 1985a. Cladistics and the land-plants: A response to Robinson. Brittonia 37:282–285.

Mishler, B. D., and S. P. Churchill. 1985b. Transition to a land flora: Phylogenetic relationships of the green algae and bryophytes. Cladistics 1:305–328.

Mishler, B. D., L. A. Lewis, M. A. Buchheim, K. S. Renzaglia, D. J. Garbary, C. F. Delwiche,

F. W. Zechman, T. S. Kantz, and R. L. Chapman. 1994. Phylogenetic relationships of the "green algae" and "bryophytes". Annals of the Missouri Botanical Garden 81:451–483.

Nixon, K. C., W. L. Crepet, D. Stevenson, and E. M. Friis. 1994. A reevaluation of seed plant phylogeny. Annals of the Missouri Botanic Garden 81:484–533.

Nixon, K. C., and J. I. Davis. 1991. Polymorphic taxa, missing values and cladistic analysis. Cladistics 7:233–241.

Olmstead, R. G., and J. A. Sweere. 1994. Combining data in phylogenetic systematics: An empirical approach using three molecular data sets in the Solanaceae. Systematic Biology 43:467–481.

Page, C. N. 1979. The diversity of ferns: An ecological perspective. Pp. 9–56 in A. F. Dyer, ed. The experimental biology of ferns. Academic Press, London.

Pankhurst, R. J. 1995. Some problems in the methodology of cladistics. Binary 7:37–41.

Patterson, C. 1981. Significance of fossils in determining evolutionary relationships. Annual Review of Ecology and Systematics 12:195–223.

Patterson, C. 1988. Homology in classical and molecular biology. Molecular Biology and Evolution 5:603–625.

Patterson, C., D. M. Williams, and C. J. Humphries. 1993. Congruence between molecular and morphological phylogenies. Annual Review of Ecology and Systematics 24:153–188.

Peer, Y., R. van der, De Baere, J. Cawenberghs, and R. De Wachter. 1990. Evolution of green plants and their relationship with other photosynthetic eukaryotes as deduced from 5S ribosomal RNA sequences. Plant Systematics and Evolution 170:85–96.

Penny, D., P. J. Lockhart, M. A. Steel, and M. D. Hendy. 1994. The role of models in reconstructing evolutionary trees. Pp. 211–230 in R. W. Scotland, D. M. Williams, and D. J. Siebert, eds. Models in phylogeny reconstruction. Oxford University Press, Oxford.

Philbrick, C. T., and D. H. Les. 1995. Systematics of North American and European *Callitriche* (Callitrichaceae). American Journal of Botany 82:156. [abstracts]

Pryer, K. M., Lutzoni, F. M., and Smith, A. R. 1994. Towards a fern phylogeny: Integrating morphology and molecules. American Journal of Botany 81:122. [abstracts]

Pryer, K. M., A. R. Smith, and J. E. Skog. 1995. Phylogenetic relationships of extant ferns based on evidence from morphology and *rbc*L sequences, American Fern Journal 85:205–282.

Raubeson, L. A., and R. K. Jansen. 1992. Chloroplast DNA evidence on the ancient evolutionary split in vascular land-plants. Science 255:1697–1699.

Raubeson, L. A., D. B. Stein, and D. S. Conant. 1994. Insights into fern evolution from mapping chloroplast genomes. American Journal of Botany 81:123. [abstracts]

Reeves, P. A., and R. G. Olmstead. 1993. Polyphyly of the Scrophulariaceae and origin of the aquatic Callitrichaceae inferred from chloroplast DNA sequences. American Journal of Botany 80:174. [abstracts]

Remy, W., P. G. Gensel, and H. A. Hass. 1993. The gametophyte generation of some early Devonian land plants. International Journal of Plant Science 154:35–58.

Remy, W., and H. A. Hass. 1991. Gametophyt aus dem Chert von Rhynie. Argumenta Palaeobotanica 8:1–27, 29–45, 69–117.

Robinson, H. 1985. Comments on the cladistic approach to the phylogeny of the "Bryophytes" by Mishler and Churchill. Brittonia 37:279–281.

Ross-Craig, S. 1973. Drawings of British plants, XXX–XXXI. Bell, London.

Rothwell, G. W. 1994. Phylogenetic relationships among ferns and gymnosperms: An overview. Journal of Plant Research 107:411–416.

Rothwell, G. W. In press. Phylogenetic radiations of ferns: A paleobotanical perspective. In R. Johns and J. Camus, eds. Pteridology in perspective. Royal Botanic Gardens, Kew.

Rothwell, G. W., and R. Serbet. 1994. Lignophyte phylogeny and the evolution of spermatophytes: A numerical cladistic analysis. Systematic Botany 19:443–482.

Rothwell, G. W., and R. A. Stockey. 1989. Fossil Ophioglossales in the Paleocene of western North America. American Journal of Botany 76:637–644.

Rothwell, G. W., and R. A. Stockey. 1994. The role of *Hydropteris pinnata* gen. et spec. nov. in reconstructing the cladistics of heterosporous ferns. American Journal of Botany 81: 479–492.

Rouffa, A. S. 1978. On phenotypic expression, morphogenetic pattern, and synangium evolution in *Psilotum*. American Journal of Botany 65:692–713.

Sheffield, E. 1994. Alternation of generations in ferns: Mechanisms and significance. Biological Reviews 69:331–343.

Smith, A. B., and D. T. J. Littlewood. 1994. Paleontological data and molecular phylogenetic analysis. Paleobiology 20:259–273.

Spencer, S. C. 1993. The evolution of vernal pool adaptation and restriction in *Navarettia* (Polemoniaceae) as inferred from anatomical, morphological, and palynological studies. American Journal of Botany (abstracts) 80:177.

Sporne, K. R. 1975. The morphology of the pteridophytes, 4th ed. Hutchinson, London.

Steenis, C. G. G. J. van. 1976. Autonomous evolution in plants: Differences in plant and animal evolution. Garden's Bulletin, Singapore 29:103–126.

Stein, D. B., D. S. Conant, M. E. Ahearn, E. T. Jordan, S. A. Kirch, M. Hasebe, K. Iwatsuki, and M. K. Tan. 1992. Structural rearrangements of the chloroplast genome provide an important phylogenetic link with ferns. Proceedings of the National Academy of Sciences, USA 89:1856–1860.

Stein, W. E., and C. B. Beck. 1987. Paraphyletic groups in phylogenetic analysis: Progymnospermopsida and Préphanérogames in alternative views of seed-plant relationships. Bulletin de la Societé Botanique Français 134:107–119.

Stewart, W. N., and G. W. Rothwell. 1993. Paleobotany and the evolution of plants, 2nd ed. Cambridge University Press, Cambridge.

Swofford, D. L. 1991. When are phylogeny estimates from molecular and morphological data incongruent? Pp. 295–333 *in* M. M. Miyamoto, and J. Cracraft, eds. Phylogenetic analysis of DNA sequences. Oxford University Press, Oxford.

Swofford, D. L. 1993. Phylogenetic Analysis Using Parsimony, version 3.1.1. Illinios Natural History Survey, Urbana, IL.

Swofford, D. L., and W. P. Maddison. 1987. Reconstructing ancestral character states under Wagner parsimony. Mathematics and Bioscience 87:199–229.

Taylor, T. N. 1988. The origin of land plants: Some answers, more questions. Taxon 37: 805–833.

Taylor, T. N., and E. L. Taylor. 1992. The biology and evolution of fossil plants. Prentice-Hall, Englewood Cliffs, New Jersey.

Waters, D. A., M. A. Buchheim, R. A. Dewey, and R. L. Chapman. 1992. Preliminary inferences of the phylogeny of bryophytes from nuclear-encoded ribosomal RNA sequences. American Journal of Botany 79:459–466.

Wiley, E. O., D. J. Siegel-Causey, D. R. Brooks, and V. A. Funk. 1991. The compleat cladist: A primer of phylogenetic procedures. Museum of Natural History, University of Kansas, Lawrence, Kansas.

Williams, D. M. 1994. Combining trees and combining data. Taxon 43:449–453.

Williams, D. M., R. W. Scotland, and S. Blackmore. 1990. Is there a direct ontogenetic criterion in systematics? Biological Journal of the Linnean Society 39:99–108.

Williams, S. E., V. A. Albert, and M. W. Chase. 1994. Relationships of Droseraceae: A cladistic analysis of *rbc*L sequence and morphological data. American Journal of Botany 81: 1027–1037.

Wolf, P. G., P. S. Soltis, and D. E. Soltis. 1994. Phylogenetic relationships of Dennstaedtioid ferns: Evidence from *rbc*L sequences. Molecular Phylogenetics and Evolution 3:383–392.

Wolfe, A. D., and C. W. DePamphilis. 1995. Systematic implications of relaxed functional constraints on the rubisco large subunit in parasitic plants of the Scrophulariaceae and Orobanchaceae. American Journal of Botany 82:172. [abstracts]

BEHAVIORAL CHARACTERS AND HOMOPLASY: PERCEPTION VERSUS PRACTICE

HEATHER C. PROCTOR
Queen's University, Kingston
Ontario, Canada

INTRODUCTION

When the study of animal behavior was first recognized as a legitimate science, ethologists included reconstruction of phylogenies among their goals (e.g., Tinbergen 1953). However, this phylogenetic orientation was short-lived. By the time Hennig's *Phylogenetic Systematics* was published in 1966, the emphasis in ethological studies had switched to questions of current utility (Burghardt and Gittleman 1990; Greene 1994). This emphasis is still dominant; in his discussion of the "evolution of behavioral ecology," Gross (1994) devotes only one sentence to the value of phylogenetic systematics. Ousted from its parental discipline, has the phylogenetic study of behavior found a berth among systematists? Seemingly not. In a review of 882 published phylogenies, Sanderson et al. (1993) found that fewer than 4% of the studies relied on behavioral characters. According to recent reviews, many

HOMOPLASY: The Recurrence of Similarity in Evolution, M. J. Sanderson and L. Hufford, eds.
131

systematists feel that behavioral characters are too variable and homoplastic to be used in cladistic studies (see discussions in Wenzel 1992; de Quieroz and Wimberger 1993; Miller and Wenzel 1995; Wenzel and Carpenter 1994). Instead, comparative studies of behavior typically involve mapping behavioral characters onto trees constructed using other, presumably more trustworthy, characters (e.g., Sillén-Tullberg 1988; Losos 1990; Langtimm and Dewsbury 1991).

Several questions arise from the rarity of phylogenies based on behavior. First, how does one define behavioral characters? Second, what are the theoretical and empirical relationships between behavior and levels of homoplasy? Third, is there a bias against behavioral characters in systematics? If so, do systematists actively avoid using behaviors because they consider them too prone to homoplasy, or are there other factors involved?

BEHAVIORS AS PHYLOGENETIC CHARACTERS: THE PERCEPTION

What Are Behavioral Characters?

At first this may seem a trivial question. Most people define behavior as muscular activity involving movement of part or all of an animal. Active behavioral characters occasionally appear in phylogenetic analyses. These include locomotion (Andersen 1979), vocalization (Crowe et al. 1992), courtship display (McLennan et al. 1988; Prum 1990), and sperm transfer behavior (Witte 1991; Proctor 1992, 1993).

However, muscular actions are not the only kinds of behavioral characters; there are also the *products* of these actions. Obvious products of behavior include the nests of fishes, birds, and insects and the constructed cases of some moth and caddisfly larvae. Morphology and sites of these structures appear as phylogenetic characters in studies of many taxa: insects (Ross 1964; Packer 1991; Wenzel 1993), fishes (Maurakis et al. 1991; McLennan 1993), amphibians (Lynch 1989), and birds (Strauch 1984). Other examples of behavioral products include spermatophores, galls, tunnels, footprints, and webs (Coddington 1986; Eberhard 1990). Even trace fossils such as dinosaur tracks can be considered a kind of petrified behavior (Eibl-Eibesfeldt 1970: 199). Such "actualized" behavior may even include products of physiological activity, such as glandular secretions or patterns of electrical discharge. Ecological attributes are also potential candidates for behavioral characters, including habitat choice, foraging sites, diet, and host–parasite associations. The last of these ecobehavioral categories has received enormous attention (e.g., plants and herbivores (Mitter et al. 1991), worms and vertebrates (Brooks and McLennan 1991, 1993)). In summary, behavior is the part of an animal's phenotype that includes its actions, products, and interactions.

Can One Delimit and Homologize Behavioral Characters?

Behavior often consists of a smooth flow of movement in which one action merges into another without clear separation into units (Drummond 1981). One reason for the rarity of active behavioral characters in phylogenetic studies may be that they are difficult to delimit (Wenzel 1992), resulting in low consistency of definitions among observers. One observer may recognize distinct actions, e.g., "mouth gape," "bow," "tail fan," while another may consider them one action: "fanning tail while bowing with open beak." This problem is exemplified in Miller's (1988) description of three authors working on oystercatchers (*Haematopus* spp.) who divided the behavioral repertoire of these birds into 12, 18, and 20 types of behavior. Not only did the number of recognized characters differ between observers, but there was only one exact correspondence among their lists.

Another difficulty in defining actions is that they must be described in context, that is, movement is always in relation to other objects. Sometimes the reference point is another part of the organism (e.g., "bill pointed to chest"), sometimes a stationary object (e.g., "wings opened parallel to ground"), and sometimes a moving object (e.g., "male chases female"). As well as spatial reference points, activities have temporal contexts. For example, wing-fluttering after bathing may be a different character than wing-fluttering after copulation. This context-dependency adds another level of complexity to identifying behavioral characters (Miller 1988). In contrast, DNA, allozyme, and morphological characters seem much more readily defined. There is likely to be little quarrel over the number of tarsal segments in a tiger beetle's foreleg, for example, and that number is not dependent on context. Like these characters, actualized behavior such as nest architecture is less likely to be interpreted differently by different observers (Wenzel 1993). On the other hand, quantitative morphological characters such as tail length or bristle number share with behavioral sequences the potential to be subjectively divided into character states. In a study of published plant phylogenies, Stevens (1991) found that authors seldom explained their rationale for subdividing quantitative characters, even though there are several delimitation techniques published (e.g., Archie 1985; Chappill 1989). Analysis of molecular data is also subjective to some extent. "Gaps" corresponding to insertions and deletions are used to align DNA sequences, but as there is little theory to guide researchers (Swofford and Olsen, 1990: 418), choice of a gap-weighting algorithm may be a matter of individual preference.

Once behaviors have been defined, can they be legitimately homologized? Some systematists (e.g., Atz 1970) are skeptical, but most students of animal behavior are open to the idea, particulary when the comparisons are among very closely related taxa (Dewsbury 1978: 253). Wenzel (1992) discusses homology in great detail and concludes that recognizing homology among behaviors is not qualitatively different from recognizing it among morphological structures; however, he admits that problems in consistently delimiting characters may make it more difficult (see discussion above). Wenzel's definition of behavioral homology is a phylogenetic one—homologous behaviors originate in a common ancestor and are similar because of descent from that ancestor (see also Nelson 1994). Regardless of whether they are morphological, genetic or behavioral, homologous characters are those that are concordant in a phylogenetic analysis while homoplastic ones fail to form a consistent pattern. As well as this *a posteriori* method of recognizing behavioral homologies, Wenzel suggests that Remane's (1952) criteria for morphological homology are often subconsciously used by ethologists when selecting characters.

Remane's criteria can be roughly translated into ethological equivalents (Eibl-Eibesfeldt 1970: 187; Wenzel 1992): the morphological criterion of *position* correlates with the ethological one of *sequence*, while those of *specific quality* (uniqueness) and *linkage by intermediate forms* are essentially the same in morphology and ethology. Wenzel (1992) suggests that the desire to create a linear flow of forms sometimes produces "ethoclines" that do not reflect the true pattern of evolution. However, creation of hypothetical intermediates is not unique to behavioral studies, as morphologists commonly design transitional forms to connect disparate but presumably homologous structures (Wiley 1981).

According to some authors (e.g., Eibl-Eibesfeldt 1970: 190) serial homology is also possible in behavior and may be a special case of connection by intermediates (Wenzel 1992). Serial homologs can exist as similar movements in different appendages (e.g., "walking" movements in arthropod mouthparts, which are leg homologs), or as similar movements of one organ that have different purposes (e.g., woodpeckers tapping to communicate and to find insects). The latter concept identifies ritualized behavior, which is the modification of a noncommunicative behavior to serve an additional communicative function (Eibl-Eibesfeldt 1970: 97), as a type of behavioral serial homology.

Although it is possible to apply Remane's criteria to behavior, Lauder (1994) questions the value of using *a priori* clues to homology for any type of character. He methodically shows how many traditional "loci" of homology (e.g., ontogeny) are not always reliable indicators of common origin and advocates the *a posteriori* cladistic approach described by Wenzel (1992).

BEHAVIOR AND HOMOPLASY

Theory: Should Behavior Be More Homoplastic Than Other Characters?

To many biologists, behavioral characters appear too labile to be useful for phylogenetic studies (de Queiroz and Wimberger 1993). Certainly, when coefficients of variation are compared between behavioral and morphological characters the CVs for behavior are often higher (Wcislo 1989; but note his query with regard to the statistical legitimacy of this comparison). Some ethologists have tried to limit "units" of behavior to those motor patterns that show little or no variability within or between individuals. However, the usefulness of these fixed or modal action patterns (FAPs and MAPs, respectively), has been questioned by other behavioral biologists (e.g., Miller 1988). West-Eberhard (1983, 1984, 1989) has emphasized how rapidly behavior, particularly that involved in social communication, diverges in comparsion with morphology. Her contention is supported by many examples of "ethospecies" that can be distinguished behaviorally but are morphologically almost identical (e.g., flash patterns in *Photuris* fireflies (Mayr 1958), song in *Empidonax* flycatchers (Peterson 1980), song in Hawaiian crickets (Otte 1989)). Wcislo (1989) focuses on how feeding behaviors, like social signals, appear to diverge phylogenetically before morphology does. His observation is also supported by literature citations of species that differ in diet but are physically indistinguishable (nest-provisioning in *Ammophila* wasps (Mayr 1958), anthropophily in black flies (Crosskey 1990), host selection in trematodes (Brooks and Wiley 1988)). The very rapid divergence of some behaviors supports the argument that behavioral characters are unlikely to be useful in reconstructing phylogenetic relationships.

However, the phenomenon of one character type changing more rapidly than others is not restricted to behavior. There are many species that are identical morphologically but are distinctive genetically (e.g., plethodontid salamanders (Larson 1989)) or chromosomally (blackflies (Crosskey 1990)). This does not mean that *all* genetic information is too labile to be used to reconstruct phylogenies. Rather, it means that one should use the type of DNA most appropriate to the phylogenetic level of interest, choosing slowly changing types like rDNA for working at high taxonomic levels and rapidly changing types like mtDNA for lower taxa and subspecific lineages. Types of behavior differ in their expected variation among taxa just as different types of DNA differ. For example, Mundinger (1979) found that flight calls of cardueline birds are autapomorphies useful for characterizing species while social and alarm calls are subfamilial synapomorphies. Some behaviors are very ancient. In reptiles, tongue protrusion for sensory purposes probably evolved in the common ancestor of squamates at least 100 Ma (de Queiroz and Wimberger 1993). Moynihan (1975) provides evidence that several social displays of cephalopods originated before the splitting of the three orders 190 Ma. In some cases, behaviors are conserved long after the applicable morphology has changed (Eibl-Eibesfeldt 1970: 192): male elk threaten each other with raised lips even though the ancestrally dagger-like canines have been reduced to tiny nubs; stump-tailed macaques attempt to perform balancing movements with tail remnants too small to be useful; and essentially hairless humans experience piloerection (a.k.a. "gooseflesh") when cold or frightened.

One cause of intraspecific variation that is unique to behavior is learning. The fact that learned behaviors can be transmitted within as well as between generations appears to make these characters inappropriate for phylogenetic purposes and may tempt some systematists to tar all behaviors with the same brush. However, the huge body of literature on phenotypic plasticity demonstrates that intraspecific variation in morphology is also a common occurrence; this fact does not make us class *all* structural characters as inappropriate fodder for cladistic study. As well, although the individual states for behaviorally and morphologically plastic characters may not be useful for phylogenetic reconstruction above the species level, both the existence of variation and the degree of plasticity itself may be good characters. West-Eberhard (1989) provides examples of species that consistently differ in whether they are polymorphic. The *ranges* of environmentally induced morphological plasticity, reaction norms, are also highly variable across species (Stearns 1989) and are subject to selection (e.g., Parejko and Dodson 1991). Similarly, behavioral reaction norms in the form of variable behavioral repertoires are species-specific characters; for example, the chipping sparrow *Spizella passerina* has only one song type while the five-striped sparrow *Aimophila quinquefasciata* can sing up to 200 types (Baptista and Gaunt 1994). Imprinting, the permanent learning that occurs in a certain sensitive period of an animal's life, is species–specific in many vertebrates (Eibl-Eibesfeldt 1970: 217). Birds can be divided into those that sing species–specific songs even when raised in social isolation and those that need to learn it from conspecifics during an imprinting period. However, even birds that learn their song are not necessarily *tabulae rasilis*, as species vary in the number of parameters that are learned versus innate. Creepers (Certhiidae), for example, learn only rhythm, syntax, and element number while Anna's hummingbird (*Calypte anna*) also learns syllable structure and frequency (Baptista and Gaunt 1994). It is even possible to use human language, which few people would argue is innate, to trace population movement and fracturing (Cavalli-Sforza et al. 1988). Finally, the very ability to learn is a species-specific character (Eibl-Eibesfeldt 1970: 217; Wcislo 1989). Bumblebees, for example, are much better at learning to recognize nectar-rich flowers than are solitary bees (Dukas and Real 1991).

How does the heritability of behavior compare with that of other characters? Relatively high heritabilities (30–50%) for continuous behavioral characters have been found for levels of geotaxis, general activity, and mating speed in *Drosophila* and general activity in rats (Manning 1975). Other activities are inherited in a discrete fashion, such as brood-cleaning behavior in honeybees (Rothenbuhler 1964), aggression in *Gryllus* crickets (Hörmann-Heck 1957), and host-searching in nematodes (Osche 1952). Hybrids sometimes display interesting mixtures of parental species' behaviors, indicating the probability of single-locus inheritance. Dilger (1962) crossed two parrots, *Agapornis roseicollis*, which tucks nesting material under its rump feathers, and *A. fischeri*, which carries this material in its bill. The F_1 hybrids tried to tuck strips of nesting material under their rump feathers but would not let go of them with their bills, a combination of characters that prevented them from constructing nests. In a literature survey of 1120 narrow sense heritability estimates, Mousseau and Roff (1987) found median heritability values of 0.26 for life-history, 0.27 for physiological, 0.32 for behavioral, and 0.53 for morphological characters.

In summary, there is no *a priori* way to determine whether a trait will be a good phylogenetic character; any observable attribute has the potential to be useful (Brooks and McLennan 1991; Lauder 1994; Miller and Wenzel 1995). Whether behavioral characters will be appropriate for a given study will depend on the types of organisms, the taxonomic level, and the type of behavior used. Some behaviors are heritable and some are not, just as morphological traits vary in the degree of genetic or environmental control. Even when behaviors are clearly learned and not heritable, the extent of the learning template can be used as a phylogenetic character (e.g., bird song (Baptista and Gaunt 1994)).

Evidence: Are Behaviors More Homoplastic Than Other Types of Characters?

de Queiroz and Wimberger (1993) compared consistency indices between behavioral and morphological characters within and between phylogenetic studies. For the 22 studies in the "within study" comparison, the mean CI for behaviors (0.841) was not significantly different from that for morphological characters (0.839). Their comparison of eight purely behavioral data sets with 36 morphological sets likewise showed no difference in CIs. de Queiroz and Wimberger (1993: 54) conclude that their data "give no justification for discriminating against behavioral characters as indicators of phylogenetic relationships." However, they did notice that purely behavioral studies tended to deal with lower taxonomic levels—genera and species—while morphological studies included more higher taxa. They felt that some may construe this pattern as evidence that behavior is useful only at low levels because it is too homoplastic for higher taxa. To counter this, de Queiroz and Wimberger suggest that detailed behavioral studies tend to focus on taxa whose behavior is easy to observe (e.g., ducks mate in plain sight on open water) and that as one moves to more distantly related groups it is unlikely that the same propitious observational environment will be present. An additional reason for this pattern is that ethological observations are seldom made with the goal of homologizing behaviors across broad groups; rather, efforts are concentrated on behaviors that potentially differentiate species (Wenzel 1992), perhaps to test hypotheses about isolating mechanisms. Finally, there are logistic difficulties in making behavioral observations across a wide range of taxa because distantly related organisms may live far apart or in habitats that require different sets of observational skills and equipment. These problems combined with time constraints may prevent any one person from undertaking broad ethological studies.

Some authors have directly compared the performance of behavioral and morphological data sets in reconstructing phylogenies of the same taxa. In these comparisons behavior has been found to be as good as or better than morphology. Prum (1990) collected an impressive array of 44 behavioral characters from the displays of male manakins (Aves: Pipridae) and compared the resultant phylogeny with that produced from 57 morphological characters of the syrinx. The behavioral tree was almost completely congruent with the morphological tree. A cladogram for stickleback fish that was derived from behavioral characters not only had a very high CI (90.3%), but produced a better resolved tree than did morphological data (McLennan et al. 1988). Wenzel (1993) found nest architecture to be more consistent than adult morphology in phylogenetic studies of social wasps.

Is there *any* evidence that one type of character is *consistently* more reliable than others? Sanderson and Donoghue (1989) compared levels of homoplasy in 42 morphological and 18 molecular data sets. They found little evidence that molecular characters are more reliable than morphology despite earlier expectations that DNA would outcompete traditional character types (Jansen and Palmer 1988; other references in Sanderson and Donoghue (1989)). Another type of comparison between molecular and morphological data is presented by Patterson et al. (1993), who examined consistencies *among* multiple trees for a given taxon. They found that disagreement among molecular trees was just as great as that among morphological trees for taxa as diverse as hominids, arthropods, angiosperms, and green algae. Again, as Miller and Wenzel (1995) suggest, there is little evidence that one type of character—morphological, molecular, or behavioral—is *a priori* more likely to reflect phylogeny than another type.

BEHAVIORAL CHARACTERS IN PHYLOGENETIC ANALYSIS: THE PRACTICE

Literature Surveys: Do Systematists Use Behavioral Characters?

In their survey of 882 phylogenetic analyses published from 1989 to 1991, Sanderson et al. (1993) found that fewer than 4% of studies relied on character types other than morphology, sequence/RFLP, or allozyme characters. As behavior had to share this 4% with other character types (e.g., biochemical, DNA–DNA hybridization), it is apparently very rarely used by systematists. Or is it? Sanderson et al. (1993) were interested in documenting the rapid increase in phylogenetic information that has occurred since the advent of molecular systematics. They point out that their survey was biased toward botanical and molecular biology journals. These two categories of literature *a priori* exclude the use of behavioral characters, unless one considers associations with pollinators to be plant behavior. As well, Sanderson et al. apparently categorized entire studies as "morphological," "DNA sequence," or "behavior," although many phylogenetic reconstructions rely on mixtures of character types. If behavioral characters are usually used in combination with other character types, behavioral studies would be underrepresented in their survey. Finally, by concentrating on very recent studies Sanderson et al. may have overlooked a larger representation of behavioral studies in older, pre-molecular literature.

To test these ideas I performed my own survey in which I looked at journals that were not by definition biased toward one type of character: *Annals of the Entomological Society of America* (1989–1993); *Canadian Entomologist* (1989–1993); *The Auk* (1989–1993); *Copeia* (1989–1993); *Evolution* (1983–1993); and *Systematic Zoology/Biology* (1966–1993). I included only studies on taxa that definitely could behave (i.e., excluding plants, fungi, and viruses) and counted all examples in which behavioral characters were included in a matrix. I counted as "behavior" any character that described actions or architectural products (e.g., nests, cases); I did not include potentially more contentious characters such as host–parasite associations, chemical secretions, or nest locations.

A total of 291 phylogenetic studies were analyzed, with representation from reptiles, amphibians, fish, mammals, birds, insects, and other invertebrate taxa. Of these studies, 247 relied solely on one type of character while 44 used more than one. Of the first kind of study only 4 (2%) were based solely on behavioral characters (Table 1, column 2). Studies based only on morphological or allozyme characters were the most common at 41 and 21%, respectively. This is similar to the results in Sanderson et al. (1993)

TABLE 1 Frequency with which Different Character Types Were Used for Phylogeny Reconstruction in the Literature Survey

Character type	All studies (N = 291)	Single character studies (N = 247)	Mixed character studies (N = 44)
Morphological	139 (48%)	101 (41%)	38 (86%)
Allozymes	65 (22%)	53 (21%)	12 (27%)
DNA restriction sites	27 (9%)	23 (9%)	4 (9%)
Immunological	24 (8%)	20 (8%)	4 (9%)
Chromosome morph	24 (8%)	14 (6%)	10 (23%)
Biochemical	19 (7%)	12 (5%)	7 (16%)
Other	19 (7%)	5 (2%)	14 (32%)
Behavioral	17 (6%)	4 (2%)	13 (29%)
DNA sequences	9 (3%)	8 (3%)	1 (2%)
Habitat	9 (3%)	0 (0%)	9 (20%)
DNA hybridization	8 (3%)	7 (3%)	1 (2%)

Note. Percentages are rounded to the nearest whole number. The number of character-type occurrences in the mixed character studies is greater than the number of studies because more than one character type occurred in each study.

except that DNA sequences were used in about 40% of their studies in comparison to the 3% in my survey of older literature.

A slightly different pattern emerges from studies with more than one character type (Table 1, column 3). Although morphology remains the most commonly used type in mixed character studies, behavioral characters increase their representation from 2 to 29%. Habitat, which never appears in single character studies, is relatively common in the mixed studies (20%). When mixed studies are considered as a proportion of all studies involving a particular character type there seem to be two categories of characters: those that often appear as sole characters (morphology, allozymes, immunological distances) and those that are more common in mixed studies (chromosomal morphology, behavior, biochemistry, habitat).

Despite the differences between my survey and that of Sanderson et al. (1993), it is apparent that behavior is rarely used to reconstruct phylogenies either alone or in combination with other character types (Table 1). Do systematists avoid using behaviors because they feel these characters are too homoplastic to be useful? To put the question more generally, why do systematists use certain characters and avoid others?

Opinion Survey: Why Systematists Avoid Certain Types of Characters

To answer this I sent questionnaires to 68 systematists located mainly in Canada and the United States. I asked them to rank character types according to the frequency with which they used them, and for those that they *did not use* I asked them to indicate the reason why. Categories of characters were morphological, behavioral, ontogenetic, allozyme, DNA sequence, biogeographical, fossil, and other. Possible reasons for not using a character type were as follows: characters not available (e.g., preserved insects do not behave); characters too costly in either time or money; characters too homoplastic; had not considered using these characters; other characters are sufficient. I also asked the respondents to specify the exact nature of each character type in order to determine the breadth of their interpretation of behavior.

Fifty-two questionnaires were returned from systematists working on taxa ranging from fungi to ungulates. Morphology was ranked by 41 systematists as the character type used most often, followed by DNA (eight respondents) and the rest of the categories at one to three respondents each (Fig. 1). Almost all respondents, 49 of 52, used morphological characters to some degree. Nearly as many (41) claimed to use biogeographical characters despite the exclamatory statements of several respondents who considered the use of biogeography in systematic research to be tautological. Behaviors were used to some degree by 28 respondents, DNA by 24, ontogeny by 22, fossils by 18, allozymes by 16, and other characters (e.g., chromosome morphology) by 17. Characters classed as behavior by respondents included mating, feeding, social displays, host types, dispersal mode, and architecture of nests, cocoons and cases. One botanist categorized pollination biology under behavior, and several mycologists placed mating compatibility, substrate preference, and symptom expression as behavior. The relatively high proportion of respondents who use behavior (28 of 52 = 54%) appears at odds with literature surveys that show behavior to be rarely used. There are two possible explanations for this descrepancy. First, many of the systematists in my survey worked for governmental institutions and museums (Agriculture Canada, Royal Ontario Museum) and their taxonomic work may be in-house rather than widely published. Second, my original distribution list was biased toward scientists and institutions I knew personally and perhaps could indicate a tendency of my own to interact with people interested in behavior. Regardless, this potential bias does not alter the validity of the most important question on the survey: if a particular character type was not used, what was the reason? For the vast majority of unused character types, the reasons sytematists avoided them were that the characters were unavailable, they were too costly, or other characters were sufficient for their purposes (Fig. 2). Only four respondents described some character types as too homoplastic, two of them in reference to DNA sequences, one to allozymes, and one to behavior. The most common reasons given for not using behavioral characters were that they were not available (perhaps because the taxon does not exhibit behavior, or the work was done from dead museum specimens) or because collecting the behavioral information would be too costly in time or money.

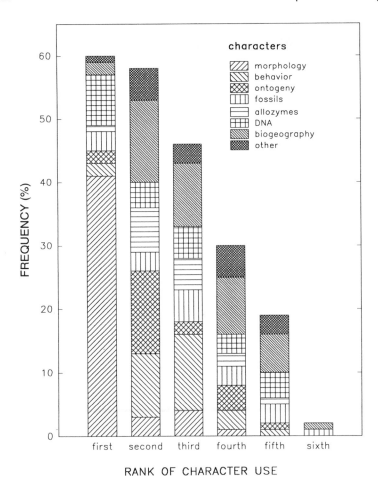

FIGURE 1   Results of opinion survey showing the frequency with which different character types were ranked from most (= first) to least often used.

The conclusion one draws from these surveys is that the rarity of behavioral characters in phylogenetic analyses is *not* because systematists view them as suspiciously labile, but because these characters are not readily available. Homoplasy was so seldom given as a reason for avoiding certain character types that one wonders if anyone uses Remane's homology criteria *a priori*. The perception that systematists actively avoid behaviors out of fear of homoplasy may be one perpetuated by the literature (e.g., de Queiroz and Wimberger 1993: 46) rather than one emerging from practice.

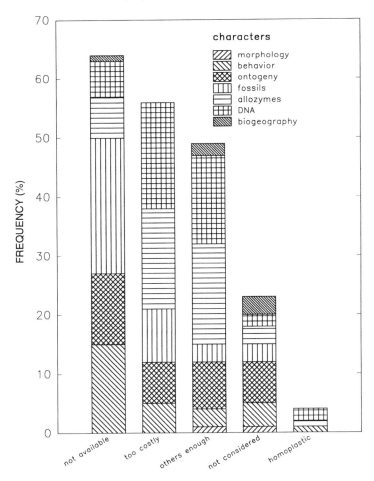

REASON FOR NOT USING CHARACTER TYPE

FIGURE 2 Results of opinion survey showing the frequency of different reasons given for *not* using particular character types.

Comparative Methods in Behavioral Ecology: Behavioral Categories, Not Characters

Despite the above discussion, readers may be forgiven if they still have the impression that behaviors are more homoplastic than other types of characters, particulary if they concentrate on ecological literature. Many behavioral ecologists interested in explaining current distributions of traits have turned to "phylogenetically sound" statistical methods to test hypotheses about adaptation (Harvey and Pagel 1991). These methods often involve correlating independent origins of the character of interest with purported causal factors (e.g., Ridley 1983; Proctor 1991). Ecological researchers are likely to define their characters very broadly for two reasons: first, because the phenomenon of interest may occur in a wide range of taxa and thus is unlikely to be similar in the narrow, phylogenetically homologous sense; second, because statistical power increases with the sample size (number of independent evolutions). Thus ecologists define their characters of interest very broadly in order to *maximize the probability of homoplasy*. Rarely are the behaviors investigated in comparative studies thought to be true homologs, rather they are suites of behaviors serving similar functions. For example, water mites are estimated to have evolved copulation 91 times in the approximately 5000 extant species (Proctor 1991). In this study, "copulation" was broadly defined as the placing of sperm by the male on or in the receiving structure of the female. It was not considered to be a single binary character but rather a complex of characters that were interpreted as homologs only when the males' sperm-transfer structures were similar in morphology (e.g., copulation using modified wing-like genital flaps; copulation using modified tarsal claws of third legs). Other such generally defined behavioral characters include "parental behavior" (Gittleman 1981), "lekking" (Höglund 1989; Prum 1994), "gregariousness" (Sillén-Tullberg 1988), "mating frequency" (Ridley 1989), "assortative mating" (Crespi 1989), "cooperative breeding" (Edwards and Naeem 1993), and "polygamy" (Wiklund and Forsberg 1991). However, in most of these latter cases, no attempt is made to break the broad functional definition of a behavioral type into the variant subsets that could indicate homologous behaviors (indeed, a priori breakdown may not be possible). The authors rely instead on optimizing the behavior onto an already extant tree to reveal the number of independent evolutions.

When one uses character states to construct a cladogram, two factors can mask independent origins of a state. First, as mentioned above, functionally defined characters are likely to show more homoplasy than structurally defined ones (Wenzel 1992; Siddal et al. 1993). Second, losses of a character state may be falsely homologized, since although there may be many independent losses of a character state, seldom are there structural or behavioral clues to this independence. There are many examples of strongly homoplastic characters in the literature whose repetitiveness may be due to the breadth of their definitions or to their involving losses rather than gains of a character. Behavioral examples include 91 gains of copulation in the 5000 species of water mites (Proctor 1991), 12 evolutions of obligatory associations with plants in the 300 species of pseudomyrmecine ants (Ward 1991), and more than 1000 evolutions of flightlessness in the Insecta (Wagner and Liebherr 1992). Morphological examples include 40 to 65 origins of light-sensing organs in the Metazoa (Salvini-Plawen and Mayr 1977), at least 50 losses of the pelvic fin in fish (Nelson 1989, 1990), 26 gains of toe fringes in 354 genera of lizards (Luke 1986), and up to 11 changes from cross- to self-pollination in the 40 species of New World *Dalechampia* (Euphorbiaceae) (Armbruster 1993).

CONCLUSIONS

Let us return to the questions posed at the beginning of the chapter. To the query "should one expect behavioral characters to be more homoplastic than other types of characters?" the answer is "no." Every property of behavioral characters has parallels among morphological and/or molecular characters. Like morphology, behavior shows ontogeny (e.g., Wenzel 1993) and plasticity and is inherited through single-locus and polygenic systems for different traits; even learning has some hard-wired properties. Undeniably, delimiting and homologizing behaviors will often prove problematic (Miller and Wenzel 1995), but homologizing morphology or molecules is also fraught with difficulty (Hillis 1994). These problems are reduced for actualized behaviors like nests and webs (Wenzel 1993), while sonographs and video technology help to make vocalizations and movements concrete and more readily comparable than the ephemeral recollections of observers. If ethologists are diligent in the detail and consistency of terminology with which they describe behaviors then they will increase the value of their work to systematists (Miller 1988).

To the question "are behavioral characters rarely used by systematists?" the answer is "yes." The reason for this lack is not that behaviors are thought to be more homoplastic, but that they are not readily obtainable for most taxa (e.g., pinned insects and study skins do not behave). When one sees behavior used in a phylogenetic context it is usually mapped onto a previously obtained tree rather than used to build the tree. Two reasons for this tendency are: (a) those who observe behaviors are usually behavioral ecologists or psychologists rather than systematists, so they may lack the skills for reconstructing phylogenies; (b) information about behavior is likely to be sparsely scattered among the taxa in a clade, with a few well-studied species (e.g., *Drosophila melanogaster* in the Drosophilidae) and many lacking any behavioral information at all (e.g., most of the other 3000 or so *Drosophila* spp.) (see also Prum (1990), with regard to manakins). These missing data make it more difficult to construct trees with behaviors than with morphological characters, for which there will always be some information in original species descriptions and museum type specimens. Perhaps animal behaviorists should emulate molecular biologists' construction of repositories for gene sequences. An "ethobank" (Greene 1994) for storage and dissemination of ethograms, illustrations, and videotapes of animal behavior would do much to promote the use of behavioral characters. Closer collaboration between ethologists and systematists in which phylogenetically important taxa were identified and studied would benefit both disciplines, and would finally put the notion of the inferiority of behavioral characters to rest.

ACKNOWLEDGMENTS

Thanks to John Wenzel for allowing me to pick his brains for minimal recompense in beer, and to the 52 systematists who responded with varying degrees of humor to my impertinent questionnaire. Comments from the editors and two anonymous referees greatly improved the manuscript. This research was supported by a Natural Sciences and Engineering Research Council of Canada operating grant.

REFERENCES

Andersen, N. M. 1979. Phylogenetic inference as applied to the study of evolutionary diversification of semiaquatic bugs (Hemiptera: Gerromorpha). Systematic Zoology 28:554–578.
Archie, J. W. 1985. Methods for coding variable morphological features for numerical taxonomic analysis. Systematic Zoology 34: 326–345.
Armbruster, W. S. 1993. Evolution of plant pollination systems: Hypothesis and tests with the neotropical vine *Dalechampia*. Evolution 47:1480–1505.
Atz, J. W. 1970. The application of the idea of homology to behavior. Pp. 53–74 *in* L.R. Aronson, E. Tobach, D.S. Lehrman, and J.S. Rosenblatt, eds. Development and evolution of behavior. Freeman, San Francisco.
Baptista, L. F., and S. L. L. Gaunt. 1994. Advances in studies of avian sound communication. The Condor 96:817–830.
Brooks, D. R., and D. A. McLennan. 1991. Phylogeny, ecology and behavior. University of Chicago Press, Chicago.
Brooks, D. R., and D. A. McLennan. 1993. Parascript: parasites and the language of evolution. Smithsonian Institution Press, Washington, DC.
Brooks, D. R., and E. O. Wiley. 1988. Evolution as entropy, 2nd ed. University of Chicago Press, Chicago.
Burghardt, G. M., and J. L. Gittleman. 1990. Comparative behavior and phylogenetic analyses: New wine, old bottles. Pp. 192–225 *in* M. Bekoff and D. Jamieson, eds. Interpretation and explanation in the study of animal behavior, Volume II. Westview Press, Boulder, Colorado.
Cavalli-Sforza, L. L., A. Piazza, P. Menozzi, and J. Mountain. 1988. Reconstruction of human evolution: Bringing together genetic, archaeological, and linguistic data. Proceedings of the National Academy of Sciences USA 85:6002–6006.
Chappill, J. A. 1989. Quantitative characters in phylogenetic analysis. Cladistics 6:217–234.
Coddington, J. A. 1986. Orb webs in "non-orb weaving" ogre-faced spiders (Araneae: Dinopidae): A question of geneology. Cladistics 2:53–67.
Crespi, B. J. 1989. Causes of assortative mating in arthropods. Animal Behaviour 38:980–1000.
Crosskey, R. W. 1990. The natural history of blackflies. Wiley, Toronto.
Crowe, T. M., E. H. Harley, M. B. Jakutowicz, J. Komen, and A. A. Crowe. 1992. Phylogenetic, taxonomic and biogeographical implications of genetic, morphological, and behavioral variation in francolins (Phasianidae: *Francolinus*). The Auk 108:24–42.

de Quieroz, A., and P. H. Wimberger. 1993. The usefulness of behaviour for phylogeny estimation: Levels of homoplasy in behavioral and morphological characters. Evolution 47: 46–60.

Dewsbury, D. A. 1978. Comparative animal behavior. McGraw-Hill, Toronto.

Dilger, W. C. 1962. The behavior of lovebirds. Scientific American 206:88–98.

Drummond, H. 1981 . The nature and description of behavior patterns. Pp. 1–34 in P. P. G. Bateson and P. H. Klopfer, eds. Perspectives in ethology: Advantages of diversity. Plenum, New York.

Dukas, R., and L. A. Real. 1991. Learning foraging tasks by bees: A comparison between social and solitary species. Animal Behaviour 42:269–276. Perspectives in ethology, Volume 4. Plenum, New York.

Eberhard, W. G. 1990. Function and phylogeny of spider webs. Annual Review of Ecology and Systematics 2:341–372.

Edwards, S. V., and S. Naeem. 1993. The phylogenetic component of cooperative breeding in perching birds. The American Naturalist 141:754–789.

Eibl-Eibesfeldt, I. 1970. Ethology: The biology of behavior. Holt, Rinehart and Winston, Toronto.

Gittleman, J. L. 1981. The phylogeny of parental care in fishes. Animal Behaviour 29:936–941.

Greene, H. W. 1994. Homology and behavioral repertoires. Pp. 369–391 in B. K. Hall, ed. Homology: The hierarchical basis of comparative biology. Academic Press, New York.

Gross, M. R. 1994. The evolution of behavioural ecology. Trends in Ecology and Evolution 9: 358–360.

Harvey, P. H., and M. D. Pagel. 1991. The comparative method in evolutionary biology. Oxford University Press, New York.

Hillis, D. M. 1994. Homology in molecular biology. Pp. 339–369 in B.K. Hall, ed. Homology: The hierarchical basis of comparative biology. Academic Press, New York.

Höglund, J. 1989. Size and plumage dimorphism in lek-breeding birds: A comparative analysis. American Naturalist 134:72–87.

Hörmann-Heck, S. v. 1957. Untersuchungen über den Erbgang einiger Vehaltenswiesen bei Grillenbastarden (*Gryllus campestris x Gryllus bimaculatus*) Zeitschrift für Tierpsychologie 14:137–183

Jansen, R. K., and J. Palmer. 1988. Phylogenetic implications of chloroplast DNA restriction site variation in the Mutisieae (Asteraceae). American Journal of Botany 75:753–766.

Langtimm, C. A., and D. A. Dewsbury. 1991. Phylogeny and evolution of rodent copulatory behaviour. Animal Behaviour 41:217–225.

Larson, A. L. 1989. The relationship between speciation and morphological evolution. Pp. 579–598 in D. Otte and J. A. Endler, eds. Speciation and its consequences. Sinauer, Sunderland, Massachusetts.

Lauder, G. V. 1994. Homology, form, and function. Pp. 151–196 in B. K. Hall, ed. Homology: The hierarchical basis of comparative biology. Academic Press, San Diego.

Losos, J. B. 1990. Ecomorphology, performance capability, and scaling of West Indian *Anolis* lizards: An evolutionary analysis. Ecological Monographs 60:369–388.

Luke, C. 1986. Convergent evolution of lizard toe fringes. Biological Journal of the Linnean Society 27:1–16.

Lynch, J. D. 1989. A review of the leptodactylid frogs of the genus *Pseudopaludicola* in northern South America. Copeia 1989:577–588.

Manning, A. 1975. Behaviour genetics and the study of behavioural evolution. Pp. 71–91 in G. Baerends, C. Beer and A. Manning, eds. Function and evolution in behaviour: Essays in honour of Professor Niko Tinbergen, F. R. S. Clarendon Press, Oxford.

Maurakis, E. G., W. S. Woolcott, and M. H. Sabaj. 1991. Reproductive-behavioral phylogenetics of *Nocomis* species-groups. American Midland Naturalist 126:103–110.

Mayr, E. 1958. Behavior and systematics. Pp. 341–362 *in* A. Roe and G. G. Simpson, eds. Behavior and evolution. Yale University Press, New Haven.

McLennan, D. A. 1993. Phylogenetic relationships in the Gasterosteidae: An updated tree based on behavioral characters with a discussion of homoplasy. Copeia 1993:318–326.

McLennan, D. A., D. R. Brooks, and J. D. McPhail. 1988. The benefits of communication between comparative ethology and phylogenetic systematics: A case study using gasterosteoid fishes. Canadian Journal of Zoology 66:2177–2190.

Miller, E. H. 1988. Description of bird behavior for comparative purposes. Pp. 347–394 *in* R. F. Johnston, ed. Current Ornithology, Vol. 5. Plenum Press, New York.

Miller, J. S., and J. W. Wenzel. 1995. Ecological characters and phylogeny. Annual Review of Entomology 40:389–415.

Mitter, C., B. Farrell, and D. J. Futuyma. 1991. Phylogenetic studies of insect-plant interactions: Insights into the genesis of diversity. Trends in Ecology and Evolution 6:290–293.

Mousseau, T. A., and D. A. Roff. 1987. Natural selection and the heritability of fitness components. Heredity 59:181–197.

Moynihan, M. 1975. Conservatism of displays and comparable stereotyped patterns among cephalopods . Pp. 276–291 *in* G. Baerends, C. Beer and A. Manning, eds. Function and evolution in behaviour: Essays in honour of Professor Niko Tinbergen, F. R. S. Clarendon Press, Oxford.

Mundinger, P. C. 1979. Call learning in the Carduelinae: Ethological and systematic considerations. Systematic Zoology 28:270–283.

Nelson, G. 1994. Homology and systematics. Pp. 101–149 *in* B. K. Hall, ed. Homology: The hierarchical basis of comparative biology. Academic Press, New York.

Nelson, J. S. 1989 and 1990. Analysis of the multiple occurrence of pelvic fin absence in extant fishes. Matsya 15 & 16:21–38.

Osche, G. 1952. Die Bedeutung der Osmoregulation und des Winkverhaltens für freilebende Nematoden. Zeitschrift für Morphologie und Ökologie der Tiere 41:54–77.

Otte, D. 1989. Speciation in Hawaiian crickets. Pp. 482–526 *in* D. Otte and J. A. Endler, eds. Speciation and its consequences. Sinauer, Sunderland, Massachusetts.

Packer, L. 1991. The evolution of social behavior and nest architecture in sweat bees of the subgenus *Evylaeus* (Hymenoptera: Halictidae): A phylogenetic approach. Behavioral Ecology and Sociobiology 29:153–160.

Parejko, K., and S. I. Dodson. 1991. The evolutionary ecology of an antipredator reaction norm: *Daphnia pulex* and *Chaoborus americanus*. Evolution 45:1665–1674.

Patterson, C., D. M. Williams, and C. J. Humphries. 1993. Congruence between molecular and morphological phylogenies. Annual Review of Ecology and Systematics 24:153–188.

Peterson, R. T. 1980. A field guide to the birds east of the Rockies, 4th ed. Houghton Mifflin, Boston.

Proctor, H. C. 1991. The evolution of copulation in water mites: A comparative test for nonreversing characters. Evolution 45:558–567.

Proctor, H. C. 1992. Sensory exploitation and the evolution of male mating behaviour: A cladistic test using water mites (Acari: Parasitengona). Animal Behaviour 44:745–752.

Proctor, H. C. 1993. Mating biology resolves trichotomy for cheliferoid pseudoscorpions (Pseudoscorpionida, Cheliferoidea). Journal of Arachnology 21:156–158.

Prum, R. O. 1990. Phylogenetic analysis of the evolution of display behavior in the Neotropical manakins (Aves: Pipridae). Ethology 84:202–231.

Prum, R. O. 1994. Phylogenetic analysis of the evolution of alternative social behavior in the manakins (Aves: Pipridae). Evolution 48: 1657–1675.

Remane, A. 1952. Die Grundlagen des Natürlichen Systems der Vergleichenden Anatomie und der Phylogenetik. Geest und Portig K. G., Leipzig.

Ridley, M. 1983. The explanation of organic diversity: The comparative method and adaptations for mating. Oxford University Press, Oxford.

Ridley, M. 1989. The timing and frequency of mating in insects. Animal Behaviour 37:535–545.

Ross, H. H. 1964. Evolution of caddisworm cases and nets. American Zoologist 4:209–220.

Rothenbuhler, W. C. 1964. Behavior genetics of nest cleaning in honeybees. IV. Responses of F1 and backcross generations to disease killed brood. American Zoologist 4:111–123.

Salvini-Plawen, L. v., and E. Mayr. 1977. On the evolution of photoreceptors and eyes. Evolutionary Biology 10:207–264.

Sanderson, M. J., B. G. Baldwin, G. Bharathan, C. S. Campbell, C. von Dohlen, D. Ferguson, J. M. Porter, M. F. Wojciechowski, and M. J. Donoghue. 1993. The growth of phylogenetic information and the need for a phylogenetic database. Systematic Biology 42:562–568.

Sanderson, M. J., and M. J. Donoghue. 1989. Patterns of variation in levels of homoplasy. Evolution 43:1781–1795.

Siddal, M. E., D. R. Brooks, and S. S. Desser. 1993. Phylogeny and the reversibility of parasitism. Evolution 47:308–312.

Sillén-Tullberg, B. 1988. Evolution of gregariousness in aposematic butterfly larvae: A phylogenetic analysis. Evolution 42:293–305.

Stearns, S. C. 1989. The evolutionary significance of phenotypic plasticity. BioScience 39: 436–445.

Stevens, P. F. 1991. Character states, morphological variation, and phylogenetic analysis: A review. Systematic Botany 16: 553–583.

Strauch, T. G., Jr. 1984. Use of homoplastic characters in compatibility analysis. Systematic Zoology 33:167–177.

Swofford, D. L., and G. J. Olsen. 1990. Phylogeny reconstruction. Pp. 411–501 *in* D. M. Hillis and C. Moritz, eds. Molecular systematics. Sinauer, Sunderland, Massachusetts.

Tinbergen, N. 1953. Social behaviour in animals, with special reference to vertebrates. Methuen, London.

Wagner, D. L., and J. K. Liebherr. 1992. Flightlessness in insects. Trends in Ecology and Evolution 7:216–220.

Ward, P. S. 1991. Phylogenetic analysis of pseudomyrmecine ants associated with domatia-bearing plants. Pp. 335–352 *in* C. R. Huxley and D. F. Cutler, eds. Ant-plant interactions. Oxford University Press, New York.

Wcislo, W. T. 1989. Behavioral environments and evolutionary change. Annual Review of Ecology and Systematics 20:137–169.

Wenzel, J. W. 1992. Behavioral homology and phylogeny. Annual Review of Ecology and Systematics 23:361–381.

Wenzel, J. W. 1993. Application of the biogenetic law to behavioral ontogeny: A test using nest architecture in paper wasps. Journal of Evolutionary Biology 6:229–247.

Wenzel, J. W., and J. M. Carpenter. 1994. Comparing methods: Adaptive traits and tests of adaptation. Pp. 79–101 *in* P. Eggleton and R. Vane-Wright, eds. Phylogenetics and ecology. The Linnean Society of London, London.

West-Eberhard, M. J. 1983. Sexual selection, social competition, and speciation. Quarterly Review of Biology 58:155–183.

West-Eberhard, M. J. 1984. Sexual selection, competitive communication and species-specific signals in insects. Pp. 283–324 *in* T. Lewis, ed. Insect communication. Academic Press, London.

West-Eberhard, M. J. 1989. Phenotypic plasticity and the origins of diversity. Annual Review of Ecology and Systematics 20:249–278.

Wiklund, C. and J. Forsberg. 1991. Sexual size dimorphism in relation to female polygamy and protandry in butterflies: A comparative study of Swedish Pieridae and Satyridae. Oikos 60:373–381.

Wiley, E. O. 1981. Phylogenetics: The theory and practice of phylogenetic systematics. John Wiley and Sons, Toronto.

Witte, H. 1991. Indirect sperm transfer in prostigmatic mites from a phylogenetic viewpoint. Pp. 107–176 *in* R. Schuster and P. W. Murphy, eds. The Acari: Reproduction, development and life-history strategies. Chapman and Hall, New York.

PART II

MEASURES OF HOMOPLASY

MEASURES OF HOMOPLASY

JAMES W. ARCHIE

California State University
Long Beach, California

INTRODUCTION

Homoplasy in the evolutionary process takes place when the origin of a character state among lineages occurs more than once, either through parallel or convergent derivation of an apomorphic state from a plesiomorphic state (not necessarily the same sequence in each lineage) or by reversal from an apomorphic to a plesiomorphic state. Such a process can be termed *evolutionary homoplasy*. In phylogenetic analysis, a transformation series among the observed character states is first hypothesized. Homoplasy is inferred by examining the distribution of character states among extant taxa and determining whether or not this distribution can be explained by single changes between plesiomorphic states and relatively apomorphic states based on the hypothesized transformation series, or whether multiple parallel or convergent changes or reversals are required. Because the actual transformations among the states are not observed, instances of evolutionary homoplasy cannot be inferred. Instead, the principle of parsimony is used to

HOMOPLASY: The Recurrence of Similarity in Evolution, M. J. Sanderson and L. Hufford, eds.
Copyright © 1996 All rights of reproduction in any form reserved.

determine the minimum number of character transformations that must be assumed or hypothesized in order to explain the distribution of character states. Whenever the minimum number of necessary steps required by the transformation series is exceeded, homoplasy is inferred to have occurred. The number of steps on most parsimonious trees beyond the minimum can be termed *cladistic* or *phylogenetic homoplasy* to differentiate it from evolutionary homoplasy. The latter will always be underestimated by cladistic homoplasy. All references to homoplasy in the following text refer explicitly to cladistic homoplasy.

Homoplasy is fundamentally a property of individual characters. However, the inference of cladistic homoplasy in a phylogenetic analysis depends on the superposition of the character on a phylogenetic tree hypothesis. The latter, in turn, is usually derived from the analysis of an entire data set, and such analyses often include the character of interest. The inference of cladistic homoplasy in such a character is then dependent on the hierarchical correlations or congruences among a set of characters that yield the hypothesis of relationships. If the characters are all perfectly consistent with a single phylogenetic hypothesis, then no homoplasy will be required for any single character in the data set. If no single hypothesis of relationships agrees with all characters, then homoplasy will be required in some number of individual characters. The amount of homoplasy inferred varies among the characters for a single tree hypothesis and, when more than a single tree hypothesis is considered, may vary for the same character between the different hypotheses. The amount of homoplasy will also vary among different character sets, either for the same or differing groups of taxa.

Whether data are morphological, behavioral, ecological, physiological, biochemical, or molecular, homoplasy has been observed to be a commonly occurring phenomenon and, for some types of data, is an expected result of whatever evolutionary process generated the distribution of character states that is observed. An analysis of character state changes on a phylogenetic hypothesis and the inference, in many instances, of homoplasious change provides incite into evolutionary processes. Homoplasy also has an effect on our ability to accurately estimate phylogenies, and the existence of homoplasy affects the confidence that we have in the phylogenies estimated or discovered from the analysis of the data. For these reasons, it is necessary to evaluate the amount of homoplasy in particular data sets and it is useful to compare data sets both for the same sets of taxa and for different sets of taxa. Measures or indices of homoplasy for both single characters and entire character sets (Farris (1989a) called measures of homoplasy for entire data sets *ensemble* measures) have been developed over the past 25 years to effectively analyze and compare homoplasy: (a) between single characters for single tree hypotheses, (b) for the same character between different hypotheses, and c) for entire data sets. This review attempts to compare both the standard measures of homoplasy that are in common use in phylogenetics and many of the more recently developed methods. What will be seen is that many of these measures have a common thread for comparison while others are quite distinct. After an analysis and comparison of their mathematical properties and interrelationships, a critical, and, hopefully, objective analysis is presented of the behavior of the various indices. A basic analysis of the relationship between homoplasy measures and phylogenetic accuracy is presented and, finally, a series of recommendations for the use of homoplasy indices is attempted. Measures of *character homoplasy* will be discussed first followed by measures of homoplasy for entire data sets or ensembles of characters.

MEASURES OF CHARACTER HOMOPLASY

Number of Extra Steps

The basic measure of the amount of character change in a character on a tree is the minimum number of steps, s, necessary to explain the distribution of the character states among the taxa. For any character there is a well defined minimum number of steps, m, that is required when no homoplasy is present based on an hypothesized transformation series. Of the actual inferred steps for a character on a specific tree hypothesis, the number of steps beyond the minimum is referred to as the number of extra steps, h. A value of $h > 0$ can be explained only by either parallel, convergent, or reverse evolution, that is, homoplasy. The existence of extra steps, $h = s - m > 0$, on a tree is, therefore, the fundamental indicator of homoplasy in a single character (or among a set of characters, see below).

It is again important to note that inferences of extra steps or homoplasy on tree hypotheses are inferences of *cladistic homoplasy* rather than *evolutionary homoplasy*. Cladistic homoplasy, although an estimate of evolutionary homoplasy, is a biased estimator and can be expected to underestimate evolutionary homoplasy. As new characters or taxa are added to a study, cladistic homoplasy levels and, more importantly, specific instances of homoplasy on tree hypotheses for a character are subject to reevaluation. Although generally expected to increase as new characters or taxa are added to a study, it is also possible for the inferred (cladistic) homoplasy, h, in a character to decrease if the minimum length tree topology changes. The location of specific character changes on the tree may also change, whether or not h changes. Clearly, however, historical, evolutionary homoplasy is not subject to modification based on the inclusion of information by the systematist.

In contrast to the minimum number of steps required for a character, Mickevich (1978) gave a formula for the maximum possible number of steps that binary characters could have on any minimum length tree. For t taxa, this quantity is equal to $t/2$ when t is an even number and $(t - 1)/2$ when t is an odd number. The maximum number of extra steps beyond the minimum, m, required for binary characters is one less than this, $h_{max} = [t/2] - 1$, where $[t/2]$ refers to the greatest integer less than the fraction $t/2$ (Mickevich 1978). The maximum number of extra steps for a specific character depends directly on the frequency of alternative character states. For binary characters, it is one less than the number of occurrences of the less frequent state. For example, if only a single taxon possesses the apomorphic state, then no homoplasy can be required for this character on any tree. If two taxa possess the apomorphic state, only a single instance of homoplasy can occur on any tree. As the frequency of the less frequent state approaches $[t/2]$, then the maximum possible homoplasy required for a character increases to h_{max}.

Although h is the fundamental measure of homoplasy, it generally is not used for evaluating relative levels of agreement with a tree among different characters specifically because h depends on both the frequency of character states among the taxa and the number of taxa. Clearly it cannot be used to compare homoplasy in separate studies containing different numbers of taxa. However, when combined together with other characters, Σs_i, or, equivalently, Σh_i, can be used to evaluate possible trees relative to one another (see below).

Character Consistency Index

Farris (1969) described the character consistency index c (or, more generally, c_i, the value of c for character i) which has been utilized in various phylogenetic analysis computer programs (Wagner78, PHYSIS, Hennig86, PAUP, MacClade, and PHYLIP) to indicate fit of a character to a tree. In parallel with the character consistency index, Kluge and Farris (1969), described the data set consistency index, CI (see below). Using Farris' (1989a) definitions, where s_i equals the number of steps for character i on a tree (minimized for that tree) and m_i equals the minimum possible number of steps for the character on any tree, then

$$c_i = \frac{m_i}{s} = \frac{m_i}{m_i + h_i} .$$

h_i in this equation equals the number of steps in character i due to homoplasy. Although this definition is general for multistate characters (where $m_i > 1$), when binary characters are used, $m_i = 1$, and the c_i equals the *reciprocal* of the number of steps per character, $c_i = 1/s_i$. For multistate characters with a_i distinct character states, c_i equals the reciprocal of the number of steps per apomorphic state, that is, $(a_i - 1)/s_i$ (for binary characters $a_i = 2$, so this is a general formula). Farris (1989a) noted that $1 - c_i$ is "the fraction of change [in a character] that must be attributed to homoplasy." The latter is now termed the homoplasy index. The character c_i varies from 1.0 when the number of steps on a tree equals the minimum and there is no homoplasy in the character and decreases as the amount of homoplasy increases. As is apparent from the formulation (i.e., $m_i + h_i$ in the denominator), as the number of homoplasy steps increases the c_i approaches 0.0 asymptomatically, but it can never equal 0.0 since the numerator m_i is greater than zero for all variable characters. The minimum possible value of c_i depends on both the frequency of the character states and the number of taxa in the study. For studies with large numbers of taxa, the maximum possible amount of homoplasy, h, for a character is larger than for studies with fewer taxa and, hence, individual c_i values tend to be smaller in studies with more taxa, t. Specifically, for binary characters, c_i will never be less than $1/[t/2]$. With these dependencies, c_i is less useful as a comparative index of character congruence than originally perceived. The c_i has generally been used only to identify characters that have high levels of homoplasy (low c_i values), but was used explicitly by PHYSIS (Farris and Mickevich 1975) as an index for character weighting (see Farris 1969, 1989a).

Character Retention Index

Farris (1989a) proposed the character retention index, r (or r_i for character i), as an improvement of the consistency index "for certain applications." In contrast to the character consistency index which compares the observed tree length (s_i or minimum steps plus homoplasy) to the steps on the tree when the character fits the tree perfectly (m_i), the retention index expresses the observed homoplasy relative to the maximum possible amount of homoplasy that the character could have on any tree,

$$r_i = \frac{s_{\max} - s_i}{s_{\max} - m_i}$$

$$= \frac{(h_{\max} + m_i) - (h_i + m_i)}{(h_{\max} + m_i) - m_i}$$

$$= \frac{h_{\max} - h_i}{h_{\max}} = 1 - \frac{h_i}{h_{\max}},$$

where s_i and m_i are as defined above and $s_{\max}$ is the maximum number of steps that the specific character can have on any tree. When there is no homoplasy on the tree (i.e., $h_i = 0$) r_i equals 1.0. In contrast, when the amount of homoplasy on the tree is as large as possible, ($h_i = h_{\max}$) the numerator $h_{\max} - h_i$ equals 0.0 and r_i will equal 0.0. r_i thus varies in the same way as c_i, but clearly varies between fixed limits that do not depend on either character state frequencies or on the number of taxa. Farris (1989a) reasoned that when $r_i = 1.0$, each character state is uniquely derived on the tree and taxa share character states due to synapomorphy. In contrast, when $r_i = 0.0$, all occurrences of the derived character states arise independently and taxa share character states due strictly to false homology or homoplasy. In the former case, all "apparent synapomorphy" in the character is retained on the tree while in the latter case, no "apparent synapomorphy" is retained on the tree; hence the name character retention index.

Rescaled Character Consistency Index

Because c_i does not vary between fixed limits, Farris (1989a) proposed rescaling c_i by first subtracting its minimum possible value and then rescaling it to vary between 0 and 1 by dividing it by its range. This entire process is equivalent to rescaling c_i by multiplying it by r_i. The product he termed the rescaled consistency index, $rc_i = r_i \times c_i$. Since r_i varies between 0 and 1, multiplying c_i by r_i accomplishes the desired rescaling. Farris (1989a) does not explicitly state why rc_i is better than the retention index, which is already properly scaled, especially as a weighting factor in weighted parsimony. However, as was done earlier with the consistency index, Farris implemented the rescaled consistency index into his computer program, Hennig86 (v. 1.0), as a character weighting scheme.

MEASURES OF DATA SET HOMOPLASY

Measures of homoplasy derived from the entire data set are more diverse than those for single characters. However, most (but not all) of these are derived from the individual character values and most can be related to a rescaling of the total amount of homoplasy among the set of characters by a combination of scalings of the individual characters.

Total Homoplasy or Number of Extra Steps

Total homoplasy (H) in a character set with N characters can be defined as the sum across all characters of the number of extra steps, h_i, present in each character,

$$H = \sum_{i=1}^{N} h_i.$$

This count of number of extra steps, first used by Camin and Sokal (1965), is the most basic measure of homoplasy on a tree for the entire data set. Using the principle of parsimony it, or more often, the total number of steps on a tree, S, is the single statistic used in selecting among possible trees for a set of taxa and a specific set of characters. Simply stated, the tree showing the minimum number of steps is chosen as the most parsimonious. The summation to determine H may be modified by weighting (multiplying) each of the h_i by some factor, for example, c_i or, more recently, rc_i, as done in Hennig86. It is apparent that because H is a sum of individual h_i values, it will increase monotonically with increasing numbers of characters. In addition, because the individual h_i are expected to increase with the number of taxa in a study and h_{max} depends on the frequency of character states, the value of H is expected to vary with both number of taxa and character state frequency. For these reasons, although H is the single direct measure of overall homoplasy, it is an inappropriate comparative index of homoplasy among data sets.

Consistency Index (Homoplasy Index)

The data set consistency index (C, or more often, CI) was first described by Kluge and Farris (1969) as a scaled measure of consistency or fit of an entire set of characters to a tree. Rather than being an average of the character consistency values, the CI is calculated by forming the ratio of the sum across all characters of the minimum number of steps, m_i, for the characters and the sum across characters of the actual number of steps, s_i, on a tree,

$$\text{CI} = \frac{\Sigma m_i}{\Sigma s_i} = \frac{\Sigma m_i}{\Sigma (m_i + h_i)}.$$

For a set of N binary characters the CI equals the reciprocal of the number of steps per character (i.e., N/S). Kluge and Farris' stated objective in scaling number of steps in this way was "to measure the deviation of the tree from a perfect fit to data." They noted specifically that CI "varies between fixed limits" and that it "is 1 if there is no convergence on the tree, and tends to 0 as the amount of convergence on the tree increases." Because of its formulation, they noted that CI "is some weighted continuous analog of 'number of extra steps'," and "can be used . . . to compare the fits of trees to different data sets" (Kluge and Farris 1969: 8). The homoplasy index, HI, is the complement of CI, HI = 1 − CI, analogous to the character homoplasy index:

$$\text{HI} = 1 - \text{CI} = 1 - \frac{\Sigma m_i}{\Sigma (m_i + h_i)}$$

$$= \frac{\Sigma m_i + \Sigma h_i - \Sigma m_i}{\Sigma (m_i + h_i)} = \frac{\Sigma h_i}{\Sigma (m_i + h_i)}.$$

For the entire data set, HI is the proportion of total character change on the tree that is due to homoplasy.

For sets of characters that are variable among taxa the numerator for CI is a positive number (minimally the number of characters if all characters are binary), and, hence, CI must be strictly greater than zero; that is, it cannot equal zero. If all characters are congruent, all $h_i = 0.0$ and CI equals 1.0. As noted above, the maximum possible values of h_i vary directly with the number of taxa in a study and also the frequency of character states. As a result, the minimum achievable value for CI is generally correlated with the number of taxa. Archie (1989a,b) and Sanderson and Donoghue (1989) found that for real data sets, observed values of CI are also negatively correlated with the number of taxa. Archie and Felsenstein (1993) demonstrated that this negative correlation is expected based on an expected mathematical relationship between the number of taxa and number of steps per character. Several authors (Archie 1989b; Sanderson and Donoghue 1989; Klassen et al. 1991) have found small negative, but nonsignificant, correlations between CI and the number of characters for real data sets. In computer simulation studies, Archie (1989b) and Archie and Felsenstein (1993) found a consistent increase in number of steps per character with an increase in number of characters (implying a negative correlation between CI and number of characters). They related this correlation to basic properties of minimum length tree optimization and to the increasing degree of character conflicts that are possible as the number of characters increases. This latter effect would not be substantial (and, as noted, has not been found to be statistically significant) in comparisons of fits of different data sets to minimum length trees, presumably because real data sets differ substantially in so many other features besides number of characters (e.g., character state frequencies; see Fig. 1 in Archie 1989a). By subsampling a single data set, however, Archie (1989b) found a consistent negative correlation between the CI and number of characters.

Retention Index (= HER_{max})

Farris (1989a) described the data set (he used the term ensemble) retention index that he had earlier implemented in his phylogenetic analysis package, Hennig86. The formula is analogous to the formula for the character retention index, r_i,

$$\text{RI} = \frac{S_{max} - S}{S_{max} - M}$$

$$= \frac{(H_{max} + M) - (H + M)}{(H_{max} + M) - M} = \frac{H_{max} - H}{H_{max}}$$

$$= \frac{\Sigma h_{max,i} - \Sigma h_i}{\Sigma h_{max,i}} = \frac{H^B_{max} - H}{H^B_{max}} = 1 - \frac{H}{H^B_{max}}$$

where S_{max} and M are the maximum number and minimum number of steps on a tree summed over all characters, respectively. Here H^B_{max} is the sum across all characters of the maximum possible homoplasy for each character. It equals the amount of homoplasy present on the totally unresolved or "bush" phylogeny. Farris (1989b) stated that RI was the complement of the distortion coefficient (Farris 1973), but, as pointed out by Archie (1990), although the character retention index, r_i, is the complement of the character distortion index, the data set distortion index is an arithmetic average of the individual character values and RI cannot be calculated from it.

Farris (1989a) proposed using RI only for evaluating the effects of different character weighting schemes, inferring that higher RI values indicate better overall fit of certain weighted character schemes. However, he also states that RI is preferable to CI "when different suites of characters . . . are compared." Since its introduction, RI has frequently been used as a general index of overall homoplasy in data sets (e.g., Nedbal et al. 1994).

Archie (1989b) independently introduced an index, the homoplasy excess ratio maximum (HER_{max}), for measuring and comparing levels of homoplasy in data sets. This latter index is identical to Farris' RI (Farris 1989b, 1991; Archie 1990). Archie (1989b) attempted to scale total homoplasy in such a way that when no homoplasy was present, the index would equal 1.0, but when homoplasy was the maximum possible for the data set it would be zero. Although Farris (1989a) clearly shows that the character retention index, r_i, will equal zero under specific conditions, he does not address whether or not ensemble RI will equal zero under some specified conditions. Clearly, if a tree exists on which all characters exhibit maximum homoplasy, then RI ($= HER_{max}$) will be zero. Farris (1991) showed that such a tree can always be found, but that it is a tree with no resolution, that is, a "bush" topology. However, Archie (1989b, 1990) showed that, under most conditions, RI will not and cannot equal zero on *minimum length trees* even when the character conflicts are as bad as possible (see the discussion on the permutation congruence index, below). This is because the totally unresolved tree on which RI equals zero will never be a minimum length tree as long as there is at least one potentially informative character in the data set. The "bush" phylogeny is generally uninteresting in phylogenetic analyses and arises only as a consensus tree of suites of minimum length or near minimum length trees. Generally, such consensus trees are not minimum length trees (Miyamoto 1985; Archie 1989b, 1990). These results suggest that the intuitive scaling of individual character homoplasy levels that gives rise to r_i is an inappropriate scaling of the combined suite of characters.

In summary, as Farris (1989a) states, both r_i and RI measure the proportion of apparent homoplasy in the data set that is retained as homoplasy on the phylogenetic tree. However, whereas r_i is scaled to vary between achievable limits on minimum length trees, RI is not so scaled.

162

Homoplasy Excess Ratio

Archie (1989b) proposed the homoplasy excess ratio statistic (HER) as an alternative to RI (= HER_{max}) to incorporate the fact that RI cannot equal zero for most data sets on minimum length trees. Recall, RI is scaled using an H^B_{max} equal to the maximum homoplasy possible across all characters on any tree topology (in most cases, the bush topology). In contrast, HER is scaled using an H_{max} (H^P_{max}) equal to the expected homoplasy found on a minimum length tree when the hierarchical character congruence is as low as expected due to random chance alone, that is, the *expected H* if the observed character states for the characters are assigned to taxa at random. The formula is directly analogous to that for RI:

$$HER = \frac{S_{max} - S}{S_{max} - M}$$

$$= \frac{\Sigma b_{max,i} - \Sigma b_i}{\Sigma b_{max,i}} = \frac{H^P_{max} - H}{H^P_{max}} = 1 - \frac{H}{H^P_{max}}.$$

H^P_{max} cannot be calculated easily. Farris (1991) pointed out that, mathematically, H^P_{max} equals the average number of steps on minimum length trees for *all possible* within-character permutations of the data matrix. Although obtaining all possible permutations for a data matrix is a straightforward (but large) problem, obtaining the minimum length trees for this set of matrices would be prohibitively time consuming.

Archie (1989b) proposed estimating H^P_{max} with the following procedure. First, randomize or permute (hence, "H^P_{max}") the states of each character separately across taxa. Each of the resulting characters has the same frequency of states as the original characters, but all hierarchical character congruences among the characters beyond what is expected due to chance alone will be destroyed. As a result, the data are expected to be random with respect to each other and to the true or any phylogeny. Second, determine the length of the minimum length tree for the randomized data. This two-step procedure is repeated a large number of times (100 repetitions was recommended). Finally, the estimate of H^P_{max} is the average amount of homoplasy or extra steps across all randomized replicates. Archie (1989b) observed that the variance in the estimate of H^P_{max} provided by the randomization procedure is quite small. As a result, in most cases the necessary number of permutations for estimating H^P_{max} with good precision should be no more than 25. Although Archie (1989b) never attempted to interpret HER with respect to the retention of apparent synapomorphy on the tree, HER clearly measures the proportion of apparent synapomorphy in the data that is retained on the tree relative to the amount that is expected due to chance alone.

HER equals zero when the hierarchical correlations among characters are as low as expected due to chance alone, but will take on negative values when the amount of homoplasy in a data set is greater than expected due to chance alone. Farris (1991) criticized HER for this, but the conditions under which HER equals zero or is less than zero are well defined. Because the scaling of HER is empirical, it takes into account not only the number of taxa and characters, but also the frequency of character states among the taxa and the properties of the data analysis technique (i.e., parsimony) in summarizing character state information. It therefore is not affected by the properties of the data set that do influence CI or RI.

Rescaled Consistency Index

Farris (1989a) suggested that, as with the character consistency index, the data set consistency index (CI) could be rescaled to vary between 0 and 1 by multiplying CI by RI to yield the rescaled consistency index, RC. However, as pointed out by Archie (1991), because neither RI nor CI can equal zero under conditions used for phylogenetic analysis, that is, when minimum length trees are used, then RC cannot equal zero on minimum length trees. This scaling of CI does not achieve its portended effect. Farris (1989a) did not explicitly propose a specific use or interpretation of RC.

Data Decisiveness

The novel concept of the decisiveness of a data set was introduced by Goloboff (1991a). He defined decisiveness as the degree to which a data set differentiates among possible trees. Data sets that provide a greater range of tree lengths among all possible trees for the taxa are more *decisive* than data sets that provide a smaller range. Goloboff demonstrated that only very specific types of data matrices are completely indecisive such that all possible bifurcating trees are equal in length. These data sets contain one each (or an exact multiple thereof) of every possible character type (i.e., every possible distribution of 0 and 1 states among taxa). Although Goloboff (1991b) states that there is no necessary relationship between decisiveness and randomness among characters, completely indecisive matrices, in fact, have greater character conflict than would be expected due to chance (they are worse than random). Indecisive matrices are so idiosyncratic that they are extremely unlikely to arise in real data sets.

Goloboff (1991a) introduced a measure of homoplasy related to the concept of the decisiveness of a data set. Le Quesne (1989) first showed that for any data set, the sum of the lengths of all possible trees and the average length calculated for the possible trees is identical regardless of the degree of congruence among the characters. This results from the fact that the total number of steps for a character on all possible trees depends only on the frequency of states in the character rather than the congruence of the character with any others in the data array (Goloboff 1991a). Using this result, Goloboff proposed an index he called data decisiveness (DD) that is formulated identically to that of RI and HER except that H_{max} (H^{DD}_{max}) is calculated from the average length of the data set on all possible trees:

$$\mathrm{DD} = \frac{\bar{S} - S}{\bar{S} - M}$$

$$= \frac{(M + H^{DD}_{max}) - (M + H)}{(M + H^{DD}_{max}) - M}$$

$$= \frac{H^{DD}_{max} - H}{H^{DD}_{max}} = 1 - \frac{H}{H^{DD}_{max}} \cdot$$

For each character (or each character with a specific frequency of character states) the number of steps on all possible trees is calculated using the formula of Carter et al. (1990). Summed across all characters, this gives the total number of steps for all characters for all possible trees. $\bar{S}$ is the average number of steps calculated by dividing the sum of the number of steps for all trees by the number of trees. To obtain H^{DD}_{max}, the minimum number of steps required for each character is subtracted from S. DD has a maximum value of 1.0 when there is no homoplasy in the data (as with RI and HER). For minimum length trees, it equals 0.0 only when the data set has one of every possible character type in the data set (or an exact multiple of this), that is, it is one of Goloboff's indecisive matrices, but otherwise is strictly positive. DD may be negative (or zero) for nonminimum length trees and is always negative for the bush topology (as long as there are informative characters in the data set).

Homoplasy Slope Ratio

The homoplasy slope ratio (Meier et al. 1991) is also a standardized measure of the total homoplasy in a data set. The expected relationship between the number of steps per character and number of taxa in data sets was first examined by Archie (1989a) and later by Archie and Felsenstein (1993). Meier et al. reexamined this relationship and derived approximate equations for the slope of the regression of number of extra steps per character on number of taxa for data with randomly assigned character states with prob(0) = prob(1) = 0.5. The estimated slope was used to standardize the slope of the regression of observed number of extra steps per character on number of taxa for real data sets. They used this ratio of the two slopes as a relative measure of homoplasy, the homoplasy slope ratio, HSR. Meier et al. also showed that the ratio of slopes is equal to the ratio of observed total homoplasy, H, divided by an *estimate* of the maximum homoplasy present on minimum length trees for their randomly generated data, $H^{R'}_{max}$:

$$\text{HSR} = \frac{\text{slope}_{\text{real}}}{\text{slope}_{\text{random}}} = \frac{H}{H^{\text{R'}}_{\text{max}}}.$$

Because HSR is not subtracted from 1, it varies inversely relative to CI, RI, and HER; that is, HSR = 0.0 for data with no homoplasy and increases to an unspecified limit as homoplasy increases.

HSR is related to the relative expected HER (REHER, Archie 1989b) in which H_{max} was calculated from the expected number of steps per character *on random trees* for random data (versus minimum length trees in HSR). For random trees, the expected number of steps was shown by Archie and Felsenstein (1993) to equal $(3t - 2)/9$, approximately, for t taxa (a more exact, but recursive, equation is given in Archie and Felsenstein 1993). The derivation of their equation is nonintuitive, but was derived from an original, exact formula for the expected number of steps on random trees for variable state data, where prob(0) = prob(1) = prob(0,1) = 1/3. The approximate and exact formulae give identical values (to three decimal places) for $t > 6$. Archie (1989b) used the result from Archie and Felsenstein for a crude scaling of total homoplasy, but recommended against its use:

$$\text{REHER} = 1 - \frac{H}{(3t - 2)/9 - n} = 1 - \frac{H}{H^{\text{R}}_{\text{max}}}.$$

Archie (1989b), Meier et al. (1991), and Archie and Felsenstein (1993, their Figs. 6 and 7) found that on minimum length trees the slope of the relationship between number of steps per character and number of taxa was lower than that expected for random trees and that as the number of characters increased, this slope also changed. Meier et al. (1991) estimated the joint relationship between number of characters and number of taxa versus the number of extra steps per character using computer simulation. This estimate was used to obtain a table of $H^{\text{R'}}_{\text{max}}$ values.

Archie (1989b) rejected REHER as an estimate of HER because of the lack of dependence of the scaling factor, H'_{max}, on the observed character data. Because of the approximate correction of HSR for the number of steps of random characters on minimum length trees, it is less subject to constraints in interpretation than is REHER, but it is less sensitive to differences in overall homoplasy caused by characteristics of the data set than is HER or RI. Meier et al.(1991) did not analyze the specific effects of increasing levels of homoplasy on HSR nor did they examine the upper limits to their index.

Permutation Congruence Index

In his discussion of the properties of the homoplasy excess ratio, Farris (1991) presented a new index that he called the permutation congruence index, K. Recall that to standardize observed homoplasy, H, HER is calculated using the *average* amount of homoplasy present on minimum length trees over all possible character permutations of the data matrix. In practice, an estimate of this average is obtained by calculating minimum length trees for a reasonable number of matrix permutations of the data. Although the expected value of HER for any random permutation of the matrix is 0.0, because the *average* from the matrix permutations is used, it is possible to obtain a value of HER less than zero (HER $<$ 0). As Farris pointed out, more than half of the possible permutations of the matrix (of which the original data is just one) may give negative values for HER if the distribution of minimum tree lengths for the matrix permutations is skewed. Farris suggested that a more reasonable standardization for H would be the *maximum* amount of homoplasy for any possible permutation of the data matrix. He called this maximum $J(X)$, but in the notation used here, it will be called H^J_{max}. The permutation congruence index, K, is calculated from this value as follows:

$$K = 1 - \frac{H}{J(X)} = 1 - \frac{H}{H^J_{max}}.$$

Because H^J_{max} is used in calculating K, it is guaranteed to have a range from 1 when no homoplasy is present in the data to 0 when the maximum possible amount of homoplasy is present for the given set of character state frequencies in the data. This scaling gives the optimal range for such a homoplasy index as reasoned by Archie (1989b). The difficulty with calculating K is that in order to determine the value for H^J_{max} for all possible permutations of the matrix X, the minimum length tree would have to be calculated for all possible permutations. Clearly, this would be insurmountable because of the shear size of the problem for analysis. Farris (1991) reasoned that for reasonably large data matrices, H^J_{max} will be close to the value of H^B_{max} used in calculating the retention index, RI, and that, as a result, RI can be used as an approximation of K. RI will be strictly *greater* than K because H^B_{max} used in its calculation is strictly greater than that used in calculating K (recall that the bush phylogeny on which the value of H^B_{max} is determined is never a minimum length tree [except for data with no possibility of synapomorphy] for any permutation of the data matrix).

Farris also reasoned that for data with small amounts of homoplasy, both K and RI will be very similar in value, presumably whether H^J_{max} and H^B_{max} are similar or not. This final conclusion seems reasonable at first; however, it is also quite peculiar. If H is small, then *all* of the relevant indices, CI, RI, HER, 1 − HSR, and K will be very similar in value, not just K and RI. Under these conditions, it would not matter which index is used. However, under general conditions, CI, RI, HER, and, presumably, K can be quite different as pointed out by Archie (1989b, in preparation).

SPECIALIZED MEASURES OF HOMOPLASY

Skewness of Tree Lengths

Fitch (1979) was the first to suggest that the distribution of tree lengths of all possible trees for a set of taxa could be used to indicate the presence of a phylogenetic signal in systematic data. For all but two of the data sets that he had examined, a highly asymmetrical distribution was present. Tree length distributions that are highly skewed (negative value of the third moment statistic for the data, $g1$) result in few trees that are similar in length to the minimal length tree and a greater range of tree lengths among all possible trees. This latter result is possibly an indication of greater data decisiveness (*sensu* Goloboff 1991a) for the data. Le Quesne (1989) found that if one of the data sets he examined, that had a strongly asymmetrical distribution of tree lengths ($g1 = -0.875$), was randomized within characters to eliminate information on relationships among taxa, the distribution becomes symmetrical ($g1 \approx 0.0$) with a substantial, concomitant decrease in the standard deviation of the tree lengths(s). He reported skewness statistics ranging from -0.220 to -0.736 for 10 other data sets. An example of this effect is presented in Fig. 1 for the *Anacyclus* data of Humphries (Swofford 1993: 10 taxa, 47 characters). For the original data (solid lines), the minimal tree length was 67 steps while for the randomized data (dotted line) the minimum tree length was 80 steps. The skewness statistics for the original and random-

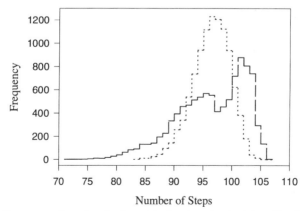

FIGURE 1   Frequency of tree lengths for original and randomized data. Tree length distributions for original *Anacyclus* data set (dashed line) and for same data set after character randomization (dotted line).

ized data are $g1 = -0.685$ and -0.042, respectively, while the standard deviations are 6.02 and 3.22.

Hillis (1991), Huelsenbeck (1991), and Hillis and Huelsenbeck (1992) formalized the use of tree-length skewness ($g1$) as an indication of the presence of phylogenetic signal in systematic data. Le Quesne (1989, his Fig. 2; see also Hillis 1991) formally demonstrated that for random phylogenetic data, the distribution of tree lengths is substantially skewed. In addition, Le Quesne (1989) and Hillis (1991) demonstrated that skewness is not qualitatively affected by tree shape. Hillis (1991) used computer simulation to generate frequency distributions for $g1$ skewness statistics for random data to allow statistical testing of $g1$ values from systematic data. His results were limited to studies with 6, 7, and 8 taxa, but many recent studies have used his critical values for general testing for phylogenetic signal.

Le Quesne's study (1989) suggests an alternative approach using character randomization. First, the tree-length distribution for a data set can be determined, either exhaustively or using a large ($n = 10,000$) random sample of possible trees. Next, the data can be randomized by character (*sensu* Archie 1989a) and the distribution of tree lengths for these randomized data determined. As previously discussed, the randomized data set is comparable to the original data in terms of character state frequencies, but the characters are expected to be random with respect to one another and to contain no information on phylogenetic relationships. These two distributions from the original and randomized data (example in Fig. 1) can be compared statistically: (a) by a Kolmogorov–Smirnov test (Sokal and Rohlf 1995), (b) by a test of the equality of the variances for the two distributions using an F-ratio test, as suggested by Le Quesne, or (c) by a direct comparison of their $g1$ skewness values using a two-sample z-test,

$$z = \frac{g1_o - g1_R}{\sqrt{\dfrac{6}{n_1} + \dfrac{6}{n_2}}},$$

where $g1_O$ is the tree length skewness for the original data and $g1_R$ is the tree-length skewness for the randomized data. Because the variance of each $g1$ statistic depends only on the sample size (n), the number of degrees of freedom for each of these sample statistics is infinite so a z-test is appropriate. Each of the suggested test comparisons between the two tree length distributions will have fairly high power because of the large sample sizes upon which the test statistics are based (although the tests are only approximate because tree lengths are not all independent). Clearly, comparisons of the original data set should be made with more than one randomization of data (its actually possible to get a randomized data set with no homoplasy). Le Quesne's study indicated that with sets of random data, large negative skewness values are possible, but the expected value is close to zero. Le Quesne also suggested that both tree length distribution skewness and variance should both be examined as uncorrelated indicators of phylogenetic signal. The randomization approach suggested here (comparable to Le Quesne's approach) is possible for studies with any number of taxa and characters and is adjusted for the effect of different frequencies of states for characters on $g1$ values.

Huelsenbeck (1991) performed a series of computer simulations of phylogenies for gene frequencies and demonstrated that data sets with large negative skewness values that are considered significant using Hillis' critical values, generally have the true tree (known from the simulation) within 3% of the range of tree lengths from the minimal length tree. For those data sets with nonsignificant $g1$'s, the closeness of the true tree to the minimum length tree was unpredictable. These results indicate that significant negative skewness can be used as an indication of substantial phylogenetic signal in the data. Huelsenbeck's study did not, however, indicate the degree to which true trees differed from the minimum length tree in terms of topology. It is possible that the set of minimum length and near-minimum length trees that include the true tree contain more than one island of trees (Maddison 1991). In addition, as pointed out by Källersjö et al. (1992), data sets with nonsignificant negative skewness statistics do not necessarily lack information on the phylogeny of the taxa and the true tree may be very similar to the minimum length tree for these data. This conclusion is apparent from Huelsenbeck's results as well (his Figs. 3, 7, 9, and 11).

F-Ratio

Farris and Mickevich (presented in the manual for the phylogenetics program PHYSIS, see Brooks et al. 1986) developed the F-ratio statistic after the f-statistic of Farris (1972) as a means of measuring homoplasy in an intertaxon character difference matrix. First, to calculate the f-statistic, two separate intertaxon difference matrices are compared: (a) a pairwise matrix of intertaxon differences in character states summed across all characters, $D(i,j)$ (the phenetic difference matrix), and (b) a matrix of inferred intertaxon differences derived from a phylogenetic tree, $P(i,j)$ (the patristic difference matrix). To obtain the patristic difference matrix, ancestral node character states are assigned using one of the techniques described by Swofford and Maddison (1987). The number of character state changes on the path between each pair of terminal nodes is then determined. The f-statistic is the sum of the pairwise differences in corresponding values from these two matrices:

$$f = \sum_{i=1}^{t-1} \sum_{j=i+1}^{t} |D(i,j) - P(i,j)| .$$

The F-ratio is calculated by dividing f by the sum of the $D(i,j)$ and multiplying the quotient by 100. f and F will equal 0 when there is no homoplasy in the data and increase as the amount of homoplasy in the data increases.

Clearly, f is unbounded because the differences between the matrices increase due to homoplasy and would be expected to increase as the number of taxa increase. F uses the sum of the $D(i,j)$ as a scaling factor, but it is unclear if F is actually bounded above (e.g., $F \leq 1$) by this scaling. The specific numerical properties of F have been examined in detail by Brooks et al. (1986) and are discussed below.

Information Divergence

Brooks et al. (1986) introduced an information theoretic statistic, D, as a measure of the fit of character state data to trees. The notation D is used to denote "divergence from equiprobability of symbol states." The measure is calculated by first determining the number of taxa sharing relative apomorphic states for each character on a directed tree,

$$H_S = - \sum_{clades} \sum_{character-states} \frac{n_{i,j}}{A} \log \frac{n_{i,j}}{A},$$

where A equals the sum of the number of taxa in each clade on the tree and $n_{i,j}$ equals the number of taxa in clade j that have the apomorphic state for character i. For each clade, only those characters that change from the plesiomorphic state to the apomorphic state on the node leading to the clade are summed. The latter restriction implies, as noted by Brooks et al., that the choice of ancestral character state reconstruction algorithm will affect the value of H_s. The divergence statistic, D, is then calculated as the difference between H_s and H_{max} (these H_s and H_{max} do not indicate homoplasy), where H_{max} is the maximum value of H_s that is achieved when all taxa have distinct states for each of the characters. Brooks et al. suggest standardizing D to have a maximum achievable value of 1.0 by dividing it by H_{max}. This standardized statistic they denoted by R. They again noted, however, that D, or the standardized value R, do not have a specific minimum value when no homoplasy is present in the data, although it will always be greater than 0. This quality of the statistic would allow the comparison of different data matrices for the comparison of "information content" that they possess for a specific tree (this is similar to Goloboff's discussion of data decisiveness), but implies that it cannot be used as a general measure of homoplasy. An additional problem is that D cannot achieve a maximum under reasonable conditions due to the choice of method for calculating H_{max}. In order for $H_s = H_{max}$, all states in the character matrix would have to be unique and all transformation series would have to be undirected. Because rooted trees are used in the calculation of H_s, a more appropriate standard for comparing H_s would be the maximum number of steps necessary to explain the distribution of the apomorphic states, that is, if all characters contained only autapomorphies. If N is the total number of apomorphic states in the character matrix, then $H_{\max}$ would equal log N. For the worst possible case, that is, the bush phylogeny, $H_s = H_{\max}$ and both D and R equal 1.0. Still, however, if no homoplasy is present on the minimum length tree, the value of D or R is unknown *a priori*. A scaling procedure similar to that used in calculating RI and HER could easily be used to ensure that this minimum would be 0, but this clearly would take the index out of a possible information–theoretic interpretation.

A COMPARISON OF HOMOPLASY MEASURES AND INDICES

Numerical Properties of Homoplasy Indices

The basic measure of cladistic homoplasy on a phylogenetic tree, h_i, is a count of the number of steps beyond the minimum number necessary to explain the distribution of character states among the taxa. At the level of evaluation of homoplasy of individual characters, the difficulty with h_i is that the maximum possible value for h_i is very dependent on both the frequency of the alternative character states and the number of taxa in the study. For this reason, it is not a useful comparative measure except in evaluating single characters on different tree topologies. The character retention index, r_i, is the most direct and interpretable scaling of h_i that allows comparisons among characters. Before a character is mapped onto a tree, hypotheses of homology of derived character states, that is, synapomorphy, are made. Once a character is mapped on a tree then r_i , with an achievable range of 0.0 to 1.0, can be interpreted directly as the proportion of the original synapomorphic states that are kept together as synapomorphies on the tree (Farris 1989a,b). The character consistency index, c_i, was the first scaled index of homoplasy that was proposed. However, c_i does not effectively scale homoplasy to vary between achievable limits. Its minimum possible value may be as high as 1/2 but no lower than $1/[t/2]$. The homoplasy index, h_i (the compliment of c_i), is interpretable as the proportion of total change on the tree due to homoplasy, but clearly must have the same limitations as c_i. Although the rescaled character consistency index, rc_i, is currently being used in weighted homoplasy, the selection of this index over the more interpretable r_i has not been clearly justified (see Farris 1989a). Goloboff (1993) recommended using $1/(h_i + 1)$ or $(k + 1)/(h_i + 1 + k)$, where k is an arbitrary concavity constant, for weighting. Extensive studies have not been conducted to determine which of the character indices should be most useful in diverse situations.

A summary of the homoplasy indices for data sets or ensembles of characters is presented in Table 1. The basic measure of homoplasy, H, is the sum

TABLE 1 Summary Comparison of Data Set Homoplasy and Character Congruence Indices

172

Author	Index name	Formula	Terms	Range
Kluge and Farris (1969)	Consistency index	$CI = M/S = M/(M+H)$	M = minimum number of steps if no homoplasy is required; S = number of steps on the tree (usually a minimum length tree); H = total steps on the tree due to homoplasy	$0 < Min^a \leqslant CI \leqslant 1.0$
Farris (1989a)	Retention index (= HER_{max} [Archie 1989b])	$RI = 1 - H/H^B_{max}$	H^B_{max} = homoplasy present on the bush phylogeny	$0 \leqslant Min^b \leqslant RI \leqslant 1.0$
Archie (1989b)	Homoplasy excess ratio	$HER = 1 - H/H^P_{max}$	H^P_{max} = average homoplasy on minimum length trees for randomized (permuted) data	$1 - H^I_{max}/H^P_{max} < 0.0^c \leqslant HER \leqslant 1.0$
Farris (1989a)	Rescaled consistency index	$RC = CI \times RI$		$0 \leqslant Min^d \leqslant RC \leqslant 1.0$
Goloboff (1991a)	Data decisiveness	$DD = 1 - H/H^{DD}_{max}$	H^{DD}_{max} = average homoplasy on all possible trees for the data	$1 - H^B_{max}/H^{DD}_{max} < 0 \leqslant Min^e \leqslant DD \leqslant 1.0$
Meier et al. (1991)	Homoplasy slope ratio	$HSR = H/H^{R'}_{max}$	$H^{R'}_{max}$ = average homoplasy on minimum length trees for randomly generated data (estimate)	$0 \leqslant HSR \leqslant Max^f$
Farris (1991)	Permutation congruence	$K = 1 - H/H^I_{max}$	H^I_{max} = maximum homoplasy on minimum length trees for all possible permutations of the data matrix	$0^g \leqslant K \leqslant 1$
Archie (1990)	Steps per character	$SC = S/N$	N = number of characters	$1 \leqslant SC \leqslant [t/2]$
Archie (this chapter)	Homoplasy per character	$HC = H/N$		$0 \leqslant HC \leqslant [t/2]$
Hillis (1991)	Tree length skewness	$g1$		$Min^h \leqslant g1 \leqslant 0 \leqslant Max^i$
Farris (1976)	F-ratio	$F = \sum\sum |D(i,j) - P(i,j)|/\sum\sum D(i,j)$	$D(i,j)$ = observed total character state differences between taxa i and j; $P(i,j)$ = observed patristic (path length) difference between taxa i and j on the minimum length tree	$0 \leqslant F < Max^i$
Brooks et al. (1986)	D	$D = H_S/H_{max}$	H_S and H_{max} are the information theory statistics for the data (see text)	$0 \leqslant D < Max^j$

[a] The minimum CI depends primarily on the number of taxa, but additionally, on the number of characters and character state frequencies. It is strictly greater than 0.0.

[b] The minimum for RI is 0.0 on the bush phylogeny, but strictly greater than this on minimum length trees. Minimum is less influenced by the number of taxa than is minimum for CI, but is still influenced by the number of characters and character state frequencies.

[c] HER equals 0.0 for data that have the same amount of homoplasy as expected due to chance alone; that is, the average homoplasy on minimum length trees for all possible character permutations of the matrix. The lowest possible value for any permutation of the matrix can be determined only by examining all permutations, but this is always greater than $1 - HI_{max}/HI^{p}_{max}$.

[d] Although both the CI and RI are bounded below by 0.0, since in practice their minima are strictly greater than 0.0 on minimum length trees, the minimum value for RC depends on both of these statistics and is indeterminate.

[e] DD can only have a value of 0.0 for Goloboff's (1991) completely indecisive matrices. The minimum is strictly greater than 0.0 for the minimum length trees for all other data sets, and likely depends on character state frequencies, but may depend on number of taxa as well. On nonminimum length trees DD may be less than 0.0 but is bounded below by the formula given.

[f] HSR equals 0.0 when there is no homoplasy. The maximum is indeterminate because the scaling factor is only an estimate of the homoplasy on minimum length trees for randomly generated data, rather than the actual character data being analyzed.

[g] Although the minimum value of K is exactly 0.0, it only takes on this value for the worst possible permutation of the characters in the data set. This amount of character conflict is substantially greater than would be expected due to chance alone and could not be expected to occur for real systematic data.

[h] g1 is an unbounded statistic that takes on positive values for right-skewed distributions and negative values for left-skewed distributions. Tree length distributions for systematic data are expected to be left skewed, so, practically, the upper limit is 0.0. The lower limit, however, is indeterminate.

[i] F equals 0.0 when there is no homoplasy in the data. The upper bound is indeterminate, but may be close to 1.0.

[j] D is bounded below by 0.0, but for any data set, the minimum will be strictly greater than 0.0 even when no homoplasy is present. The upper limit is indeterminate.

across characters of the individual h_i values. Again, however, H is too dependent on characteristics of the data set (numbers of characters and taxa, character transformation series and polarity decisions, weighting schemes, and frequency of alternative character states) to be a useful comparative index of the level of homoplasy between data sets. The indices that have been proposed for comparative purposes fall into three basic categories. First is the consistency index, CI, that scales total homoplasy, H, obliquely by comparing observed character change, that is, homoplasy plus minimum necessary change, to *minimum necessary* change. Second are those indices that are a direct scaling of total homoplasy, H, to some *maximum possible* homoplasy for the data. These include the retention index, homoplasy excess ratio, permutation congruence index, homoplasy slope ratio, and data decisiveness. The similarity of these indexes is apparent from their common formula:

$$I_H = 1 - \frac{H}{H_{max}}.$$

The sole difference between them (except for the homoplasy slope ratio which is not subtracted from 1) is the scaling factor used, H_{max}. To allow consistent comparisons, in the discussion below, I will use HSR* = 1 − HSR for the Meier et al. (1991) index so it varies in a consistent fashion with the other indices. The third class of indices are similar only in that they do not use total homoplasy, H, directly in their calculation. Instead, they attempt to evaluate the relationships between character fit to trees in distinct fashions. These include the F-ratio, the information divergence statistic, and tree-length skewness.

The consistency index, CI, is qualitatively distinct from the other homoplasy indices because it compares observed homoplasy to the minimum number of steps for the characters rather than the maximum. The interpretation of HI, the compliment of CI, is that it represents the proportion of character change on a tree that is due to homoplasy. Although a seemingly reasonable interpretation of CI is that it is the proportion of character change on the tree due to synapomorphy, this interpretation is inappropriate since even on a bush phylogeny that has no synapomorphy, the CI will be greater than zero. Similarly, the value of CI is affected by the presence of characters that cannot influence the topology of the tree, that is, characters with single autapomorphies (Brooks et al. 1986; Farris 1989a; Meier et al. 1991; Klassen et al. 1991). Also, because the minimum possible value of CI on minimum length trees is correlated with the number of taxa and characters, both CI and HI (even though heuristically useful) are inappropriately scaled to permit meaningful comparisons between studies using different sets of characters or different taxa.

The series of indices that scale total homoplasy by a maximum are somewhat confusing and difficult to compare. Clearly, there are different opinions about what the maximum amount of homoplasy can be for a data set.

Hence, these indices are perhaps best compared by evaluating under what specific conditions each index will take on specific values, particularly extreme values. First, all of these take on a value of 1.0 when no homoplasy is present in the data. The retention index uses the largest H_{max} among all of the indices and the largest possible for the data and consequently will have the highest value (further from 0.0) among the series of indices for any data set. This was the basis for calling it HER_{max} (Archie 1989b). RI will always be greater than zero for minimum length trees and always equals zero on the completely unresolved (bush) phylogeny. For minimum length trees, however, the minimum value of RI is dependent to some extent on characteristics of the data set; in particular, the number of characters affects RI and, to a lesser extent, character state frequencies and number of taxa (Archie 1989b). In part, these dependencies appear to be a result of one of the basic properties of minimum length trees in phylogenetics: since there are an extremely large number of possible trees for even a small number of taxa, even data sets with high homoplasy levels can be summarized effectively and the inferred cladistic homoplasy is consistently an underestimate of the amount of evolutionary homoplasy (Archie and Felsenstein 1993). The retention index, RI, is not standardized for this feature. For real data sets, Archie (1989b) found values of RI ranging from 0.53 to 0.94 while in most cases in the literature RI is above 0.5. If the same data sets (from 8 to 44 taxa) analyzed by Archie (1989a,b) are randomized by character, the RI values range from 0.17 to 0.58. Although the RI values for randomized data were uncorrelated with the number of taxa, presumably they depend on both number of characters and character state frequencies.

The permutation congruence index, K, is related to RI and is also scaled such that it is guaranteed to be greater than or equal to zero. K will equal zero on minimum length trees when the character conflicts in the specific data set are as poor as possible. Under these conditions, however, the character incongruences are substantially greater than expected due to chance. Also, because the scaling factor for K (i.e., H^J_{max}) depends directly on character state frequency and number of taxa, interpretation of specific values of K is unclear. For example, for two data sets with character incongruence equal to that expected due to chance alone but with different character state frequencies or different numbers of taxa, the values of K could differ substantially. The behavior of K under such contrasting conditions has not been investigated in detail. Finally, the actual calculation of K is an extremely computer-intensive operation (more so than calculating HER); hence, its implementation is very unlikely. Farris' (1991) plea to use RI as an estimate of K when there is little homoplasy in the data is not very useful.

For the homoplasy slope ratio, the scaling uses an *estimate* of the maximum amount of homoplasy expected on *minimum length trees* by chance alone. This contrasts with REHER (Archie 1989b) that uses the analytically determined, maximum homoplasy expected on *random trees*. Archie and

Felsenstein (1993) showed that the homoplasy on random trees overestimated that expected on minimum length trees (see also Archie 1989b). The scaling used by HSR, however, does not correct for character state frequency differences common in real data sets that several authors have shown can affect the possible level of character conflict (e.g., Meacham 1981). As a result, for truly random two-state character data with unequal frequencies of the two states, as found in real data sets (Archie 1989a), the minimum possible value of HSR* may be either greater or less than zero (recall, HSR* = 1.0 − HSR). Because HSR* is scaled to an estimate of the average expected homoplasy in idealized data sets, HSR* may take on negative values. The minimum value is unknown. Meier et al. (1991) reported HSR values ranging from 0.03 to 0.46 for real data sets (HSR* = 0.97 to 0.54). For the data of Bremer (1988) that Archie (1989c) found did not differ from random (HER = 0.046 and −0.0084), the HSR* values are 0.555 (t=9)and 0.143 (t=6). This result would seem to indicate (a) that the minimum value for HSR* may generally be well above 0.0 and (b) that it varies substantially for data with similar levels of character incongruence. Meier et al. (1991) noted that HSR and HER are only moderately correlated. Presumably this is a result of the variability of H^p_{max} used in HER related to differences in character state frequencies in the data, whereas HSR* does not account for this. Meier et al. also found that, for a series of real data sets, HSR was uncorrelated with the number of taxa, t, but for the same data sets Archie (1989b) found that HER was correlated with t. There are alternative interpretations of these results. Meier et al. (1991) claimed that the correlation with t for these data sets demonstrated an undesirable property of HER. However, it has been shown theoretically that the value of HER is uncorrelated with the number of taxa (Archie 1989b; Farris 1991). The correlation of HER with t for the data sets examined by Archie (1989b) is likely the result of an overall increase in "average" homoplasy with increasing t. In contrast, because HSR is not corrected for character state frequency, its relation to actual homoplasy is only approximate and the lack of correlation of HSR with t is likely the result of this insensitivity. Meier et al. did not present an analysis of the effect of increasing t for a constant or increasing probability of character change on either index.

Data decisiveness (DD) was proposed as a measure of the degree to which possible trees for a set of taxa differed in length based on the observed character data. As observed by Fitch (1979) and Le Quesne (1989), there is generally a greater range in tree lengths in structured character data than in random or randomized character data. This can be measured either as the standard deviation of the tree length distribution, as proposed by Le Quesne (1989) or by DD, an index that measures how much the homoplasy in the minimum length tree is less than the average homoplasy for the entire tree-length distribution. As the mean length for the distribution is independent of the character congruence in the data (Le Quesne 1989), DD will increase as the range and standard deviation of tree lengths increases. Because the tree length distribution becomes more skewed with a longer left tail as data congruence increases (Huelsenbeck 1991), DD appears to be an appropriate description of the distribution. As suggested above for comparing an original tree-length distribution with that of randomized data, a test of significance of DD could be devised using character randomization (similar to the procedure described for testing minimum tree lengths; Archie 1989a). For example, a distribution of sample DD values for randomized data could be obtained by carrying out a sequence of character randomizations followed by determinations of sample DD values. This can be done fairly quickly for randomized data using exhaustive (branch-and-bound) searches for problems with few taxa or, for larger data sets, heuristic searches. If the original DD value is greater than $(1 - \alpha)\%$ of the DD values from the randomized data, then the original character data shows a statistically significant amount of structure.

DD is scaled so that it equals 1.0 when there is no homoplasy in the data. However, it equals 0.0 for minimum length trees only when all possible trees have exactly the same length; that is, when the data are totally "indecisive" (Goloboff 1991a). This will be the case only for Goloboff's indecisive matrices in which every one of the possible characters for the taxa are present an equal number of times. This condition seems extremely unlikely; no example of character sets that could even be considered a random sample from this type of matrix has been identified. All of the data matrices examined by Archie (1989a) deviated from this distribution substantially and this fact was the motivation for using character randomization to evaluate observed homoplasy. Because phylogenetic data are not random samples from indecisive matrices, the extent to which the original DD_O exceeds DD_R for randomized data may be a more useful statistic than DD itself. Thus

$$\Delta DD = DD_O - DD_R = \frac{\overline{S} - S}{\overline{S} - M} - \frac{\overline{S} - S_R}{\overline{S} - M}.$$

$$= \frac{S_R - S}{\overline{S} - M}$$

This difference can be further standardized with a measure of variability, for example, the standard deviation of the DD_R values. Because the values of DD, and, hence, ΔDD, are dependent on the skewness of the tree-length distribution, these indices should be correlated with $g1$ for the data. This possibility has not yet been investigated.

The homoplasy excess ratio (HER) was proposed as an alternative to the consistency index after the observed and expected properties of the latter statistic were elucidated (Archie, 1989a,b; Archie and Felsenstein 1993). The objective in proposing this index was that the scaling of total homoplasy, H_{max}, should incorporate: (a) the properties of the specific data set and

(b) the properties of the character summarization abilities of minimum length trees. In addition, the H_{max} value chosen to scale total homoplasy yields specific values under conditions that were relevant to phylogenetic systematics: HER equals 1.0 when no homoplasy is present and 0.0 when the character conflicts in the data are as large as would be expected due to chance alone. The absolute minimum value of HER is negative (Farris 1991) and can be determined by replacing H in its equation with H^J_{max} the denominator for the permutation congruence index. Because H^J_{max} is expected to vary with the characteristics of the data set, this absolute minimum value for HER will vary with these same characteristics. However, this property does not affect the interpretation of HER relative to conditions for which it was developed. Specifically, if the character congruences are equal to those expected due to chance alone, HER will have a value of zero. To the degree that conflicts are less than this, HER will be positive. If conflicts are greater than expected due to chance then HER will be negative, but, because the scaling of the value below zero depends on properties of the data set, critical comparisons should not be made for negative values. Data sets having negative HER values should not be relied upon to possess phylogenetic information in any case (see below).

Of the three indirect measures of homoplasy, the F-ratio has been in use the longest, although it has appeared only sporadically in the literature. Brooks et al. (1986) presented a careful analysis of the F-ratio. They demonstrated that the value of F is influenced by the choice of technique used for ancestor character state reconstruction. This is because character state changes are counted multiple times in the calculation of F. For example, any change occurring on a branch leading to some terminal taxon is included in the difference calculation every time the taxon is compared to the other taxa. In contrast, if the changes are placed further down the tree, then they are not counted as frequently (this effect can be demonstrated using the ACCTRAN or DELTRAN options in PAUP). F is also affected by the presence of autapomorphous characters; increasing the number of autapomorphies increases the F-ratio (recall, this also increases CI). Autapomorphies fit all trees perfectly and can have no homoplasy. Finally, Brooks et al. (1986) found that nonminimum length trees may have lower F-ratios than minimum length trees. The use of F cannot be recommended because of these properties. Finally, although the F-ratio has been used in the past to choose among multiple minimum length trees, there does not seem to be a well justified motivation for this procedure.

The information divergence statistic, D, was an attempt to use information theory to evaluate homoplasy. However, D is difficult to interpret as a measure of homoplasy, particularly because its maximum and minimum limits are uncertain and depend directly on the characteristics of the data set. It has found little use in the systematic literature.

In contrast to these first two indirect homoplasy measures, tree-length skewness has become commonly used, not as a comparative measure of homoplasy, but as a general measure of data set quality. Skewness values that are determined to be statistically significant and negative using Hillis' (1991) critical values are taken to indicate that the data set contains useful phylogenetic information (for examples, see Arévalo et al. 1994; Nedbal et al. 1994; Morales and Melnick 1994; Spotila et al. 1994). Nonsignificant values *may* be interpreted as lacking phylogenetic signal, but, generally (and perhaps appropriately), phylogenetic relationships for these data are interpreted but usually with some caution. This latter approach seems reasonable. Based on computer simulation studies, Huelsenbeck (1991) found that large and statistically significant negative values of $g1$ indicate that the tree hypothesis is more likely to resemble the true tree than if there is a nonsignificant $g1$ value, but, even if the skewness statistic is nonsignificant, the true tree may still be equal or similar to the minimum length tree.

Examples can be used to illustrate some of the difficulties with interpreting tree-length skewness statistics quantitatively. Källersjö et al. (1992) presented an example (replicated in Table 2) using two fabricated data sets to illustrate several points. Their first example demonstrates that skewness can be insensitive to overall support for a tree. For either of their sample data sets (Table 2 (A and B)), $g1$ can be found by an exhaustive search of all possible bifurcating trees ($g1 = -0.288$ and -0.959, respectively). If the entire character sets are replicated, doubling the number of characters supporting the minimum length trees, the same $g1$ statistics are found. It is also the case that if any number of the characters in Table 2(A) is deleted, the same $g1$ statistic is obtained. In contrast, it is also easy to demonstrate that tree-length skewness is sensitive to the number of characters supporting specific nodes on the tree for data sets providing more tree resolution (the data sets of Table 2 (A and B) provide limited tree resolution). For the data set in Table 2(C), $g1 = -0.5403$ for the entire tree length distribution. If two replicates of the entire data set are used in the analysis, $g1$ remains the same. However, if support for the node containing the first two taxa is increased by duplicating character 1 either two or three times, $g1$ decreases, respectively, to -0.5761 and -0.6385 (remember, larger magnitude negative $g1$ values imply greater skewness). If a homoplasious character with apomorphous (1) states in taxa 1 and 3 is added to the original set, $g1$ decreases from the value for the original data set to -0.5445 (tree length = 10). This seems inappropriate but, in this case, the characters conflict only with each other and there are two equally parsimonious trees that are congruent with all other characters. If a character with apomorphies in taxa 1 and 4 is added, however, $g1$ increases to -0.5257, while if the character has apomorphies in taxa 1 and 5, $g1$ increases slightly to -0.5216. Although tree length again equals 10 for both of these cases (and most homoplasy statistics would take

TABLE 2   Sample Data Sets Used to Evaluate the Properties of Tree Length Skewness (g1)

Taxon	A Characters 1234567890	B Characters 1234567890	C Characters 12345678
1	1111100000	1100000000	11110000
2	1111100000	0011000000	11110000
3	1111100000	0000110000	01110000
4	1111100000	0000001100	00110000
5	1111100000	0000000011	00010000
6	0000011111	1000000001	00001111
7	0000011111	0110000000	00001111
8	0000011111	0001100000	00000111
9	0000011111	0000011000	00000011
10	0000011111	0000000110	00000001

Note. The first two data sets (A and B) have 10 characters each and are reproduced from Källersjö et al. (1993). Data set three (C) has 8 characters.

on single values for all three cases), the characters are incongruent with two and three other characters, respectively. Homoplasy alone is not equivalent to degree of character incongruence in this sense, but tree length skewness reflects both features. The Källersjö et al. (1992) data sets (Table 2 (A and B)) demonstrate another difficulty in interpretation of tree-length skewness. Whereas the characters in Table 2 (A) are all congruent with a single tree topology, the characters in Table 2 (B) show substantial incongruence and support two different tree topologies of equal length that are, however, more resolved than that from Table 2 (A). In spite of this, the $g1$ value for Table 2 (B) is substantially larger in magnitude (smaller negative number) than that for Table 2 (A). Skewness therefore reflects not only lack of incongruence, but, in some manner, diversity of support for clades on a tree. However, the data in Table 2 (C) support a single fully resolved topology and have an intermediate $g1$ value.

The behavior of tree length skewness is enigmatic: (a) skewness does in some ways, but not all ways, measure overall support for the tree, (b) skewness is in some ways related to support for specific nodes on the tree, (c) skewness appears somewhat correlated with homoplasy, (d) skewness is in some ways related to degree of incongruence among the characters, and (e) skewness in some ways measures diversity of support for nodes. The primary reservation in the use of tree-length skewness is that its value does not seem to be a direct measure of data set reliability for comparative purposes. The large differences in skewness values for the data sets in Table 2, identify the nature of this difficulty; clearly there are differences in tree support for these three data sets, but whether the magnitude of tree-length skewness captures this difference is questionable. The relation of tree-length skewness to homoplasy is indirect, at best, although it appears to incorporate additional information on character congruence beyond what is contained solely in tree length.

Homoplasy and Tree Estimation

Few investigators actually discuss the implications of either high or low CI and RI values (or any other homoplasy statistic) relative to expected tree accuracy. In fact, it is not apparent from the literature how CI and RI values should be used (e.g., Farris 1989a, 1991; Archie 1989b). While the statistical significance of $g1$ values is reported (using Hillis' 1991 critical values), it is apparent that quantitative evaluation of $g1$ as a measure of phylogenetic signal is inappropriate (Källersjö et al. 1993), and most authors do not use it this way (e.g., Arévalo et al. 1994). Although there have been many studies documenting and arguing the relationship between rates and equality of character change and the accuracy of estimated trees (e.g., Saitou and Imanishi 1989; Tateno et al. 1994; Kuhner and Felsenstein 1994; Charleston et al. 1994; Huelsenbeck 1995), few if any of these studies have addressed the question of how the expected reliability of the resulting tree hypothesis should be evaluated for real data sets in relation to inferred character change complexity, that is, cladistic homoplasy: what is the relationship between *cladistic homoplasy* measured on trees using any of the indices described here and tree accuracy. I investigated this question with a simple computer simulation using a program, DNATREE, written by myself. Random trees were generated by starting with a single lineage. A fixed probability of branching was assigned to all extant lineages at successive time intervals. DNA sequences of specified length were mapped onto these growing trees and allowed to evolve along with the tree with each site having an equal probability of change in each time interval. A simple Jukes-Cantor model of equal probability of change among the four nucleotides was used. Clearly, all of these conditions are unrealistic but the results illustrate an important phenomenon that requires further investigation using more realistic parameters such as those used by so many simulation studies. Increasing levels of homoplasy were injected into the evolutionary process by increasing the probability of change in different simulations. The simulations were stopped at the stage when lineage $t + 1$ was about to be produced. The simulations yielded DNA sequences for $t = 10, 20,$ and 25 terminal taxa (results for $t = 10$ are shown). The data were analyzed with PAUP. The resulting trees were compared to the known phylogeny of the data using a strict consensus method (CI_1, Rohlf 1982). For each data set, CI, RI, and HER were calculated. A total of 10 data sets were used for each change probability setting. A summary of the results for CI and HER are plotted in Fig. 2. In Fig. 2A, the horizontal axis represents both the mutation rate and, concomitantly, the

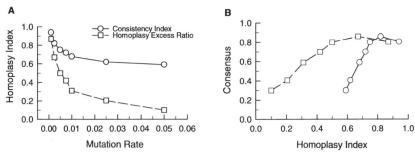

FIGURE 2 Results of computer simulation study of relationship between known tree and estimated tree. (A) Graph of consistency index and homoplasy excess ratio in relation to nucleotide substitution (mutation) rate. The latter is directly proportional to total evolutionary homoplasy. (B) Graph of relationship between consensus of estimated and true tree (Rohlf's CI_1) with minimum length tree and two measures of cladistic homoplasy (CI and HER) for the data in A.

amount of homoplasy in the data because of the evolutionary model used for character change. On the vertical axis is the mean homoplasy index value for the data, either CI or HER. As expected, as rate of change increases, both CI and HER decrease. Whereas HER decreases toward 0.0, CI decreases to a minimum of around 0.6 (this minimum decreases for $t > 10$). In Fig. 2B, the mean consensus between the estimated and true tree is plotted against the calculated CI or HER values. As the consistency index decreases, the mean consensus decreases. This same effect is apparent for HER. However, because of the increased range of values for HER, the graph is much more spread out than that for CI. More importantly, it is apparent that as CI decreases from around 0.85 to its minimum there is a precipitous drop in mean consensus. In contrast, there is a uniform decline in mean consensus as HER decreases toward 0.0 (HER is approximately linearly related to the log of mutation rate). These figures plot the mean values from numerous replications and the results from the individual replicates are quite varied. However, the implications of Fig. 2 are quite clear. First, cladistic homoplasy is correlated with accuracy of the tree estimate. Second, a relatively small change in CI is associated with a large change in expected accuracy. Because of the scatter (cloud) of values from the original data, accuracy cannot be clearly estimated over a substantial percentage ($>60\%$) of the achievable range of CI. At the top end of the scale, accuracy cannot be predicted (although its generally high) using CI. Finally, because of its expanded range, the value of HER is a much more effective indicator of expected accuracy of the tree estimate. Clearly, the relationship between homoplasy and accuracy will be shown to be a complex one, but the results from this simulation bode well the importance of evaluating data set homoplasy levels. These simulations must be expanded to include a wider array of parameter values and homoplasy statistics, but the results are apparent. One of the important reasons for evaluating cladistic homoplasy is that it is a direct indicator of expected accuracy of the inferred trees. In this regard, HER is a much better indicator of cladistic homoplasy and expected accuracy than is CI.

Recommendations: A Sense of Purpose

The specific reasons for interest in measuring homoplasy in characters or in a data set include the following: (a) to permit weighting of characters in phylogenetic analysis inversely relative to cladistic homoplasy, (b) to examine patterns of evolution and coevolution between specific characters, (c) to make comparisons among different classes of characters with regards to the relative patterns of synapomorphy and homoplasy, and (d) to evaluate the relative quality of the data as a measure of the expected quality, presumably equated to accuracy, of the tree estimate. At present, character level homoplasy measures are used primarily for weighting schemes in phylogenetic analysis. Few if any detailed studies using computer simulations have been carried out to determine whether particular weighting schemes improve tree accuracy (for one example, see Fitch and Ye 1991). Additional studies are clearly needed.

Detailed studies of homoplasy levels in characters have been made possible with graphical tools provided in phylogenetics computer programs. These are particularly well developed in MacClade (Maddison and Maddison 1992). However, although numerous investigators are beginning to analyze evolutionary patterns of various types of characters (see Mickevich and Weller 1990; Brooks and MacClennan 1991; Wenzel 1992 for reviews), few generalizations have appeared. The character retention index (r_i) and homoplasy index ($1 - c_i$) are the most interpretable indices for this level of analysis, although for single characters, simple examinations of the character transformations on the tree appear adequate (e.g., Carpenter 1989).

Various studies (Mickevich 1978; Sanderson and Donoghue 1989; de Queiroz and Wimberger 1993; Sanderson and Donoghue 1996) have compared homoplasy levels in different classes of characters using the CI as a measure of homoplasy to determine whether homoplasy is more prevalent in specific types of characters or groups of taxa. Although no substantial differences between classes of characters or classes of taxa have been found, the problems with the CI may have nullified the possibility of detecting true differences in any of these studies. For example, the resampling studies of a single data set carried out by Archie (1989b) identified correlations (N versus HER) not yet found in comparisons of real data sets using CI, presumably for the same reasons. The studies comparing classes of characters should be repeated using an index such as HER which is specifically adjusted for data set differences.

Data set quality is currently evaluated using several different methods, including homoplasy indices: in most cladistic studies, the CI and RI are reported while a large percentage of studies also report Felsenstein's (1985) bootstrap values and an increasing number of studies report $g1$ values for tree-length distributions. All of these presumably are used as indicators of both data set quality and the confidence that investigators place in the phylogenetic hypothesis. The application of the bootstrap to phylogenetic analysis is currently receiving substantial research, primarily because of the difficulty in interpreting specific bootstrap values on phylogenies (Huelsenbeck 1995; Li and Zharkikh 1995; Archie, in preparation). Expanded studies are necessary on the relationship between homoplasy measures, accuracy, and repeatability in phylogenetic analysis. The results from the preliminary analyses presented here indicate that accuracy is related to level of cladistic homoplasy. Whereas CI is not especially effective as an indicator of homoplasy level, because of its scaling features, HER varies appropriately with accuracy.

The fundamental premise that has been presented here for the use of data set homoplasy measures is that it is desirable to make comparisons among diverse data sets, for example, to evaluate whether different types of data tend to show more or less homoplasy or to evaluate the quality of the data for phylogenetic inference. It may be, however, that a general description of the level of homoplasy is desired which captures the fact that, for example, with increasing numbers of taxa or characters, the level of homoplasy tends to increase. For such studies, H is the fundamental measure of homoplasy, but H just feels too raw. The CI has been used as the simplest scaling of H, since its development some 25 years ago, and many investigators are comfortable with it as a more general indicator of homoplasy. For all the reasons presented above (and in the literature cited), however, its difficult to understand this level of comfort. The compliment of CI, that is, HI is more directly interpretable, but clearly also has a scaling problem. For binary characters, the CI is the reciprocal of the average number of steps per character, SC = S/N = 1/CI, while for nonbinary characters, it is the reciprocal of the average number of steps per apomorphic state (SC is used for both types of characters). The reciprocal of CI, therefore, contains the information in CI but in a form directly interpretable in terms of character change on the tree (Archie 1990) and lacks the pretense of being scaled between fixed limits. When there is no homoplasy in the data, SC equals 1.0 and as homoplasy increases, because, for example, of increasing character conflicts or increasing numbers of taxa, SC increases basically without bound (it is actually bounded by $[t/2]$, but t can be very large). If autapomorphous characters are included in a data set, SC will be closer to 1.0 than if they are not included. Archie (1989a,b) examined 28 data sets and found SC ranging from 1.06 to 4.71 while if these same data sets were randomized, SC ranged from 1.39 to 6.40.

An alternative, related statistic, is the average amount of homoplasy per character, HC = H/N = $(S-N)/N$ = SC − 1. HC has a minimum value of 0.0. As a pure descriptor of character change on the tree, either of these indices would seem to be more useful than CI as far as interpretability. The fact that neither of these direct measures of homoplasy has been adopted by the fields of systematics or evolutionary biology is an oddity.

More than one measure of homoplasy may be desirable or appropriate for comparative purposes. Whereas certain indices are easy to calculate and others require more computer-intensive operations, this factor alone should not determine which index is used (e.g., Meier et al. 1991; Farris 1991; Källersjö et al. 1992). CI, RI, and HSR are easy to calculate, but their scaling seems not completely adequate for general comparative purposes. CI suffers from too many difficulties to be useful and HSR is too crude an approximation of, possibly, a useful comparison (once standardized for data set properties, it is equivalent to HER). In contrast, HER is standardized for properties of the data and is appropriately scaled to maximize interpretability. Its drawback is that it requires some computer-intensive operations for its calculation. However, in proportion to the time taken to gather systematic data, the time required to calculate HER is minimal. Archie (1989b) recommended using RI to estimate HER by a rescaling using a regression analysis equation, but the quality of this estimate should be refined. The desired interpretation of RI as the level of synapomorphy retained on the tree (Farris 1989a) has the difficulty that simply due to chance alone, a large percentage of apparent synapomorphy will be retained on a minimum length tree. HER is the exact rescaling of RI to incorporate this fact. The two indices related to tree-length skewness, $g1$ and DD (and ΔDD) require intermediate amounts of computational effort. The relationship between these latter statistics remains to be investigated, but clearly, the lack of a significant $g1$ or DD alone is not sufficient to rule out useful information in a set of data. For interpretation of evolutionary change on a tree, it appears that SC or HC should be the most useful, particularly for evolutionary process studies.

ACKNOWLEDGMENTS

I thank Michael Sanderson for inviting me to contribute to this volume and to him and an outside reviewer for careful evaluation of an earlier version of the MS. Preparation of the manuscript forced me to make a needed thorough review and comparison of homoplasy indices in a more objective framework than I had previously attempted. I also thank numerous colleagues for discussions on homoplasy indices throughout the past 8 years, particularly Joe Felsenstein, Wayne Maddison, the University of Hawaii Evoluncheon Group, and members of the Poquot Mafia, especially Chris Simon and Dan Faith. This research has been sponsored in part by grants from the National Science Foundation (BSR 85-00141, BSR89-18042, and DEB91-19091) and California State University at Long Beach.

REFERENCES

Archie, J. W. 1989a. A randomization test for phylogenetic information in systematic data. Systematic Zoology 38:219–252.

Archie, J. W. 1989b. Homoplasy excess ratios: New indices for measuring levels of homoplasy in phylogenetic systematics and a critique of the consistency index. Systematic Zoology 38:235–269.

Archie, J. W. 1989c. Phylogenies of plant families: A demonstration of phylogenetic randomness in DNA sequence data derived from proteins. Evolution 43:1796–1800.

Archie, J. W. 1990. Homoplasy excess statistics and retention indices: A reply to Farris. Systematic Zoology 39:169–174.

Archie, J. W., and J. Felsenstein. 1993. The number of evolutionary steps on random and minimum length trees for random evolutionary data. Theoretical Population Biology 43: 52–79.

Arévalo, E., S. K. Davis, and J. W. Sites, Jr. 1994. Mitochondrial DNA sequence divergence and phylogenetic relationships among eight chromosome races of the *Sceloporus grammicus* complex (Phrynosomatidae) in central Mexico. Systematic Zoology 43:387–418.

Bremer, K. 1988. The limits of amino acid sequence data in angiosperm phylogenetic reconstruction. Evolution 42:795–803.

Brooks, D. R., and D. A. McLennan. 1991. Phylogeny, ecology and behavior. Univ. of Chicago Press, Chicago.

Brooks, D., R. T. O'Grady, and E. O. Wiley. 1986. A measure of the information content of phylogenetic trees, and its use as an optimality criterion. Systematic Zoology 35:571–581.

Camin, J. H., and R. R. Sokal. 1965. A method for deducing branching sequences in phylogeny. Evolution 19:311–326.

Carpenter, J. M. 1989. Testing scenarios: Wasp social behavior. Cladistics 5:131–144.

Carter, M., M. Hendy, D. Penny, L. Szekely, and N. Wormald. 1990. On the distribution of lengths of evolutionary trees. S.I.A.M Journal of Discrete Mathematics. 3:38–47.

Charleston, M. A., M. D. Hendy, and D. Penny. 1994. The effects of sequence length, tree topology, and number of taxa on the performance of phylogenetic methods. Journal of Computational Biology 1:133–151.

de Queiroz, A., and P. H. Wimberger. 1993. The usefulness of behavior for phylogeny estimation: Levels of homoplasy in behavioral and morphological characters. Evolution 47:46–60.

Farris, J. S. 1969. A successive approximations approach to character weighting. Systematic Zoology 18:374–385.

Farris, J. S. 1972. Estimating phylogenetic trees from distance matrices. American Naturalist 106:645–668.

Farris, J. S. 1973. On comparing the shapes of taxonomic trees. Systematic Zoology. 22:50–54.

Farris, J. S. 1989a. The retention index and rescaled consistency index. Cladistics 5:417–419.

Farris, J. S. 1989b. The retention index and homoplasy excess. Systematic Zoology 38:406–407.

Farris, J. S. 1991. Excess homoplasy ratios. Cladistics 7:81–91.

Farris, J. S., and M. F. Mickevich. 1975. PHYSIS. Computer program published by the authors.

Felsenstein, J. 1985. Confidence limits on phylogenies: An approach using the bootstrap. Evolution 39:783–791.

Fitch, W. M. 1979. Cautionary remarks on using gene expression events in parsimony procedures. Systematic Zoology 28:375–379.

Fitch, W. M., and J. Ye. 1991. Weighted parsimony: Does it work? Pp. 147–154 *in* M. M. Miyamoto and J. Cracraft, eds. Phylogenetic analysis of DNA sequences. Oxford, New York.

Goloboff, P. 1991a. Homoplasy and the choice among cladograms. Cladistics 7:215–232.

Goloboff, P. 1991b. Random data, homoplasy and information. Cladistics 7:395–406.

Goloboff, P. 1993. Estimating character weights during tree search. Cladistics 9:83–91.

Hillis, D. M. 1991. Discriminating between phylogenetic signal and random noise in DNA sequences. Pp. 278–294 *in* M. M. Miyamoto and J. Cracraft, eds. Phylogenetic analysis of DNA sequences. Oxford Univ. Press, New York.

Hillis, D. M., and J. Huelsenbeck. 1992. Signal, noise, and reliability in molecular phylogenetic analyses. Journal of Heredity 83:189–195.

Huelsenbeck, J. P. 1991. Tree-length distribution skewness: An indicator of phylogenetic information. Systematic Zoology 40:257–270.

Huelsenbeck, J. P. 1995. Performance of phylogenetic methods in simulation. Systematic Biology 44:17–48.

Kållersjö, M., J. S. Farris, A. G. Kluge, and C. Bult. 1992. Skewness and permutation. Cladistics 8:275–287.

Klassen, G. J., R. D. Mooi, and A. Locke. 1991. Consistency indices and random data. Systematic Zoology 40:446–457.

Kluge, A., and J. S. Farris. 1969. Quantitative phyletics and the evolution of anurans. Systematic Zoology 18:1–32.

Kuhner, M. K., and J. Felsenstein. 1994. A simulation comparison of phylogeny algorithms under equal and unequal evolutionary rates. Molecular Biology and Evolution 11:459–468.

Le Quesne, W. 1989. Frequency distributions of lengths of possible networks from a data matrix. Cladistics 5:395–407.

Li, W.-H., and A. Zharkikh. 1995. Statistical tests of DNA phylogenies. Systematic Biology 44:49–63.

Maddison, D. R. 1991. The discovery and importance of multiple islands of most-parsimonious trees. Systematic Zoology 40:315–328.

Maddison, W. P., and D. R. Maddison. 1992. MacClade: Analysis of phylogeny and character evolution, version 3. Sinauer, Sunderland, MA.

Meacham, C. A. 1981. A probability measure for character compatibility. Mathematical Biosciences 57:1–18.

Meier, R., P. Kores, and S. Darwin. 1991. Homoplasy slope ratio: A better measurement of observed homoplasy in cladistic analyses. Systematic Zoology 40:74–88.

Mickevich, M. F. 1978. Taxonomic congruence. Systematic Zoology 27:143–158.

Mickevich, M. F., and S. J. Weller. 1990. Evolutionary character analysis: Tracing character change on a cladogram. Cladistics 6:137–170.

Miyamoto, M. M. 1985. Consensus cladograms and general classifications. Cladistics 1:186.

Morales, J. C., and D. J. Melnick. 1994. Molecular systematics of the living rhinoceros. Molecular Phylogenetics and Evolution 3:128–134.

Nedbal, M. A., M. W. Allard, and R. L. Honeycutt. 1994. Molecular systematics of hystricognath rodents: Evidence from the mitochondrial 12S rRNA gene. Molecular Phylogenetics and Evolution 3:206–220.

Rohlf, F. J. 1982. Consensus indices for comparing classifications. Mathematical Biosciences 59:131–144.

Saitou, N., and T. Imanishi. 1989. Relative efficiencies of the Fitch- Margoliash, maximum-parsimony, maximum-likelihood, minimum-evolution, and neighbor-joining methods of phylogenetic tree construction in obtaining the correct tree. Molecular Biology and Evolution 6:514–525.

Sanderson, M. J., and M. J. Donoghue. 1989. Patterns of variation in levels of homoplasy. Evolution 43:1781–1795.

Sanderson, M. J., and M. J. Donoghue. 1996. The relationship between homoplasy and confidence in a phylogenetic tree. Pp. XX–XX *in* M. J. Sanderson and L. Hufford, eds. Homoplasy and the evolutionary process. Academic Press, New York.

Sokal, R. R., and F. J. Rohlf. 1995. Biometry, 3rd ed. W. H. Freeman, San Francisco.

Spotila, L. D., N. F. Kaufer, E. Therior, K. M. Ryan, D. Penick, and J. R. Spotila. 1994. Sequence analysis of the ZFY and *Sox* genes in the turtle, *Chelydra serpentina*. Molecular Phylogenetics and Evolution 3:1–9.

Swofford, D. 1993. PAUP: Phylogenetic analysis using parsimony, version 3.1. Computer program distributed by the Illinois Natural History Survey, Champaign, Illinois.

Swofford, D., and W. Maddison. 1987. Reconstructing ancestral character states under Wagner parsimony. Mathematical Biosciences 87:195–229.

Tateno, Y., N. Takezaki, and M. Nei. 1994. Relative efficiencies of the maximum-likelihood, neighbor-joining, and maximum-parsimony methods when substitution rate varies with site. Molecular Biology and Evolution 11:261–277.

Wenzel, J. W. 1992. Behavioral homology and phylogeny. Annual Review of Ecology and Systematics 23:361–381.

THE MEASUREMENT OF HOMOPLASY: A STOCHASTIC VIEW

JOSEPH T. CHANG AND JUNHYONG KIM
Yale University
New Haven, Connecticut

DEFINITION OF HOMOPLASY

Analogue: A part or organ in one animal which has the same function as another part or organ in a different animal.

Homologue: The same organ in different animals under every variety of form and function.
—Richard Owen, 1843 (from Panchen 1994)

According to Richard Owen's original definitions, homology and analogy (cf., Panchen 1994) are not mutually exclusive concepts. The term homoplasy is used to describe the case in which a feature in one organism is analogous but not homologous to a feature in another organism (Panchen 1994). (Strictly speaking, the term analogy might be seen to involve some notion of function; however, in current usage similarity seems to be substituted for analogy.) However, a variety of homology concepts have been proposed (e.g., Patterson 1982, 1988; Wagner 1989; reviewed in Donoghue 1992) and the definition of homoplasy changes depending on which homology concept is adopted. In systematics, both the transformational homology

HOMOPLASY: The Recurrence of Similarity in Evolution, M. J. Sanderson and L. Hufford, eds.
Copyright © 1996 All rights of reproduction in any form reserved.

view—that structures are homologous if they can be traced to a condition that originated in a common ancestor—and the taxic view—structures are homologous if they are a shared derived state for a monophyletic group— seem to be popular (Donoghue 1992). For both of these views, multiple independent origin of structures, as opposed to single common origin, im- plies homoplasy. Determining homoplasy is considerably more difficult using the transformational view since different character states are allowed to be derived from a common state with the possibility of homology. Under the taxic view put forth by Patterson (1982), homoplasy occurs when character states fail the three proposed tests: similarity (general identity), congruence (single origin demarking a monophyletic group), and conjunction (more ac- curately nonconjunction; not found in same organism). Under the taxic def- inition, only identical (in some general sense; say, forelimbs) attributes that can be traced to a single origin and shared among all members of a mono- phyletic group are homologous. Therefore, determination of homology or homoplasy is simplified.

The importance and the utility of different concepts of homology de- pends on the biological focus of the researcher, whether it is systematics, functional, developmental, etc., and it is not our purpose to espouse one concept over another. In this paper, we take an evolutionary approach of defining homoplasy as identity of parts or structure between organisms that is due to chance and not by descent. This definition is very similar to the taxic definition except, as can be seen below, we attempt to take into account the phyletic evolutionary process, unlike the original taxic concept (Patter- son 1982). The structure is convenient because we do not try to deal with more fundamental notions such as "identity," "structure," "part," etc., but assume that these are given. We return to some of these concepts later in the discussion.

THE USE OF MEASURES OF HOMOPLASY IN SYSTEMATICS

In comparison with the general biological interest of the concepts of homol- ogy and homoplasy, the role of these concepts in systematics seems to be more specialized: for example, their use as "falsifiers" of phylogenetic hy- pothesis (Farris 1983). In particular, the concept of homoplasy or some measurement of the "degree of homoplasy" seems to be widely used in sys- tematics as a measure of the reliability of a phylogenetic estimate obtained from a collection of characters. A more minor, but related, use of the degree of homoplasy found in the literature is as a general measure of evolutionary plasticity of a character or a data set (Sanderson and Donoghue 1989). We will discuss each of these two uses in turn.

Using a measure of the degree of homoplasy as a surrogate measure for reliability of an estimated phylogeny appears early in the literature following the development of the maximum parsimony method of phylogenetic estimation. For example, the character consistency index was first described by Kluge and Farris (1969). However, the connection between the reliability of an estimated phylogeny and the amount of homoplasy is not entirely clear; it seems to arise from a shift of homoplasy as an optimality criterion for maximum parsimony to a goodness-of-fit criterion. This shift can be seen in the general framework for the maximum parsimony method as articulated by Farris (1983). Given a phylogenetic tree and a character with a distribution of states at the tips of the tree, the maximum parsimony algorithm determines the minimum number of originations of the various states of the character given the hypothesis of that particular tree. Any origination beyond the minimum (one less the number of states in the character) is considered an ad hoc hypothesis of homoplasy for the character. Ad hoc, because the hypothesis of homoplasy is implied by the tree and not other external biological criteria.

The maximum parsimony framework is based on the idea of the most parsimonious tree as the best estimate of the phylogeny. Farris (1983), in fact, states that the desirability of the tree with the least amount of "ad hoc homoplasy" is independent of the actual amount of homoplastic evolution found in nature: ". . . no degree of abundance of homoplasy is by itself sufficient to defend the choice of a less parsimonious genealogy over a more parsimonious one." From this argument alone it is not clear why homoplasy might also relate to some sense of reliability. For this, Farris' next sentence seems relevant: "That abundance can diminish only the *strength of preference* for the parsimonious arrangement . . ." [emphasis added]. It is perhaps from this view that measures of homoplasy are also seen as measures of accuracy or reliability. However, this view does not provide well grounded arguments as to why the quantitative measure of homoplasy should relate to the quantitative degree of reliability.

The use of homoplasy levels (measured using maximum parsimony) as a measure of evolutionary plasticity is more intuitive. The parsimony length correlates with the number of originations of a character state which then correlates with some notion of evolutionary plasticity. Empirical evidence for this was found by Sanderson and Donoghue (1989) who showed that the consistency index (Farris 1989) was negatively correlated with the number of taxa in actual data sets. They noted that this negative correlation is consistent with a model of character evolution in which the probability of character-state change increases with the total number of branches in a tree (and consequently the total amount of evolution). This in itself shows only that the consistency index is related to the total amount of evolutionary time in a given group of organisms. Whether this also implies evolutionary plasticity or not depends on what we mean by plasticity. That is, the propensity of a character to change through phyletic evolution might be measured in terms of per unit time or simply the total amount of evolution in all relevant taxa. Either way, it seems that what we would really like is an estimate of the rate of evolution since quantities related to evolutionary plasticity can be derived from such rate estimates. Measures using maximum parsimony procedures do not perform adequately in this regard since change is measured in units of cladogenic events.

These two common uses of measures of homoplasy deviate from the original notion of homoplasy, even from the taxic definition of homoplasy. A given character state shared between two or more organisms is or is not homologous (or homoplastic). A quantitative measure of homoplasy for an individual character does not seem appropriate given the definitions. If we have a data set, we might characterize the total level of homoplasy in the data set by the percentage of the characters deemed homoplastic, however, most measures of homoplasy are not simple counts of putative homoplastic characters. An alternative view is to quantify the probability that a shared character state is homologous (or homoplastic) and use this value as a measure of homoplasy. (Patterson (1982) has also suggested some kind of probability computation for his congruence test of homology.) As demonstrated below, some well known indices of homoplasy derived from the maximum parsimony procedure (see Archie, this volume, for a review) are not direct estimators of levels of homoplasy in this stochastic sense.

If a quantitative measure of homoplasy relates to the probability that a character is homoplastic, where does the uncertainty come from? One possibility is that a determination of homology and homoplasy may be uncertain because of possible errors in the putative ancestral relationships. A shared derived state might be considered homoplastic because the taxa sharing that state are not considered monophyletic, yet if the taxa are in fact monophyletic the character state could be homologous. Another possibility is that even if there were no errors in the tree, there can still be uncertainty in the homology statement. For example, two taxa might be monophyletic and share a derived state, yet the state might be identical because of unseen independent originations during the phyletic evolution of each lineage. The maximum parsimony method counts origination events in units of cladogenesis and therefore cannot determine such phyletic events. Construction of an appropriate measure of homoplasy then requires a determination of uncertainty in the estimated tree and a stochastic model of the phyletic character change process. Given such quantities we can measure the probability that character states are identical, but not by descent.

Before continuing the above exposition, we should examine the first kind of uncertainty a little further. We noted that we might declare a character state homoplastic because according to the estimated tree the character state is not a shared derived state of a monophyletic group—which might be mistaken. However, this assessment really depends on a strictly taxic notion of homology/homoplasy. For example, consider a four taxa rooted tree {{A,B},{C,D}} with the character state assignment {{1,0},{1,0}}. According to the taxic view, the character state {1} shared between taxa A and C is homoplastic because it is not a shared derived state of the four taxa. However, why should the states at B and D influence what we think about the shared states at A and C? From a process point of view, the states at taxa B and D would influence how we think about the character change process between A and C only if we were to make some homogenity assumption about the overall process of character change (both states 1 and 0) over the whole tree. Otherwise, the evolutionary connection between the character states at A and C depends only on the path between A and C. This means that mistakes in our estimates of branching order of the trees should not influence our determination of homology/homoplasy except in an indirect manner. Therefore, unless we are concerned with the strict taxic view of homology/homoplasy, the first kind of uncertainty, the tree estimate, should not directly affect quantifying the probability of homoplasy.

However, this still leaves uncertainty of the second kind, that caused by multiple state changes along phyletic history. This leads us to introduce a measure of homoplasy that depends on the path length between any two (or more) taxa sharing a character state. Namely, we quantify homoplasy by the following conditional probability: given that two randomly chosen taxa are identical for a particular character state, what is the probability that they are identical by chance (i.e., *not* by descent)? Roughly speaking, this quantity will be affected by distance between pairs but not so much by the branching order of the taxa. In the following we make this notion more precise and suggest an estimator for this probability under a simple model of evolution. We suggest that this estimator is an index of homoplasy that is faithful to the above definition of homoplasy. We contrast our index with two other indices, the retention index (Farris 1989) and HER (Archie 1989) with some toy examples and discuss possible interpretations.

A PROBABILISTIC MEASURE OF HOMOPLASY

The equality $X(t) = X(u)$ means that the state of character X is identical for taxa t and u. Let us also write $X(t) \equiv X(u)$ to mean that $X(t)$ and $X(u)$ are identical by descent (ibd). That is, $X(t) \equiv X(u)$ if the character X does not experience any changes of state on the path that joins t and u; of course $X(t) \equiv X(u)$ implies $X(t) = X(u)$. The idea of our homoplasy measure is this. For a given character, suppose we choose a random pair of taxa and find that they share the same character state. We are interested in the probability that this happened "by accident." Accordingly, we define a homoplasy measure

H to be the conditional probability that, given that two randomly chosen taxa share the same state, they are not ibd. That is, letting random taxa T and U be chosen uniformly over the $n(n-1)$ pairs of distinct taxa and independently of the evolutionary process,

$$1 - H = P\{X(T) \equiv X(U)|X(T) = X(U)\}$$

$$= \frac{P\{X(T) \equiv X(U)\}}{P\{X(T) = X(U)\}}$$

$$= \frac{\sum_{t,u} P\{X(T) \equiv X(U)|T = t, U = u\}P\{T = t, U = u\}}{\sum_{t,u} P\{X(T) = X(U)|T = t, U = u\}P\{T = t, U = u\}}.$$

However, $P\{T=t, U=u\} = 1/n(n-1)$ for all $t \neq u$. Also, by the assumed independence of T and U from the X process,

$$P\{X(T) \equiv X(U) \mid T=t, U=u\} =$$

$$P\{X(t) \equiv X(u) \mid T = t, U = u\} = P\{X(t) \equiv X(u)\},$$

and, similarly, $P\{X(T) = X(U) \mid T=t, U=u\} = P\{X(t) = X(u)\}$. Thus,

$$H = 1 - \frac{\sum_{t,u} P\{X(t) \equiv X(u)\}}{\sum_{t,u} P\{X(t) = X(u)\}}.$$

The denominator involves only joint distributions of character values at terminal nodes and can therefore be estimated directly from the data in an obvious way. In the numerator, for simplicity let us consider models of the type considered by Cavender (1978), in which characters are assumed to be binary and to switch values at the times of a Poisson point process along the branches of the tree. For such models there is a simple, well-known relationship between probabilities of identity and probabilities of identity by descent. Letting $p(t,u)$ denote the probability $P\{X(t) \neq X(u)\}$, the expected number of changes, v, on the path joining t and u is given by

$$v = \frac{-1}{2} \log(1 - 2p(t,u)).$$

The probability of identity by descent is the probability of no state changes on a path of length v, which is e^{-v}. That is,

$$P\{X(t) \equiv X(u)\} = e^{-v(t,u)} = \sqrt{1 - 2p(t,u)}.$$

Therefore,

$$H = 1 - \frac{\sum_{t,u} \sqrt{1 - 2p(t,u)}}{\sum_{t,u} 1 - p(t,u)}.$$

Estimation

Having expressed H in terms of the joint probabilities $p(t,u)$ on terminal taxa, which can be estimated simply by the corresponding fractions in the sample, a straightforward "plug-in" estimator suggests itself. It is natural to estimate $p(t,u)$ by the fraction

$$\hat{p}(t,u) = \frac{1}{n} \sum_{i=1}^{n} I\{X^i(t) \neq X^i(u)\},$$

where $I\{X^i(t) \neq X^i(u)\}$ denotes an indicator random variable that is 1 if $X^i(t) \neq X^i(u)$ and 0 otherwise. The index i runs through the characters in a data set. This suggests estimating H by

$$\hat{H} = 1 - \frac{\sum_{t,u} \sqrt{1 - 2\hat{p}(t,u)}}{\sum_{t,u} 1 - \hat{p}(t,u)}.$$

Examples

In this section we illustrate the behavior of our homoplasy index (H), the retention index (R), and homoplasy excess ratio (HER) by evaluating these indices on some simple models. Both R and HER are computed from the maximum parsimony length of characters in a data set. Briefly, $R = 1 - H_o/H_{max}$, where H_o is the excess in total length of the data set for the most parsimonious tree and H_{max} is the maximum excess length of the data set possible under any tree (this can be found by computing the length over a star phylogeny). Similarly, HER $= 1 - H_o/H_{per}$, where H_o is again the excess parsimony length while H_{per} is the expected excess length of the data set under random permutation of the character vectors. The randomization is carried out by taking each character and permuting across the taxa, thereby destroying the phyletic information. An excellent review of these two indices and other such measures is given in the chapter by J. Archie in this volume. We note that we often abuse language in the examples below when we say "expected" or "$R =$", etc. Our evaluations of these indices under the models given below will be exact for the standard "mythical infinite data set" generated by that model. The evaluations will be close approximations with high probability if we observed large numbers of characters.

Example 1: Star Phylogeny

Suppose we have a star phylogeny with equal branch lengths with the parameter $p = P(X(\text{root}) \neq X(\text{tip}))$. For concreteness we will consider the case of a four taxa tree (Fig. 1a). We then have 16 possible binary state combinations at the tips of this tree: (0,0,0,0), (0,0,0,1), and so on to (1,1,1,1). For a star phylogeny, the probability of each kind of character pattern will depend only on whether there is no change, one change, etc., and will follow a binomial distribution. We will compute the limiting behavior of R, HER, and our H under this model as the number of observed characters grows to infinity.

Assuming a marginal probability of 1/2 for each state at each node, the binary complement of each character pattern have equal probabilities (e.g., $P\{(0,0,0,0)\} = P\{(1,1,1,1)\}$) and the Wagner parsimony algorithm treats them in the same manner, so we consider a character pattern and its binary complement as a class and use $[0,0,0,0]$ to mean the set $\{(0,0,0,0), (1,1,1,1)\}$ and so on. So we have

$$P\{[0,0,0,0]\} = (1 - p)^4 + p^4,$$

$$P\{[0,0,0,1]\} = P\{[0,0,1,0]\} = P\{[0,1,0,0]\} = P\{[1,0,0,0]\}$$

$$= p(1 - p)^3 + p^3(1 - p),$$

$$P\{[0,0,1,1]\} = P\{[0,1,0,1]\} = P\{[0,1,1,0]\} = 2p^2(1 - p)^2.$$

For both R and HER, only the informative patterns $[0,0,1,1]$, $[0,1,0,1]$, $[0,1,1,0]$ have nonzero excess length. Therefore, we need to consider only these three classes in the computations. There are three possible fully re-

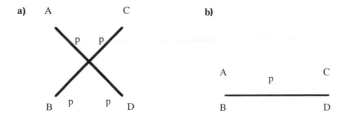

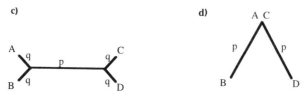

FIGURE 1 Models of four taxa trees for the examples discussed in the main text. (a) A star phylogeny with identical probability of change, p, per branch. (b) An extremely informative tree with zero terminal branches and an internal branch with probability of change p. (c) Same as model (b) but with finite positive terminal branches (probability of change q). (d) An extreme version of an inconsistent four taxa tree.

solved topologies for four taxa and since the probabilities of all informative character patterns are the same, the expected length for a character generated by a star phylogeny is the same for any of the three topologies. Therefore, arbitrarily choosing one topology, the excess length is zero for one informative pattern, say [0,0,1,1], and the excess lengths are one for the two other patterns. So we have the expected excess length

$$H_0 = 0 \cdot \{2p^2(1-p)^2\} + 1 \cdot \{2p^2(1-p)^2\} + 1 \cdot \{2p^2(1-p)^2\}.$$

For the expected length, H_{max}, measured on a star phylogeny, we simply replace the first term in the above sum with $1 \cdot \{2p^2(1 - p)^2\}$ since all characters with two zeros and two ones have excess length one over a star phylogeny. Using these two quantities to compute R, we arrive at

$$R = 1 - \frac{4p^2(1 - p)^2}{6p^2(1 - p)^2} = 1 - \frac{2}{3} = \frac{1}{3}.$$

To compute HER, we need the probability of each character pattern under the suggested permutation. However, we can immediately see that this probability depends only on the marginal frequency of ones and zeros in a character pattern which is exactly equivalent to the probability of the original characters. Therefore, $H_0 = H_{per}$ and HER = 0 for all values of p.

To compute our H value, we need to compute $p(t,u)$ for all pairs of taxa. In this case we have six pairs, (A,B), (A,C), (A,D), (B,C), (B,D), and (C,D) respectively. An enumeration of the character patterns will show that $p(t,u) = 2p(1-p)$ for all pairs of taxa. Therefore, we have

$$H = 1 - \frac{6\sqrt{1 - 4p(1 - p)}}{6(1 - 2p(1 - p))} = 1 - \frac{\sqrt{1 - 4p(1 - p)}}{1 - 2p(1 - p)}.$$

For example, we have $H = 0$ when $p = 0$, and $H = 1$ when $p = 1/2$.

With a star phylogeny it seems reasonable for a measure of homoplasy to be an increasing function of the length of the branches (p value). That is, according to the meaning of the term homoplasy, if p is small then there will be little homoplasy—only a small fraction of the resemblance among the taxa is due to separate originating events. Similarly, if p is near 1/2, each branch will have many state changes with high probability—most of the resemblance is due to separate origination events. However, in this example, both R and HER take constant values regardless of the length of the branches. The definitions of R and HER have the interpretation that low values imply high homoplasy and vice versa. In this example HER has a constant value, zero, with the implication that all data sets generated by a star phylogeny have maximum levels of homoplasy. In this sense, neither R nor HER measures homoplasy.

If we move from the definition of homoplasy to the idea of a homoplasy measure as an indication of reliability or informativeness of data, the values of HER makes some sense. We could view data from a star model as having no phylogenetic information with respect to the hierarchical organization of the taxa. (Of course, different views can be taken depending on whether we wish to consider polytomous trees as a possible phylogeny.) R, on the other hand, takes a curious value of 1/3 which does not seem to have an intuitive interpretation.

Example 2: Extremely Informative Tree

Consider a tree with four taxa A, B, C, and D having the topology $((AB)(CD))$, a long internal branch with parameter p, and very short branches to the terminal nodes (Fig. 1b). In fact, to simplify matters, suppose that the length of all branches other than the internal branch is zero. Then we have $P\{[0,0,0,0]\} = 1 - p$, $P\{[0,0,1,1]\} = p$, and all other patterns have probability zero.

For computation of R, we have $H_o = 0 \cdot p$ and $H_{max} = 1 \cdot p$. Therefore, $R = 1$ regardless of the value of p.

With HER, permutation of the pattern $[0,0,1,1]$ produces two other patterns $[0,1,0,1]$ and $[0,1,1,0]$ and each of these three patterns then has the probability $p/3$. Therefore, $H_{per} = 2 \cdot p/3$ but $H_o = 0$ and HER $= 1$.

For our measure, H, we again examine each character pattern and the implied joint probability $p(t,u)$ and we find $p(A,B) = 0$, $p(A,C) = p(A,D) = p(B,C) = p(B,D) = p$, and $p(C,D) = 0$. So we have

$$H = 1 - \frac{2 + 4\sqrt{1 - 2p}}{2 + 4(1 - p)},$$

and $H = 0$ when $p = 0$ and $H = 1/2$ when $p = 1/2$.

In this example, again neither R nor HER takes the length of the internal branch into account; both suggest no homoplasy. This would be reasonable for this example from the point of view of reliability since the length of all other branches are zero. On the other hand, large p implies that the internal branch is very long. In such cases, any character that takes identical states on both sides of the internal branch is likely to be due to chance reversals, suggesting a large probability of homoplasy. This is not reflected in R or HER.

Our H is a function of the length of the internal branch and takes the value 1/2 when $p = 1/2$. The interpretation of this value requires recalling the definition of our measure. Among the six pairs for comparison of the four taxa, the four comparisons across the internal branch ((A,C), (A,D), (B,C), and (B,D) pairs) are homoplastic with probability 1 since $p = 1/2$ corresponds to an infinitely long branch. The two comparisons (A,B) and (C,D) are homologous with probability one since zero-length branches separate (A,B) and (C,D). However, we are conditioning on the event that the chosen pair of taxa are identical. For the four comparisons across the internal branch the probability that they are identical is 1/2 since the probability of the character pattern [0,0,1,1] is 1/2. For the (A,B) and (C,D) pairs, the probability of identity is one since [0,0,0,0] and [0,0,1,1] imply identity for these two pairs. We can loosely reason that there are four homoplastic pairs and two homologous pairs but we are half as likely to choose the homoplastic pairs since they are identical only half of the time, therefore, the probability of homoplasy is 1/2.

We continue this example by assuming this time that the internal branch has probability of change p, while the four other terminal branches have probability of change q (Fig. 1c). After enumeration of the character probabilities and a great deal of algebra we obtain

$$R = 1 - \frac{4(1 - q)^2 q^2}{p - 4pq + 6q^2 4pq^2 - 12q^3 + 6q^4}$$

$$HER = \frac{p(1 - 2q)^2}{p - 4pq + 6q^2 4pq^2 - 12q^3 + 6q^4}$$

$$H = 1 + \frac{(1 - 2q)(1 + 2\sqrt{(1 - 2p)})}{8pq^2 - 6q^2 - 8pq + 6q + 2p - 3}.$$

Although these equations are not easily interpretable, several features can be noted. First, when q goes to 1/2, both $1 - HER$ and H goes to 1, but R goes to 1/3. This is similar in result to the star phylogeny case. Figure 2 shows a plot of the three indices as a function of q value (the value of R and HER are plotted as $1-R$ and $1-HER$ to allow consistent interpretation with H). The plot is drawn for two values of p, 1/4 and 1/2 (solid lines and dotted lines, respectively). In this plot, both R and HER have an unusual behavior: for fixed values of q, levels of indicated homoplasy are higher when p is smaller. This again makes sense in terms of reliability but not levels of homoplasy. Both R and HER show an S-shaped curve as a function of q, while H approaches its maximum more simply. The value of H also starts at different levels depending on the value of p. Both of these observations indicate that H is more sensitive to the total branch length of the tree while R and HER seem to be more indicative of the relative length of the terminal branches and the internal branch.

Example 3: Inconsistent Model

Figure 1d shows the usual example of inconsistency of parsimony after Felsenstein (1978), with true tree topology ((AB),(CD)). Consider the limiting case with only the long branches having positive length and the short branches having zero branch length. The possible patterns and their probabilities are: $P([0,0,0,0]) = (1-p)^2$, $P([0,1,0,1]) = p^2$, $P([0,0,0,1]) = p(1-p)$, and $P([0,1,0,0]) = p(1-p)$. The only phylogenetically informative character pattern is [0,1,0,1] whose excess parsimony length is zero (*measured over*

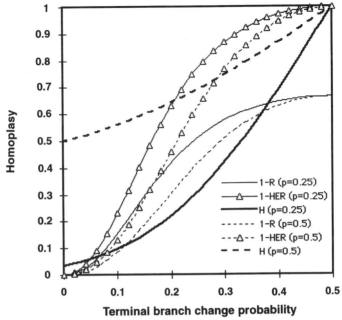

FIGURE 2   Values of homoplasy indices R (thin lines), HER (marked with triangles), and H (thick lines) computed for the four taxa model shown in Fig. 1c. The values are computed as a function of q for two fixed values of p, $p = 1/4$ (solid line) and $p = 1/2$ (dotted line).

the wrong tree). Therefore, both R and HER take the value one for this model. Some computations show that H is

$$H = 1 - \frac{1}{2}\left(\frac{1 + 4\sqrt{1 - 2p} + (1 - 2p)}{1 + (1 - p)(2 - p)}\right),$$

and $H = 0$ when $p = 0$ and $H = 5/7$ when $p = 1/2$.

For this model, R and HER will say that there is no homoplasy, or at least very little homoplasy. However, according to the literal meaning of this term, this example is plagued by homoplasy; for example, with probability p^2, we have $X(A)=X(C)\neq X(B)=X(D)$, in which case the character X is necessarily homoplastic. On the other hand, in terms of reliability of the inference, this situation is the prototypical, notorious example of a difficult problem. So it seems worrisome that a purported measure of reliability would report no sign of trouble here: while it is very unlikely that the inference is correct, the homoplasy measures R and HER indicate that the inference is reliable. Here R and HER measure neither homoplasy nor reliability. Our measure H, while a function of p, takes a maximum of 5/7. This value, although it may seem curious at first, can be interpreted in a manner similar to the previous example.

DISCUSSION

Homoplasy is "not homology." However, "not homology" has different implications depending on the notion of homology one adopts. According to the taxic view, homology is defined as derived states shared among monophyletic groups. Shared derived states of multiple evolutionary origin is homoplasy by this definition. In this paper we have attempted to formulate a probabilistic measure of homoplasy using the question "for a randomly chosen pair of taxa, given that their character states are identical, what is the probability that they are identical not by descent but by chance?" Measures of homoplasy based on the maximum parsimony length of characters are poor estimators of such probabilities because they ignore the possibility of multiple events within phyletic lineages. We presented one possible estimator for this probability based on a simple Poisson model of evolution for binary characters. However, we have not fully developed the possible family of stochastic measures of homoplasy from our framework. For example, two possible directions for extensions are the treatment of more complicated models of phyletic evolution and the inclusion of the phylogenetic tree in the quantification. The latter might involve also quantifying the uncertainty of the estimated tree as alluded to earlier in this paper. More importantly, we emphasize the desirability of starting out with an explicit probabilistic framework before developing estimators or indices.

We have repeatedly noted the difficulty of using measures of homoplasy for inferring reliability or evolutionary plasticity. In fact, more recently, explicitly statistical measures of reliability such as the bootstrap (Felsenstein 1985) are widely used. However, the problem of assessing reliability is not easy. For instance, in the inconsistency example of Felsenstein, the standard use of the bootstrap in conjunction with parsimony will also fail to indicate any problem; an inconsistent estimator will not be rescued by this use of the bootstrap. Yet although the results given by the bootstrap might be improved by applying it with a consistent estimator, there is no such hope for improvement with homoplasy measures that are tied to the parsimony framework. It simply may not be reasonable to expect the reliability evaluation to be done by the same goodness-of-fit measure that is optimized by the parsimony criterion.

The situation seems similar in the area of quantifying evolutionary plasticity. Evolutionary plasticity really relates to the rates of evolution and more direct estimators of such rates or hypotheses tests have been proposed (e.g.,

Sanderson 1993). This leaves the question: what utility is there for a quantitative measure of homoplasy *per se*? We have tried to make sense of such measures by giving a probabilistic interpretation. While this allows consistent interpretation of numerical values, the utility of those values is not clear. Even granted that we have independent, identically distributed, characters such as DNA sequences (perhaps), and we ask "for this gene what is the probability of homoplasy of a nucleotide?", it still seems unlikely that we would be interested in the answer if we restrict ourselves to something like the taxic view of homology. We would be interested in the possible homology of nucleotides if we were trying to align two sequences. However, the sense of homology for this problem is different from that of the taxic view. In the alignment case, we are asking whether two structures (nucleotides) are related by descent regardless of whether they are identical or not. This question seems to fall within the transformational or the biological view of homology.

It seems to us, then, that most questions that have been posed in terms of homoplasy under the maximum parsimony framework are actually questions of reliability of the estimated tree or rates of evolution. Such questions should be addressed directly in those terms and not be confounded with the complex issues surrounding homology, analogy, and homoplasy. It also seems that homology and homoplasy questions *per se* are really about whether different structures in different organisms are comparable or not and, given that they are comparable, what evolutionary processes link the structures (Wagner 1994). These questions are more fundamental than questions of independent origins or monophyly. They seem best addressed from more pluralistic views.

REFERENCES

Archie, J. W. 1989. Homoplasy excess ratios: New indices for measuring levels of homoplasy in phylogenetic systematics and a critique of the consistency index. Systematic Zoology 38: 235–269.
Archie, J. W. 1996. Measures of homoplasy. Pp. XXXX *in* M. J. Sanderson and L. Hufford, eds. Homoplasy and the evolutionary process. Academic Press, San Diego.
Cavender, J. A. 1978. Taxonomy with confidence. Mathematical Biosciences 40:271–280.
Donoghue, M. J. 1992. Homology. Pp. 170–179 *in* E. F. Keller and E. A. Lloyd, eds. Keywords in evolutionary biology. Harvard Univ. Press, Cambridge.
Farris, J. S. 1983. The logical basis of phylogenetic analysis. Pp. 7–36 *in* N. Platnick and V. Funk, eds. Advances in cladistics, vol. II. Columbia Univ. Press, New York.
Farris, J. S. 1989. The retention index and rescaled consistency index. Cladistics 5: 417–419.
Felsenstein, J. 1978. Cases in which parsimony or compatibility methods will be positively misleading. Systematic Zoology 27:401–410.
Felsenstein, J. 1985. Confidence limits on phylogenies: An approach using the bootstrap. Evolution 39:783–791.
Kluge, A., and J. S. Farris. 1969. Quantitative phyletics and the evolution of anurans. Systematic Zoology 18:1–32.

Panchen, A. L. 1994. Richard Owen and the concept of homology. Pp. 22–57 *in* B. K. Hall, ed. Homology: The hierarchical basis of comparative biology. Academic Press, New York.

Patterson, C. 1982. Morphological characters and homology. Pp. 21–74 *in* K. A. Joysey and A. E. Friday, eds. Problems of phylogenetic reconstruction. Academic Press, New York.

Patterson, C. 1988. Homology in classical and molecular biology. Molecular Biology and Evolution 5:603–625.

Sanderson, M. J. 1993. Reversibility in evolution: A maximum likelihood approach to character gain/loss bias in phylogenies. Evolution 47:236–252.

Sanderson, M. J., and M. J. Donoghue. 1989. Patterns of variation in levels of homoplasy. Evolution 43:1781–1795.

Wagner, G. P. 1989. The biological homology concept. Annual Review of Ecology and Systematics 20:51–69.

Wagner, G. P. 1994. Homology and the mechanisms of development. Pp. 22–57 *in* B. K. Hall, eds. Homology: The hierarchical basis of comparative biology. Academic Press, New York.

PART III

GENERATION OF HOMOPLASY

COMPLEXITY AND HOMOPLASY

DANIEL W. McSHEA[1]

University of Michigan
Ann Arbor, Michigan

INTRODUCTION

Consider the two objects in Fig. 1. Which is more complex? Object A might seem more complex because it contains more parts; or A might seem simpler, because its parts are all the same size and disconnected from each other, while B's vary in size and are nested within each other. On the other hand, the arrangement of parts in A might look more complex because it is less regular, less predictable, while in B the parts are neatly lined up along the midline of the object. The problem here is that complexity has various meanings. Both objects are complex, but in different senses.

Complexity has been the subject of a number of empirical and theoretical studies in recent years. At issue in some, for example, has been the existence of a general trend in complexity (McShea 1991, 1993; Valentine et al. 1993); the relationship between complexity and such factors as ecological special-

[1]Present address: *Zoology Department, Duke University, Box 90325, Durham, NC 27708-0325.*

HOMOPLASY: *The Recurrence of Similarity in Evolution*, M. J. Sanderson and L. Hufford, eds.

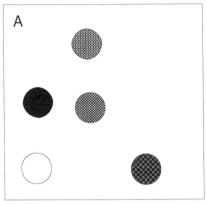 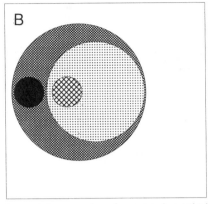

FIGURE 1   Two complex objects (squares), each containing a number of parts (circles). Which is more complex?

ization (Flessa et al. 1975), morphological variability (Lauder 1981), environmental variability (Hughes and Jackson 1990), and species longevity (Boyajian and Lutz 1992); and the probability of evolutionary homoplasy (Sanderson 1993; see also Gould 1970; MacBeth 1980; Laurent 1983). However, the significance of much of this interesting work, and its relation to other issues in evolutionary biology, has been at least partly obscured by the ambiguities and multiple meanings of the word "complexity."

Regarding trends, for example, many share the impression that organisms have become more complex, on average, over the history of life. But more complex in what sense? Number of parts may have increased but irregularity of configuration may not, or vice versa. Nestedness of parts may have increased, but their numbers may have decreased, or vice versa. Is it complexity of structure or complexity of function that is thought to increase? We cannot begin to study trends until the various senses of complexity have been distinguished.

This is likewise true for complexity and homoplasy. Consider the powerful logic connecting complexity and evolutionary irreversibility, the logic of Dollo's Law (Gould 1970; Gould and Robinson 1994). A complex structure such as an organism, or a complex part of one, that has been significantly modified in evolution never returns precisely to its former state—to the ancestral condition—because the changes along the way have been too many, and the resulting pathway too improbable to be retraced (Sanderson 1993 and references therein). Functional reversal is possible and simple morphological patterns may recur—dolphins have regained the fusiform shape of their fish ancestors—but if the transformation has been sufficiently complex, differences will be found in the details. Precise reversal is possible in principle, but exceedingly improbable in practice.

The logic is sound, but whether or not it is applicable depends on which sense of complexity is intended. One sort of complex evolutionary transformation might be a series of additions of independent steps to the developmental program of an organism. In this case, for reversal-in-detail to occur, each would have to be removed (although not necessarily in the same order). The greater the number of independent additions, that is, the greater the complexity of the transformation, the more improbable that complete reversal will occur. On the other hand, if the complexity is built up hierarchically, so that developmental steps added later are dependent on those added earlier, reversal may be a simple matter of eliminating the early steps. Take the lower supports out from under a wing of the developmental superstructure, and it collapses, restoring level ground. This example does not contradict Dollo's Law; the law remains valid, but only within a restricted domain defined by a particular sense of complexity.

In this chapter, I explore the various senses of complexity and develop a scheme for organizing and understanding them. Then I attempt to show, in a preliminary way, the manner in which each sense might be related to homoplasy, in particular, to the ease or probability of homoplasy. The relationship is important not only for the study of the development of complex structure (Wake and Roth 1989), but also for systematics, where homoplasy is mainly considered an obstacle to the discernment of evolutionary relationships. One assumption of some cladistic methodologies is that homoplasy is improbable (see discussion in Sober 1988), relative to other sorts of transformation, and that therefore the best-supported phylogeny is the one with the fewest homoplastic events. One purpose here is to analyze and evaluate this assumption in the case of complex transformations. For example, one apparent implication of Dollo's Law is that the assumption is especially likely to be valid for complex structures; as we shall see, the reasoning is correct for some senses of complexity, but not others.

COMPLEXITY

No single understanding of complexity has been proposed that encompasses all nuances and context-specific meanings. However, a common theme has emerged from a number of theoretical treatments in biology (Hinegardner and Engelberg 1983; Kampis and Csányi 1987; Katz 1986, 1988; Wicken 1987; McShea 1991; Slobodkin 1992) and is implicit in recent empirical studies (Cisne 1974; Boyajian and Lutz 1992; McShea 1992, 1993; Valentine et al. 1993). Complexity is heterogeneity. Most generally, the more diverse a system is, the more numerous its component parts and interactions and the more irregular their configuration, the more complex it is. However, as Fig. 1 dramatizes, complexity is itself a complex term, encompassing a number of different senses or types. I have identified eight types in all, based on three distinctions.

Three Distinctions

Differentiation and Configuration

Differentiation has to do with numbers of parts (or interactions; see below) and the differences among them. For systems in which parts are discrete and types easily distinguished, differentiational complexity is simply the number of different types (Valentine et al. 1993). Where variation among parts is continuous and types intergrade, complexity is some function of degree of differentiation, such as the variance or a variance analog (McShea 1992, 1993). Configurational complexity is irregularity of arrangement. For objects in which the parts are arranged in a linear series, irregularity of configuration might be a function of degree of differentiation among adjacent parts, perhaps the number of changes in part-type along the series.

To make the contrast clearer: a parade is typically highly differentiated, consisting of many differently dressed individuals, but configurationally simple, in that individuals typically march in regular rows. More abstractly, in Fig. 2, A is more differentiated, because it contains five different parts (shades), while B has only four. On the other hand, the arrangement of parts in A is more regular, so B is more complex configurationally.

Objects and Processes

In the present context, an object is a larger-scale entity which is composed of smaller-scale parts, and a process is a larger-scale activity or operation which is composed of smaller-scale interactions. Thus, object complexity refers to differentiation or irregularity among the physical parts of a system, and process complexity to the same properties of the interactions among the parts. It is the parts that interact, but their interactions can be considered in the abstract, without reference to the parts themselves. Indeed, there is no necessary correspondence between the complexity of objects and processes. In Fig. 2, C and D have the same object differentiation, because they consist of precisely the same parts, but they differ in the number of distinct interactions among parts, shown by the arrows. D has more arrows and thus greater process differentiation.

Organismal complexity has at least four components: molecular, developmental, morphological, and physiological. Molecular and morphological complexity refer to the heterogeneity of objects. For DNA molecules, the parts might be genes or nucleotides, and for whole organisms, cells or organs. Developmental and physiological complexity refer to processes. For development, the component interactions might be morphogenetic events, and for physiology they might be metabolic steps. Only developmental and

211

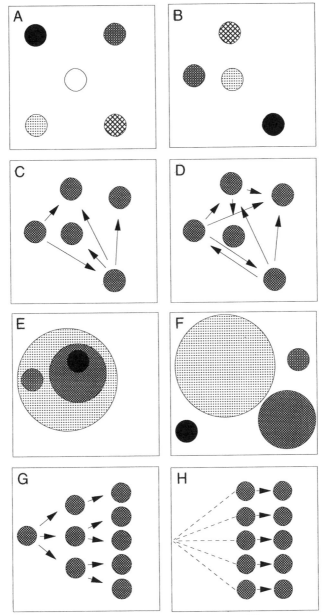

FIGURE 2 Systems of parts and interactions, illustrating the differences among the various senses or types of complexity. (A and B) Differentiational versus configurational complexity; (C and D) object versus process complexity; (E and F) hierarchical versus nonhierarchical object complexity; (G and H) hierarchical versus nonhierarchical process complexity.

morphological complexity will be considered here, although the logic of the argument can be easily adapted for molecules and physiology.

Hierarchical and Nonhierarchical Complexity

Nonhierarchical complexity refers to the number and configuration of parts at a given spatial scale, or interactions at a given spatial and temporal scale. For either parts or interactions, complexity is thus a scale-relative property; for example, the surface of a marble is smooth and featureless (i.e., simple) at the usual scale of human perception but rough and irregular (i.e., complex) at the molecular scale. Nonhierarchical complexity also varies with scalar *range*; in Fig. 2, E might be considered to have four parts at a single scale, or two parts at one scale and one at each of two other scales, depending on how the scalar spectrum is carved. Finally, scalar levels are independent to some extent (Allen and Starr 1982; Salthe 1985), and thus no object or process has a true or essential complexity. In other words, no scalar level has priority over any other, at least *a priori*. In particular, complexity at the genetic level is no more the "true" complexity of an organism than complexity at the scale of its cells or organs.

For objects, hierarchical complexity refers to the number of levels of nestedness of parts within parts. A classic object hierarchy in biology is the nested series: species, populations, individuals, organs or organ systems, cells, and so on. Upper level entities physically contain the lower and constrain their behavior to some extent (Salthe 1985). Such hierarchies have been called scalar (Salthe 1993), ecological (Eldredge and Salthe 1984), or structural hierarchies (O'Neil et al. 1986). In Fig. 2, E and F have the same parts, but E is hierarchically more complex, because its parts are physically nested within each other to some extent.

For processes, hierarchical complexity is the number of levels in a causal specification hierarchy (Salthe 1993). An army chain of command is such a hierarchy, with the highest ranking officers issuing the most general orders, causing the lower ranks to give more specific orders. Likewise, development is partly a causal specification hierarchy, with the more general features of the bauplan specified early and initiating interactions leading to elaborations of those features (Riedl 1978; Wimsatt 1986; Arthur 1988; Salthe 1993).

In Fig. 2, G and H have nearly the same parts, but in G the command structure is hierarchical, while in H it is not. (The dashed lines in H are just a reminder that, in organisms at least, developmental pathways may tend to converge to one or a small number of initiating events, and to the extent that this occurs, an ontogeny as a whole must be at least partly hierarchical. Still, at shorter time scales, interactions may proceed independently, in parallel, as shown in the figure.)

Notice that physical nestedness of objects in a system is not crucial to the hierarchical relationship of its interactions (as in G). In an army, commands of increasing specificity flow from the highest levels, generals, to the lowest, privates, but privates are not physically nested within generals. Nor is a one-to-many relation between early- and late-occurring interactions crucial. The increasing specificity of interactions (orders) along the chain of command would still be hierarchical even if every level were occupied by just a single individual.

Eight Types

The three distinctions—differentiation versus configuration, object versus process, and hierarchical versus nonhierarchical—are independent, and thus together, in their various combinations, define eight independent senses, or types, of complexity for every system. So for differentiation, every system would have a: (1) hierarchical process complexity; (2) nonhierarchical process complexity; (3) hierarchical object complexity; and (4) nonhierarchical object complexity. The list would be the same for configurational complexity (bringing the total to eight).

The configurational types are not listed, because configurational complexity will not be addressed here. Assessing irregularity of parts and interactions is more difficult—even intuitively—than assessing their differentiation, and appropriate metrics have not been developed for the most part (cf. Yagil 1985; McShea 1992). In Fig. 2, the interactions in D seem configurationally more complex than those in C, but it is difficult to specify why or how.

Difficulties

Complexity and Functionality

The scheme outlined above is purely structural in that the complexity of a system is a function only of its parts (its physical structure) or of its interactions (its dynamical structure). In either case, complexity is independent of function. So, for example, a live organism and a dead one of the same species would have about the same object complexity, if the number and configuration of their parts were about the same. In common usage, however, complexity is closely connected with functionality, and to many this structural view will seem somewhat thin or narrow.

The complaint is reasonable. However, one problem with integrating some notion of functionality into the structural view is that compound concepts are awkward in practice. Heterogeneity is difficult enough to quantify in organisms, and workable measures of complexity—even qualitative measures—would be doubly difficult to devise if complexity were simultaneously a measure of heterogeneity and functionality. One virtue of a purer or narrower view of complexity is that measurement requires knowledge of structure and nothing else.

More important for present purposes is the fact that function undoubtedly has a key role in the origination and maintenance of complex structure and may have a significant effect on the probability of homoplasy in certain complex structures (see below). And in order to examine the relationship between structure and function, the two must be kept conceptually distinct. In other words, if functionality and structural heterogeneity were collapsed somehow into a single variable—which we would then call complexity— investigation of their relationship using this variable would be impossible.

Overall Complexity

Is Fig. 1B more complex than Fig. 1A *overall?* The question has no easy answer, because the eight types of complexity are conceptually independent. We encounter the same problem with other compound measures, such as size: a balloon might be larger than a cannonball in volume but smaller in mass. Which is larger "overall?" It is hard to know how to even approach the question, and thus hard to imagine how a meaningful notion of overall complexity could be devised.

Complexity and Randomness

Three kinds of randomness are relevant here. First, systems can be random in the sense that a group of systems of the same type all seem the same. One random number string looks like another; one compost heap looks like another. However, randomness in this sense is relative to interests and use. To a sea gull, every compost heap may look very different (R. Thomas in preparation). The narrow view avoids this subjectivity. Each compost heap is considered unique and its complexity is the heterogeneity of its specific, unique structure, whether or not that structure is considered significant in some context.

Second, if complexity is heterogeneity, then it might seem reasonable to also equate it with entropy, another sort of randomness (Wicken 1987). A car wreck is entropic in that a large number of different microstates (possible compositions and part-configurations) are equivalent to, or correspond to, the same macrostate (the same car wreck). In other words, entropy is a relationship between microstates and macrostate. On the other hand, a car wreck is complex if it is especially heterogeneous, regardless of how many alternative compositions and configurations would be equivalent. Complexity is a property of a single microstate (Wicken 1987), of one specific composition and configuration, not a relation between microstates and macrostate. Thus a complex system may also be entropic, but complexity is not the same as entropy.

Finally, Crutchfield and Young (1989; see also Crutchfield 1991) have argued that systems contain both a "regular" and a "random" component, and that complexity is the differentiation of the regular portion only. In their view, random refers to the unique features of systems, such as the precise number of cells in an individual, while regular refers to shared features, such as the number of cell types. The intent is to restrict complexity to features that are "rule-based" (Gell-Mann 1994). This approach is consistent with the narrow view, if part-types are understood as first-order regularities. For example, a decision that two cells are the same type can be construed as a decision that their similarities are rule-based and their differences not.

The above view is also consistent with certain information-theoretic approaches, including the notion that the complexity of a system is the length of the shortest complete description of it (Löfgren 1977), provided that only the regularities of the system are considered (Papentin 1980, 1982). This approach seems to require that an organism be understood as a kind of message; then, its complexity would be just the length of the shortest symbol string that constitutes or conveys the message. Number of different parts is the number of different symbols required and thus would be a kind of zeroth-order measure of complexity.

Other apparent difficulties—such as the dependence of complexity measure on descriptive frame, and disparities among the notions of complexity used in differing research programs—are discussed in McShea (1996).

HOMOPLASY

The issue here is the relative probability, in qualitative terms, of homoplasy in complex evolutionary transformations of organismal development and morphology. Probability is evaluated for the four differentiational types of complexity. Treatment of the problem is analytic; this is mainly an attempt to sketch the logical implications for homoplasy of the conceptual scheme outlined above. As will be seen, the logic is not impervious to particulars. For example, a particular selection regime can render a probable evolutionary reversal improbable, and vice versa (see examples in Bull and Charnov 1985). Some of these confounding particulars are discussed.

Before proceeding, a clarification may help prevent confusion later on: the issue is not the probability of transformations leading to homoplasy in complex organisms as opposed to simple organisms. The complexity of the organisms themselves is undoubtedly an important factor in homoplasy, but as will be seen, the logic developed here is relevant only to the complexity of the homoplastic *transformations*, and not of the organisms transformed.

Types of Homoplasy

Like complexity, homoplasy has a number of senses or types: reversal, parallelism, and convergence (Wake 1991). Here I adopt (and adapt for present purposes) Hennig's understanding of these terms (Hennig 1966; cf. Haas and Simpson 1946): reversal is the return in a single lineage of a developmental trajectory or a morphological structure to its ancestral condition, as in the transition series A → B → C → B → A, where A, B, and C are morphologies or developmental trajectories in an ancestor–descendant series or lineage. Parallelism is the independent production, simultaneous or not, in two or more lineages of similar trajectories or morphologies from similar starting points by similar morphological or developmental routes: A → B → C and A → B → C. Convergence is the production, again simultaneous or not, in two or more lineages of similarity from different starting points and by different routes: A → B → C and D → E → C.

Only reversal and parallelism can be treated in the current framework. In these two, the same transformations are repeated, either in the same direction or the reverse. If we can assume for reversal that the forward and backward probabilities of a transformation are about the same (although see below), and for parallelism that probabilities are about the same in both lineages, then it is possible to say something *a priori* about joint probabilities of occurrence, at least qualitatively. In convergence, however, the two lineages undergo very different transformations and we have no information at all *a priori* about the probability of their arriving at the same place, morphologically or developmentally.

Morphological or developmental transformations can be classified as additions, deletions, or modifications of parts or interactions (O'Grady 1985). Abstractly, an addition might be represented as A → AB, a deletion as AB → A, and a modification as A → B. Only additions and deletions can be treated in the present framework. For these, the complexity of the transformation can be understood as the number of parts (in morphology) or interactions (in development) added or lost. For modifications, the complexity of the transformation presumably lies at a lower hierarchical level, in the gears or inner workings, so to speak, of A in the modification A → B. In a modification, these lower-level workings are not specified, and probabilities cannot be evaluated. Where they are specified, the problem may reduce to a case of addition or deletion at the lower level (although see Alberch 1985), and the analysis can proceed at that level.

Relative Probabilities

For each of the four types of differentiational complexity, I will consider the relative difficulty or relative probability of reversal of addition, reversal of deletion, parallel addition, and parallel deletion. Table 1 summarizes the findings.

Hierarchical Developmental Complexity: Reversal

Figure 3A shows an idealized process hierarchy in which the later interactions are caused by, and are dependent upon, the earlier interactions. The circles represent morphological features present at various stages in the development of an organism and the arrows represent developmental steps,

TABLE 1 *Relative Difficulty of Reversal and Parallelism for Additions and Deletions of Complex Processes*

		Hierarchical	*Nonhierarchical*
Reversal			
Complex addition	A→AB→A	Easy	Moderately difficult
Complex deletion	AB→A→AB	Difficult (except by suppression of a suppressor; see text)	Difficult
Parallelism			
Complex addition	A→AB; A→AB	Difficult	Difficult
Complex deletion	AB→A; AB→A	Easy	Moderately difficult

perhaps inductions (but not necessarily). Imagine that in the ancestral condition in an evolving lineage, only the structure represented by the darkly shaded circle is present, but in the course of evolution, steps (arrows) are added as shown, with additional intermediate and terminal structures resulting as well (lightly shaded circles). The suggestion has been made that the build-up of such hierarchical dependencies may characterize developmental evolution generally, at least at higher taxonomic levels (Riedl 1978; Wimsatt 1986; Arthur 1988; Kauffman 1993).

Reversal of such a complex addition is relatively easy and therefore probable. Removal of the first step nips the developmental flowering at the stem. Possible examples include cases in which large or prominent complex structures have been lost entirely (Arnold et al. 1989), such as the loss of the elaborate pluteal larval stage in some sea urchin species (Raff 1987) and the loss of the entire adult stage in some salamanders (Wake 1991).

Reversal of a deletion would be much more difficult. Imagine now an ancestor starting with a complex developmental trajectory such as that in Fig. 3A, and suppose that this trajectory is lost in evolution, perhaps by the action of a suppressor of some sort acting on step 1. Restoring the ancestral condition might be as simple as restoring step 1, which could in principle be as simple as suppressing the suppressor (Laurent 1983). On the other hand, if a significant amount of time has passed, other developmental modifications will have occurred, some of which may have altered the latent later steps and structures. In other words, the hierarchical developmental cascade may have deteriorated, or its components may have been diverted to other purposes, during the time it was out of service and it may not operate as before. In that case, reversal would require reassembly of the complex developmental trajectory from scratch, exactly the sort of transformation that is

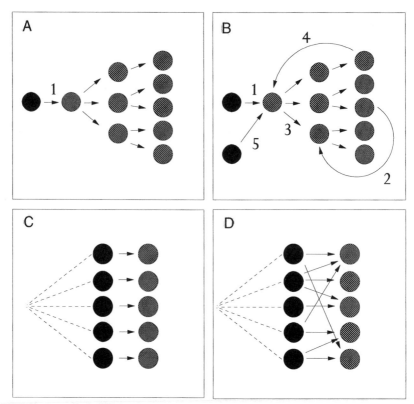

FIGURE 3   Hierarchically (A and B) and nonhierarchically (C and D) complex processes. For the discussion of reversals of additions and of parallel deletions, the darkly shaded circles are the ancestral condition and the arrows and lightly shaded circles have been added in evolution. (For reversals of deletions and parallel additions, the extended system is the ancestral condition.) See text. Notice that there is no requirement that the darkly shaded ancestral structures are adult features and that the added interactions and parts are terminal additions. In other words, the additions could just as well be interpolations within development.

pronounced prohibitively improbable by the logic of Dollo's Law. Still, latent competencies are known to persist sometimes in developmental programs (Lande 1978).

Caveats

Some possible difficulties and exceptions need to be explored. First, natural selection acting early in development may significantly affect probabilities of change. One reason is that additions of developmental steps are likely to become functionally integrated (Saunders and Ho 1976) over time and to become increasingly difficult to remove without causing gross interference in development. To the extent that such integration occurs, selection can be expected to oppose reversals of complex hierarchical additions.

Selection acting later in ontogeny, reflecting to a greater extent the organism's interaction with the outside world, is more problematic. Added complex objects and processes may well be functional later in ontogeny—may even become essential—and in such cases selection would be expected to strongly oppose reversal (i.e., deletion). Indeed, the argument has been made that additions of parts ought to be favored over deletions, on average, because additional parts increase the division of labor among parts and thus increase efficiency (Bonner 1988). Further, additions of interactions, manifest as increased degrees of developmental independence, ought to be favored over deletions because the added interactions increase evolutionary versatility (Vermeij 1973). On the other hand, it is arguable that selection should tend to rationalize and streamline development and structure by eliminating parts and interactions where possible (Rensch 1960), and that, on average, additions tend to interfere with existing mechanisms (Castrodeza 1978) more than they increase efficiency or versatility.

The arguments on both sides seem plausible, and numerous examples can be found to build a case either way, but at present neither generalization has sufficient empirical support (cf. Mabee 1993). Pending resolution, it seems reasonable to assume no net selective bias in favor of either additions or deletions, and therefore no effect on average on probability of reversal or parallelism.

Second, this analysis is clearly structuralist, in that it emphasizes the structural relations among parts and interactions and their historical constraints (Lauder 1981), but the logic does not take into account arguments by structuralists and others (e.g., Kauffman 1993; Thomas and Reif 1993; Wake 1991), which predict homoplasy as the expected consequence of diversification within highly restricted developmental or morphological spaces of possibilities. Such restrictions could result from some combination of structural or functional constraints or may simply be generic properties of the underlying evolutionary dynamic. These arguments complement rather than contradict the logic developed here. For example, given the discussion above for the difficulty of reversal of hierarchical deletions, we can see that any structural constraint tending to increase the probability of reversal of a hierarchical deletion must be sufficiently strong or precise to force reversal down a rather long and ramifying pathway (Fig. 3A). In other words, the present logic provides a set of background expectations which arguments in particular cases for structural (and other) constraints must accommodate.

Third, the perfect one-way flow of causation shown in Fig. 3A is probably not realistic for the development of most complex structures. At least some and possibly many feedback loops are likely to have been added in the course of evolution, if they were not present originally. To the extent that this occurs, the hierarchical structure of the causal network is reduced. For example, in Fig. 3B, the addition of step 2 creates an alternate pathway which bypasses the hierarchical control originally present in step 3. However, not all added pathways have this effect; the addition of step 4 leaves step 1 in its former controlling role. (Step 1 could be bypassed, but only if a pathway external to the system, such as step 5, were introduced.) The addition of such steps invalidates the conclusions drawn here to the extent that it occurs; but such confounding effects will often be empirically detectable, either in the correlation structure of interactions and parts or, ideally but less easily, in direct studies of actual developmental pathways.

Finally, and more generally, a limitless number of different mechanisms and routes to complexity can be imagined. Parts and interactions can be added or lost by coupling or decoupling of existing parts and interactions, as well as by outright addition or deletion. Deletions can occur heterochronically, or by changes in spatial relationships among interactions, as well as by simple inactivation. And so on. What is more, endless complications of the idealized cases in Figs. 2 and 3 can be imagined. Feedback loops and lateral links can be added or lost, and hierarchical control can be added or removed. Here only the logic of the simplest ideal cases (e.g., Fig. 3A) is examined.

Hierarchical Developmental Complexity: Parallelism

The remaining two hierarchical cases, parallel addition and parallel deletion need little discussion, because the logic is almost the same as for their respective analogs, reversal of a deletion and reversal of an addition. A hierarchically complex cascade added in one lineage is unlikely to be gained in another lineage. As for reversal of a deletion, the logic of Dollo's Law all but forbids it. Unlike reversal of a deletion, however, no easy alternative route, such as removal of developmental repressor, exists. Regarding the final hierarchical category, parallel deletions, a complex cascade lost in one lineage could easily be lost in another, because both require a simple interruption or heterochronic change at one or more early steps. The same caveats discussed above apply in these analogous cases as well.

Nonhierarchical Developmental Complexity

Figure 3C shows the nonhierarchical case in which the ancestral condition presents a number of developmentally independent (see above) structures, the darkly shaded circles. Suppose that in the course of evolution, a complex of interactions and parts are added, shown by the arrows and the lightly shaded circles.

Reversal in this case would be fairly difficult or improbable. Because the interactions (arrows) involved in the complex were added independently, it is reasonable to suppose that they would have to be lost independently. The probability of reversal would be the product of the probabilities of each independent deletion. The more interactions, i.e., the more complex the addition, the greater the improbability of complete loss.

Again, some possible exceptions need to be examined. Deletion of only one or two interactions could compromise the function of the complex. True reversal requires more than loss of function—it requires total deletion of the added developmental package, but a deletion of function could predispose the organism to further developmental deletions. Deprived of their selective advantage, the accumulation of developmental errors could eliminate the remaining steps fairly quickly.

Also, developmental integration occurring in an added complex of interactions could (partly) invalidate the above reasoning. For example, although the end members (lightly shaded circles) may have been added independently, developmental cross-links and mutual dependencies may be added in the course of evolution (Fig. 3D). Loss of independence complicates the situation. To the extent that the system becomes partly hierarchical, reversal to the ancestral condition becomes more probable. A possible example is the evolution of the vertebrate eye, which may have arisen stepwise in a series of independent additions of interactions (which in turn correspond to parts such as photoreceptors, lens, etc.) (Arnold et al. 1989). Considerable integration has taken place, however, and much of the original independence has apparently been lost (Arnold et al. 1989), which increases the ease with which the eye can be substantially reduced and ultimately lost (e.g., in cave species).

The logic for nonhierarchical reversal of additions also applies to reversal of nonhierarchical deletions. Regaining a complex of lost interactions and parts is difficult, again because of the improbability of multiple independent events. For nonhierarchical developmental reversal, the logic of Dollo's Law applies with full force. It is worth pointing out here an asymmetry between additions and deletions that may already be obvious. Deletions ought to be more probable on average than additions, for no other reason than it is easier to destroy than to create. Thus, reversal of complex nonhierarchical additions and deletions are both difficult, but reversal of an addition probably somewhat less difficult.

Finally, parallel additions are logically equivalent to reversals of deletions, and parallel deletions equivalent to reversals of additions.

Morphological Complexity

Figure 2E shows a hierarchical object, consisting of nested parts. Imagine that such a complex structure has been added to an organism in the course of evolution. Interestingly, the probability of reversal may not depend on the route by which the structure was assembled, i.e., on whether it was "built up" (with successively larger parts emerging out of interactions among the smaller), "built down," or assembled piecemeal (from parts fabricated independently of each other). The parts in a hierarchical structure are physically contained within some larger entity and thus spatially localized within the organism, which may tend to make their removal fairly simple. Thus, reversal of a hierarchical morphological addition should be relatively easy, as should parallel deletions. (Of course, to the extent that the parts of a nonhierarchical structure are also spatially localized, the same argument applies.) A leaf or a tail might be simply excised developmentally.

The logic for other homoplasies in morphological hierarchies—reversal of deletions and parallel additions—is the same as for developmental homoplasy and requires no special discussion. This is also true for nonhierarchical morphological transformations, and, again, the above caveats apply.

SUMMARY

Eight independent types of complexity can be identified, based on three distinctions (Fig. 2): differentiational versus configurational complexity; object versus process complexity (which in biology could be morphological versus developmental complexity); and hierarchical versus nonhierarchical complexity. The eight types correspond to the eight possible combinations of these: differentiational hierarchical object complexity, differentiational hierarchical process complexity, and so on. In this analysis, only the differentiational types are considered, and the focus is on development.

At issue is the ease or probability of homoplasy in complex evolutionary transformations. Homoplasy encompasses three types of transformation: reversal, parallelism, and convergence. Each type can take the form of an addition, a deletion, or a modification of a complex developmental process or morphological structure. Only parallelism and reversal of additions and deletions can be addressed in the present framework.

For each type of complexity and for each type of transformation, structural considerations suggest a logic with implications for homoplasy. In general terms, the probability of complex reversal or parallelism ought to depend on type of complexity, especially hierarchical versus nonhierarchical (Table 1). More specifically, deletions occurring in parallel or reversals of additions ought to be easier than addition in parallel or reversal of deletions, especially in hierarchical structures. Nonhierarchical reversals and parallel changes ought to be difficult, but somewhat less difficult for deletions occurring in parallel and for reversals of additions.

This analysis offers a scheme for thinking about complexity and homoplasy, one that may help clarify future discussion. The findings are not conclusions about the probability of homoplasy in actual cases. Rather, they show the logical implications of the scheme, that is, the necessary consequences of the geometric relationships among parts and interactions that are implicit in the scheme. In effect, what is provided is a standard for comparison with what is known in particular cases about evolutionary change in the complexity of development and morphology.

ACKNOWLEDGMENTS

I thank B. Hallgrimsson, D. Ritchie, and M. Sanderson for careful readings of the manuscript, J. Valentine and C. May for a preview of a very interesting paper on hierarchies (in preparation), and R. Thomas for a preview of a thought-provoking paper (from which the felicitous sea gull example was taken) on complexity (also in preparation).

REFERENCES

Alberch, P. 1985. Problems with the interpretation of developmental sequences. Systematic Zoology 34:46–58.

Allen, T. F. H., and T. B. Starr. 1982. Hierarchy: Perspectives for ecological complexity. Univ. of Chicago Press, Chicago.

Arnold, S. J., P. Alberch, V. Csányi, R. C. Dawkins, S. B. Emerson, B. Fritzsch, T. J. Horder, J. Maynard Smith, M. J. Starck, E. S. Vrba, G. P. Wagner, and D. B. Wake. 1989. How do complex organisms evolve? Pp. 403–433 in D. B. Wake and G. Roth, eds. Complex organismal functions: Integration and evolution in vertebrates. Wiley, New York.

Arthur, W. 1988. A theory of the evolution of development. Wiley, New York.

Bonner, J. T. 1988. The evolution of complexity. Princeton Univ. Press, Princeton, New Jersey.

Boyajian, G., and T. Lutz. 1992. Evolution of biological complexity and its relation to taxonomic longevity in the Ammonoidea. Geology 20:983–986.

Bull, J. J., and E. L. Charnov. 1985. On irreversible evolution. Evolution 39:1149–1155.

Castrodeza, C. 1978. Evolution, complexity, and fitness. Journal of Theoretical Biology 71:469–471.

Cisne, J. L. 1974. Evolution of the world fauna of free-living arthropods. Evolution 28:337–366.

Crutchfield, J. P. 1991. Knowledge and meaning ...chaos and complexity. Working paper 91-09-035. Santa Fe Institute, Santa Fe, New Mexico.

Crutchfield, J. P., and K. Young. 1989. Inferring statistical complexity. Physical Review Letters 63:105–108.

Eldredge, N., and S. N. Salthe. 1984. Hierarchy and evolution. Oxford Surveys in Evolutionary Biology 1:184–208.

Flessa, K. W., K. V. Powers, and J. L. Cisne. 1975. Specialization and evolutionary longevity in the Arthropoda. Paleobiology 1:71–81.

Gell-Mann, M. 1994. The quark and the jaguar. W. H. Freeman, New York.

Gould, S. J. 1970. Dollo on Dollo's Law: Irreversibility and the status of evolutionary laws. Journal of the History of Biology 3:189–212.

Gould, S. J., and B. A. Robinson. 1994. The promotion and prevention of recoiling in a maximally snaillike vermetid gastropod: A case study for the centenary of Dollo's Law. Paleobiology 20:368–390.

Haas, O., and G. G. Simpson. 1946. Analysis of some phylogenetic terms, with attempts at redefinition. Proceedings of the American Philosophical Society 90:319–347.

Hennig, W. 1966. Phylogenetic systematics. Univ. of Illinois Press, Urbana, Illinois.

Hinegardner, R., and J. Engelberg. 1983. Biological complexity. Journal of Theoretical Biology 104:7–20.

Hughes, D. J., and J. B. C. Jackson. 1990. Do constant environments promote complexity of form? The distribution of bryozoan polymorphisms as a test of hypotheses. Evolution 44:889–905.

Kampis, G., and V. Csányi. 1987. Notes on order and complexity. Journal of Theoretical Biology 124:111–121.

Katz, M. J. 1986. Templets and the explanation of complex patterns. Cambridge Univ. Press, Cambridge.

Katz, M. J. 1988. Pattern biology and the complex architectures of life. Longwood Academic, Wolfeboro, NH.

Kauffman, S. A. 1993. The origins of order. Oxford Univ. Press, New York.

Lande, R. 1978. Evolutionary mechanisms of limb loss in tetrapods. Evolution 32:73–92.

Lauder, G. V. 1981. Form and function: Structural analysis in evolutionary morphology. Paleobiology 7:430–442.

Laurent, R. F. 1983. Irreversibility: A comment on MacBeth's interpretations. Systematic Zoology 32:75.

Löfgren, L. 1977. Complexity of description of systems: A foundational study. International Journal of General Systems 3:197–214.

Mabee, P. M. 1993. Phylogenetic interpretation of ontogenetic change: Sorting out the actual and artefactual in an empirical case study of centrarchid fishes. Zoological Journal of the Linnean Society 107:175–291.

MacBeth, N. 1980. Reflections on irreversibility. Systematic Zoology 29:402–404.

McShea, D. W. 1991. Complexity and evolution: What everybody knows. Biology and Philosophy 6:303–324.

McShea, D. W. 1992. A metric for the study of evolutionary trends in the complexity of serial structures. Biological Journal of the Linnean Society 45:39–55.

McShea, D. W. 1993. Evolutionary change in the morphological complexity of the mammalian vertebral column. Evolution 47:730–740.

McShea, D. W. 1996. Metazoan complexity and evolution: Is there a trend? Evolution 50:477–492.

O'Grady, R. T. 1985. Ontogenetic sequences and the phylogenetics of parasitic flatworm life cycles. Cladistics 1:159–170.

O'Neil, R. V., D. L. DeAngelis, J. B. Waide, and T. F. H. Allen. 1986. A hierarchical concept of ecosystems. Princeton Univ. Press, Princeton.

Papentin, F. 1980. On order and complexity I: General considerations. Journal of Theoretical Biology 87:421–456.

Papentin, F. 1982. On order and complexity II: Application to chemical and biological structures. Journal of Theoretical Biology 95:225–245.

Raff, R. A. 1987. Constraint, flexibility, and phylogenetic history in the evolution of direct development in sea urchins. Developmental Biology 119:6–19.

Rensch, B. 1960. Evolution above the species level. Columbia Univ. Press, New York.

Riedl, R. 1978. Order in living organisms: A systems analysis of evolution. Wiley, New York.

Salthe, S. N. 1985. Evolving hierarchical systems. Columbia Univ. Press, New York.

Salthe, S. N. 1993. Development and evolution. Massachusetts Institute of Technology Press, Cambridge, MA.

Sanderson, M. J. 1993. Reversibility in evolution: A maximum likelihood approach to character gain/loss bias in phylogenies. Evolution 47:236–252.

Saunders, P. T., and M. W. Ho. 1976. On the increase in complexity in evolution. Journal of Theoretical Biology 63:375–384.

Slobodkin, L. B. 1992. Simplicity and complexity in games of the intellect. Harvard Univ. Press, Cambridge, MA.

Sober, E. 1988. Reconstructing the past. Massachusetts Institute of Technology Press, Cambridge, MA.

Thomas, R. D. K., and W.-E. Reif. 1993. The skeleton space: A finite set of organic designs. Evolution 47:341–360.

Valentine, J. W., A. G. Collins, and C. P. Meyer. 1993. Morphological complexity increase in metazoans. Paleobiology 20:131–142.

Vermeij, G. J. 1973. Biological versatility and earth history. Proceedings of the National Academy of Sciences, USA 70:1936–1938.

Wagner, G. P. 1995. The biological role of homologues: A building block hypothesis. N. Jb. Geol. Palaont. Abh. 195:279–288.

Wake, D. B. 1991. Homoplasy: The result of natural selection, or evidence of design limitations? American Naturalist 138:543–567.

Wake, D. B., and G. Roth. 1989. The linkage between ontogeny and phylogeny in the evolution of complex systems. Pp. 361–377 *in* D. B. Wake and G. Roth, eds. Complex organismal functions: Integration and evolution in vertebrates. Wiley, New York.

Wicken, J. S. 1987. Evolution, thermodynamics, and information. Oxford Univ. Press, New York.

Wimsatt, W. C. 1986. Developmental constraints, generative entrenchment, and the innate-acquired distinction. Pp. 185–208 *in* W. Bechtel, ed. Integrating scientific disciplines. Dordrecht, Boston.

Yagil, G. 1985. On the structural complexity of simple biosystems. Journal of Theoretical Biology 112:1–23.

EXAPTATION, ADAPTATION, AND HOMOPLASY: EVOLUTION OF ECOLOGICAL TRAITS IN DALECHAMPIA VINES

W. SCOTT ARMBRUSTER
University of Alaska
Fairbanks, Alaska

INTRODUCTION

Homoplasy comprises evolutionary convergence, reversals, and parallelisms causing multiple appearance of the same character state in a phylogeny. It is treated as error or noise in the reconstruction of phylogeny, because it obscures phylogenetic patterns. Implicit in the parsimony procedure that forms the basis of most phylogenetic analyses is the assumption that homoplasy is relatively uncommon. Yet homoplasy may sometimes be very common. As a result, systematists often select characters that minimize homoplasy in the data set; "good" characters exhibit low homoplasy, and "bad" characters exhibit high homoplasy. Hence levels of homoplasy are probably underestimated in most systematic studies.

Homoplastic characters are interesting in their own right, however. They may constitute the majority of evolutionary change during the course of evolution and diversification of lineages. The ubiquity of homoplasy and its

HOMOPLASY: The Recurrence of Similarity in Evolution, M. J. Sanderson and L. Hufford, eds.

role in diversification indicate that homoplasy and the evolutionary processes that generate it are important and deserve direct study (Sanderson 1991; Wake 1991).

A fundamental question that emerges from consideration of homoplasy is, "what biological factors influence the level of homoplasy of different traits"? Putting aside homoplasy caused by environmental effects, there remain numerous possible causes of homoplastic evolution of genetically determined traits (Wake 1991). These include the role of chance events (e.g., drift in more or less selectively neutral traits), genetic and developmental systems (e.g., regulatory genes and developmental patterns; see Wake 1991), natural selection (e.g., similar selective pressures leading to parallelism or convergence), and the interaction of chance and selection (e.g., exaptation [*sensu* Gould and Vrba 1982; = preadaptation]). What is the relative importance of these different sources of homoplasy? What do they tell us about the general course of evolution? The study of homoplasy can help us understand the influence of internal "constaints" on phenotype production versus external selective pressures acting on those phenotypes to create the patterns of diversity we see today (see Mayr 1960).

The goal of this contribution is to examine homoplastic evolution from an ecological perspective. In particular, I will consider the evolution of plant–animal relationships. An initial question is whether plant–animal interactions are evolutionarily labile or conservative, and if labile, how important is homoplasy? The complexity of the interactions between plants and animals might suggest that they would be evolutionarily conservative. Recent analyses of three different plant–pollinator systems (*Dalechampia* [Euphorbiaceae], *Lapeirousia* [Iridaceae], *Disa* [Orchidaceae]), however, indicate that plant–animal interactions may often be evolutionarily labile and exhibit considerable homoplasy (Armbruster 1993; Goldblatt et al. 1995; Johnson 1995). The first system forms the basis of the present contribution. I attempt to address the question: what biological properties and processes contribute to high homoplasy of ecological characters?

STUDY SYSTEM

Most of the discussion below is drawn from my studies of the interactions between insects and members of the plant genus *Dalechampia* (Euphorbiaceae). In this section I briefly review the systematics and natural history of the genus.

Dalechampia comprises about 120 extant species and occurs throughout the lowland tropics except in Australia and Oceania. The greatest diversity occurs in the neotropics, but Madagascar forms a secondary center of diversity.

Most species of *Dalechampia* are twining vines or lianas, although a few species are shrubs. The flowers are unisexual, but usually 4–13 staminate flowers and three pistillate flowers are united into a functionally bisexual blossom (pseudanthium). The clusters of staminate and pistillate flowers are subtended by two, usually large, showy bracts. In most species a large cluster of secretory bractlets is associated with the staminate flowers (Webster and Webster 1972). This "gland" secretes triterpenoid resin or monoterpene fragrances as rewards to resin-collecting, female bees (*Hypanthidium* [Megachilidae], Euglossini [Apidae]) or fragrance-collecting, male, euglossine bees, respectively. In the process of collecting the reward the bees contact the staminate and pistillate flowers and effect pollination (Armbruster 1984, 1988a; Armbruster et al. 1989). A few species appear to offer only pollen as a reward for pollen-collecting bees or pollen-feeding beetles (Armbruster et al., 1993).

Dalechampia vines are attacked by a variety of generalist and specialist herbivorous insects (and also appear to be popular with elephants in Africa). The flowers are fed on by generalists such as tettigoniids (Orthoptera) and meloids (Coleoptera) and the specialists *Dynamine* and *Neptidopsis* (Lepidoptera: Nymphalidae). The leaves are eaten by a few generalists such as tettigoniids and acridids (Orthoptera) and by several specialists, including *Syphraea* (Coleoptera: Chrysomelidae), *Hamadryas*, *Ectima*, and *Byblia* (Lepidoptera: Nymphalidae; Armbruster 1994).

MATERIALS AND METHODS

Observations on pollinators and herbivores of ca. 60 *Dalechampia* species were made at field sites in Mexico, Belize, Costa Rica, Panama, Venezuela, Ecuador, Peru, Brazil, Tanzania, South Africa, and Madagascar. Dates of study and exact field location can be found in Armbruster (1988b), Armbruster and Mziray (1987), and Armbruster et al. (1993). Floral and leaf secretions were analyzed using thin-layer chromotography, flash chromotography, HPLC, integrated gas chromotography–mass spectrometry, and NMR (see Armbruster et al. 1989, 1995 for details).

Morphological studies were conducted on plant material collected in the field and on herbarium material. These data were analyzed cladistically using PAUP 3.1.1 (Swofford 1993) on a subset of taxa (all ecotypes and sibling species were dropped from the cladistic analysis and added to the tree later) to estimate the phylogenetic relationships among species (see Armbruster (1993) for details). Ecological traits were then mapped onto the trees under the parsimony assumption (assumption of minimal evolutionary change) using MacClade 3.0 (Maddison and Maddison 1992) to reconstruct the historical sequence of evolutionary change in the ecological traits (see Armbruster

(1993) for details). The occurrence of ecological homoplasy was identified by examining this historical hypothesis. The biological basis of the homoplasy was inferred from the historical reconstruction or from independent ecological and biochemical data.

EVOLUTION OF PLANT–POLLINATOR INTERACTIONS

Figure 1 depicts the inferred historical course of evolution of plant–pollinator relationships in New World species of *Dalechampia*. The general conclusions from this result are that the ancestral *Dalechampia* (at least in the New World) were pollinated by fragrance-collecting male euglossine bees or pollen-collecting female (or worker) bees (this part of the reconstruction is not strongly supported; see Armbruster 1993). Pollination by resin-collecting

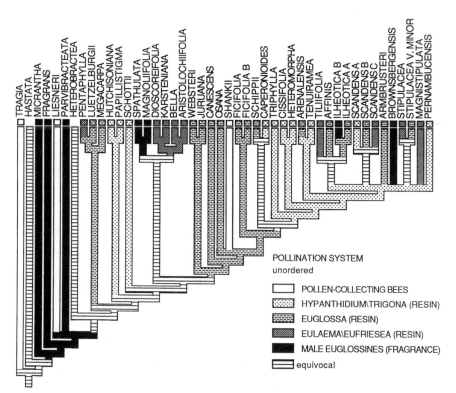

POLLINATION SYSTEM
unordered

☐ POLLEN-COLLECTING BEES
▨ HYPANTHIDIUM\TRIGONA (RESIN)
▨ EUGLOSSA (RESIN)
▨ EULAEMA\EUFRIESEA (RESIN)
■ MALE EUGLOSSINES (FRAGRANCE)
▢ equivocal

FIGURE 1   Strict consensus tree (from 321 maximally parsimonious trees) showing the best supported phylogenetic relationships among New World *Dalechampia* species. Mapped onto this cladogram is the pollination ecology of the terminal (extant) taxa; pollination ecology of ancestors (nodes) is inferred by parsimony.

bees originated only once (at least in the New World) fairly early in the evolution of the genus; most extant species are members of the initial resin-reward lineage. Pollination by fragrance-collecting male euglossine bees shows a surprising degree of parallelism (or reversal): only seven species have been shown to be pollinated by male euglossines, yet they represent three to four independent origins of the pollination system. Pollination by pollen-collecting bees has apparently originated twice, either representing a reversal or a parallelism. There have been repeated reversals and parallelisms in the evolution of pollination by small (*Hypanthidium*) and large (*Eulaema*) resin-collecting, female bees.

A similar analysis of Old World species (with a sample of New World species) is depicted in Fig. 2. From this result we can infer that pollination by pollen-feeding insects has evolved independently a third time in the lineage containing all the Malagasy species.

EVOLUTION OF DEFENSE SYSTEMS

Mapping antiherbivore defense systems onto the same phylogenetic tree shows a pattern of escalating defense through the evolution of multiple defensive systems. The first defense system to evolve appears to have been defense of staminate flowers from florivores by secretion of resin from subtending staminate bractlets (Fig. 3). The defensive function of these secretions has been demonstrated experimentally; they mechanically inhibit feeding by herbivorous insects (Armbruster et al. 1995). The next line of floral defense to evolve was the nocturnal closure of involucral bracts. This may have evolved independently in two lineages (Fig. 4). The defensive significance of nocturnal closure has also been shown experimentally (Armbruster and Mziray 1987). Another line of defense that evolved was deployment of the same triterpene resins by the sepals of the pistillate flowers as is deployed by staminate bractlets. This apparently evolved twice also (Fig. 5). A final line of defense is protection of developing fruits by the involucral bracts becoming persistent and closing around the pistillate flowers as the fruits develop. This has also evolved independently in two lineages (Fig. 6).

EVOLUTION OF NOVELTY AND PARALLELISM IS ASSOCIATED WITH EXAPTATIONS

If the reconstruction of the evolutionary history of interactions between *Dalechampia* and insects is correct, it permits identification of the conditions associated with parallel evolution of ecological traits. In most cases,

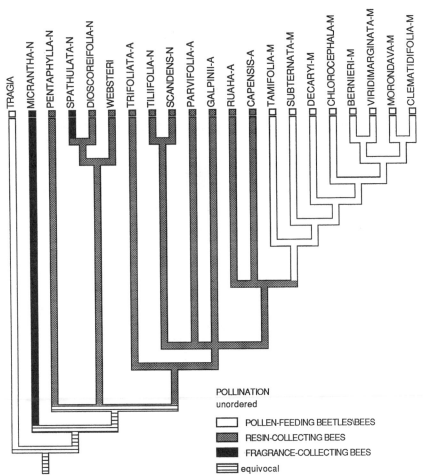

FIGURE 2   Strict consensus tree (from 40 maximally parsimonious trees) showing the best supported phylogenetic relationships among Old World and a sample of New World *Dalecham-pia* species. Mapped onto this cladogram is the pollination ecology of the terminal (extant) taxa; pollination ecology of ancesters (nodes) is inferred by parsimony. Abbreviations after specific epithets indicate geographic distribution of species: N = neotropical; A = Africa; M = Madagascar.

parallelisms are associated with preaptations (preadaptations) already being in place.

Origin of the Resin Reward

Pollination by resin-collecting bees is very rare in flowering plants. It has originated at least twice independently: once in *Dalechampia* and once in

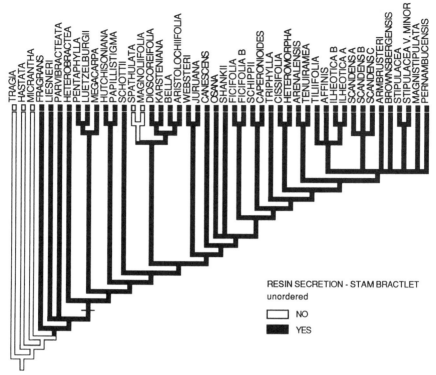

FIGURE 3   Strict consensus tree (modified from Fig. 1) with resin secretion from bractlets in staminate cymule mapped onto the terminal (extant) taxa. The condition of ancestors is inferred by parsimony. The horizontal slash mark indicates the inferred origin of pollination by resin-collecting bees.

Clusia (Clusiaceae). This broad-scale parallelism is associated with the preaptive secretion of resin (or latex) for defense functions in *Dalechampia* and probably also *Clusia*. The ancestral members of the lineage that gave rise to the first resin-reward species appear to have defended their staminate flowers with resin (Fig 3; Armbruster et al. 1995). This resin was secreted by the bractlets that later became the resin gland. Hence the origin of pollination by resin-collecting bees probably occurred by the incidental recruitment of bees to blossoms to collect the defense resins and the "accidental" pollination of the flowers. This generated selective pressures leading to the increase in resin production, amalgamation of the bractlets into a gland-like structure, and other modifications of the blossom morphology. Thus the origin of pollination by resin-collecting bees was predicated on a preapation: resin secretion for defense of flowers. A similar situation appears to hold in *Clusia* (Armbruster and Loquvam unpublished data).

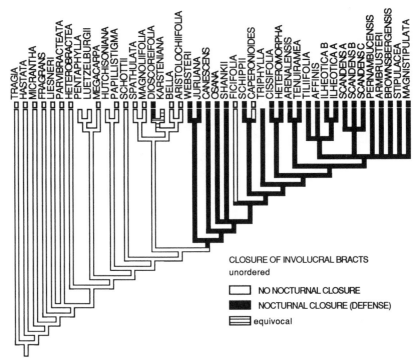

FIGURE 4   Strict consensus tree (modified from Fig. 1) with nocturnal closure of enveloping involucral bracts mapped onto the terminal (extant) taxa. The condition of ancestors is inferred by parsimony.

Origins of Fragrance Rewards

The most striking parallelism (or reversals, if male-euglossine pollination is ancestral) in pollination ecology of *Dalechampia* is the repeated evolution of pollination by fragrance-collecting male euglossine bees. The historical analysis shows separate origins in three to four lineages, yet there are only some seven species involved. How can such extensive homplasy be occurring? With historical analysis we can infer some of the conditions that prevailed at each independent origin.

The shift from pollination by resin (or pollen)-collecting bees to pollination by fragrance-collecting bees in all cases appears to have occurred by exaptation. Two different preaptations have been identified. The first is secretion of advertisement fragrances by the stigmatic surface of the pistillate flowers. The blossoms of most species of *Dalechampia* emit a sweet, flowery fragrance, not as a reward, but as a means of advertising and providing an associative learning cue for the resin- or pollen-collecting pollinators. In most cases it is the stigmas that secrete the major fragrance components. This was

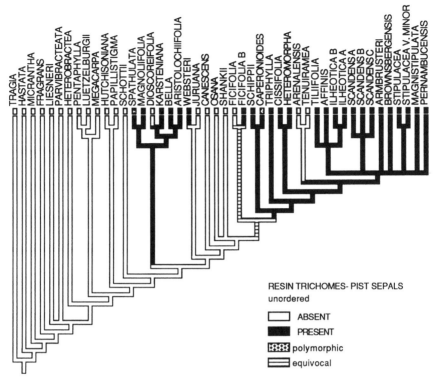

RESIN TRICHOMES- PIST SEPALS
unordered

☐ ABSENT
■ PRESENT
▨ polymorphic
▤ equivocal

FIGURE 5 Strict consensus tree (modified from Fig. 1) with resin secretion by the pistillate sepals mapped onto the terminal (extant) taxa. The condition of ancestors is inferred by parsimony.

apparently true of the ancestor of the lineage leading to *D. micrantha* and *D. fragrans* (which are male-euglossine pollinated), the ancestor of the lineage leading to *D. brownsbergensis*, and the apparently separate ancestor of *D. ilheotica* (the latter two species also being male-euglossine pollinated) (see Fig. 1). In all three lineages, the origin of male-euglossine pollination appears to have depended on fragrances already being secreted for other reasons.

The second preaptation leading to male-euglossine pollination is the secretion of triterpene resin rewards. The ancestor of the lineage containing *D. spathulata* and *D. magnoliifolia* (both male-euglossine pollinated) almost certainly secreted a resin reward. The origin of male-euglossine pollination appears to have occurred by the loss of an enzyme in the triterpene (resin) biosynthetic pathway, causing the synthesis of an intermediate product, monoterpenes, instead of resin, from the identical gland structure. Monoterpenes are volatile, fragrant compounds, and many are attractive to male euglossines. This mutation affecting the triterpene synthesis pathway

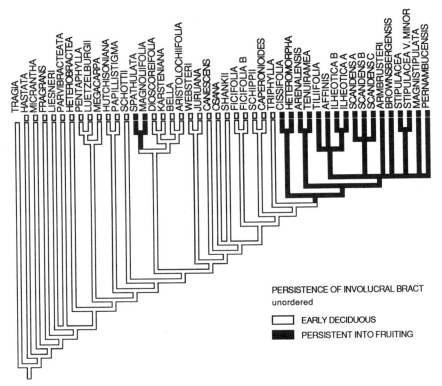

FIGURE 6   Strict consensus tree (modified from Fig. 1) with persistence of involucral bracts into the fruiting period mapped onto the terminal (extant) taxa. The condition of ancestors is inferred by parsimony.

probably led to an immediate shift from pollination by resin-collecting bees to pollination by fragrance-collecting bees. Hence the origin of male-euglossine pollination in this lineage was also predicated on a preaptation: synthesis of triterpene resins.

Origins of Pollen Rewards

Pollination by pollen-collecting bees and/or pollen-feeding beetles has evolved at least twice and perhaps three times in *Dalechampia*. The Malagasy species are the most diverse and highly adapted of the pollen-reward species; they form one monophyletic lineage apparently derived from resin-reward ancestors from Africa (Armbuster et al. 1993; Armbruster and Baldwin unpublished data). In the neotropics there are two pollen-reward species; each represents an independent origin of the pollen-reward system.

If the pollen-reward system in *Dalechampia* is ancestral, then the origins of pollination by pollen-collecting bees in Madagascar and one or both of the neotropical lineages represent reversals. If pollination by male euglossines is ancestral, they represent parallelism. In either case, the homoplasy would appear to have been promoted by the fact that the blossoms of all species produce pollen. Hence pollen production for fertilization purposes represents a preaptation permitting the origins of pollination by pollen-collecting bees. *Dalechampia* is, of course, not unique in this respect. This preaptation is found in all plants with perfect flowers or pseudanthia.

It is somewhat surprising that the switch to pollination by pollen-collecting bees has not occurred more often in *Dalechampia* (and other groups), considering the ubiquity of the preaptation. Perhaps the fitness cost of paying for service with pollen (which contain the male gametes) is relatively high. Clearly the switch to pollination by pollen-feeding insects is influenced by other factors than just pollen being present. Nevertheless, switches to pollination by pollen-feeding insects would be rarer if pollen were not produced by most flowers and pollen were not eaten by many animals.

Origins of Defense Systems

The first line of floral defense to evolve appears to have been secretion of resin by the bractlets that surround the staminate flowers. Subsequently, additional protection to the flowers (staminate and pistillate) was provided by the evolution of nocturnal closure of the involucral bracts. This may have evolved separately in two lineages (although the cost in steps to get one origin is not great and closure may have originated only once). In either case, there was again a preaptation in place: the involucral bracts had already evolved large size (large enough to envelop the flowers if closed) apparently to attract the attention of pollinators (an advertisement system). Thus the evolution of the plant's defense system against florivores was influenced by an adaptation for pollination.

A third line of floral defense has evolved apparently to defend the pistillate flowers: the same kind of triterpene resins as are deployed by the staminate bractlets are secreted at the tips of lobes of the pistillate sepals. This occurs in many of the more advanced species and appears to have evolved twice or possibly three times, representing another parallelism. A preaptation appears to be a likely cause of the parallelism. The existence of the genes for the biosynthetic machinery for resin synthesis (initially expressed only in the tissues of the staminate cymule) surely facilitated the origin of this new defense system for pistillate flowers.

The fruits of many of the advanced species in the genus appear to be defended by another system. The involucral bracts persist after anthesis and close around the developing fruits. This has evolved independently in two lineages. This parallelism again appears to be a consequence of exaptation:

the large enveloping bracts were already present and serving both advertisement and defense functions for the receptive flowers. Only minor modifications were needed for the bracts to assume a role of protecting the developing fruits and seeds.

NATURAL SELECTION
GENERATES HOMOPLASY

Natural selection appears to have generated homoplasy many times in *Dalechampia*. The role of natural selection is implicit in the previous discussion of exaptation. I wish to discuss two additional examples. One involves changes in pollination by various species of resin-collecting bees, and the other involves the shift to pollination by pollen-feeding insects and the colonization of Madagascar.

Reversals and Parallelisms in Evolution of
Resin-Reward Species

The historical analysis (Fig. 1) indicates that shifts from pollination by small megachilid bees to pollination by large euglossine bees and vice versa have occurred several times. There is also evidence for one to several reversals: adaptation for pollination by larger bees, shifting to pollination by smaller bees, and back to pollination by larger bees. These shifts all appear to be related to natural selection acting on the amount of resin offered as the reward, the size of the flowers, and geometry of the blossom (Armbruster 1990, 1993). Natural selection for shifts in pollinators is often generated by "competitive" interactions with sympatric congeneric species. The result is a character displacement pattern that can be manifested relatively rapidly, even within species (Armbruster 1985, 1986).

Rapidity of evolution and frequency of reversal and parallelism seems to be related to the strength of selection and the responsiveness of the genetic system to selection. In this case selection is apparently quite strong because pollen grains being lost to alien stigmas are genetic copies not making it into the next generation. Also, some reduction of seed set may result from interspecific pollination (Armbruster and Herzig 1984). The characters that control pollination by resin-collecting bees include: (1) the amount of resin secreted (which is usually related to the size of the resin gland and determines the size of the largest floral visitors), (2) the separation between the gland and the anthers (which determines the size of the smallest visitors to pick up pollen on their bodies), and (3) the separation between the gland and the stigmas (which determines the size of the smallest pollen-bearing visitors to deposit pollen on stigmas; Armbruster (1988a, 1990)). These three characters appear to have multigenic (quantitative) inheritance and exhibit considerable genetic variation within species and probably populations (Armbruster 1985, 1991). Such natural genetic variation in these ecologically important characters permits quick response to the strong selective pressures generated by the local biotic environments and hence leads to repeated homoplasy.

Evolutionary Reversals Associated with Shifts to Pollen Rewards

A notable evolutionary reversal affecting a large number of floral traits can be observed in the shift from pollination by resin-collecting bees to pollination by pollen-collecting bees. One such shift is associated with the colonization of Madagascar from Africa and the other has occurred in the neotropics.

Cladistic analyses of morphological characters of Old and New World species (Armbruster et al. 1993) in combination with preliminary ITS sequence data (Baldwin and Armbruster unpublished data) indicate that *Dalechampia* colonized Madagascar from Africa, and that the colonist(s) offered resin rewards to attract pollinators. Madagascar is particularly depauperate in resin-collecting bees, however (C. D. Michener personal communication; see Armbruster et al. 1993). This appears to have generated a strong selective pressure against pollination by resin-collecting bees and for pollination by insects that are attracted by pollen, including pollen-feeding beetles and pollen-collecting bees. Selective "fine-tuning" of the pollen-reward blossoms apparently included selection for increased pollen production and perhaps decreased resin production. The response to this selective pressure appears to have involved regulatory genes. Instead of simply increasing pollen or stamen number per flower, the Malagasy species have greatly increased the number of staminate flowers over their African relatives. This appears to have been accomplished by evolutionary reversal to a primitive inflorescence arrangement: five fertile pleiochasial arms, each subtended by several resin-iferous bractlets and a free involucellar bract, and no resin gland. This condition is seen in the most primitive South American species. It is not as improbable as it at first seems, however, because the resin gland is derived from the sterile fourth, or fourth and fifth arms of the staminate pleiochasium. Hence the coding for the primitive arrangement of flowers, bracts, and bractlets may have been preserved in the genome as suppressed information, and this chunk of DNA may have been reactivated when a mutation altered a regulatory gene. Producing male flowers on the sterile arms (in place of a resin gland that was no longer needed) was apparently the "easiest" way to substantially increase pollen production, and this condition became fixed early in the evolution of the Malagasy lineage.

As improbable as this series of reversals in character evolution may seem, there is good evidence that the same thing also occurred in one New World lineage. *Dalechampia shankii* is pollinated by pollen collecting bees, yet cladistic analysis indicates that it is recently derived from a resin-reward ancestor (Fig. 1). Again, selection for increased pollen production appears to have caused the number of staminate flowers to increase by reactivation of suppressed genetic information. The effect is a 33% increase in flower (and pollen) number by production of staminate flowers by the previously sterile, fourth arm in the staminate pleiochasium.

Concomitant with the reversals increasing staminate–flower number in both the Malagasy and neotropical lineages is the reversal to the primitive state of two other characters. The resiniferous bractlets have a dispersed arrangement and the resin gland is absent. However, in neither case has resin synthesis and secretion been completely lost.

CONCLUSIONS

It is clear from the historical analysis of *Dalechampia* evolution that homoplasy is common in characters with ecological significance. The generality of this conclusion is suggested by similar results in two independent studies of two monocot genera (Goldblatt et al. 1995; Johnson 1995). Three reasons seem to be responsible for this: (1) Strong selective pressures can easily reverse the direction of evolution of traits exhibiting additive genetic variation (especially quantitative traits). (2) Many character reversals may occur through changes in regulatory genes that either suppress previously expressed genetic information or control development. Selection favoring a shift toward a preexisting ("primitive") character state (e.g., larger number of flowers) may favor simply turning off the regulatory gene suppressing genetic information. Alternatively, a shift to a preexisting character state may be effected by suppression of later developmental stages (paedomorphy). Both will cause sudden, simultaneous reversals in several to many characters. (3) Parallelisms may occur commonly in lineages in which preaptations are already in place. For example, in *Dalechampia*, shifts to pollination by fragrance-collecting bees have occurred several times because fragrances were already being secreted by flowers for advertisement "purposes," or as precursors for other compounds (i.e., change through "reductive homoplasy"). Developmental patterns that permit or promote paedomorphic changes can also be interpreted as exaptations. Thus it appears that natural selection interacts with the "internal" factors that influence the generation of novel phenotypes, and they act together to generate "biases" in the pattern of diversification.

Another conclusion emerges from this study. It appears that characters of ecological importance are often of little systematic utility. In *Dalechampia*, features that strongly influence the pollination ecology of a population (e.g., quantitative traits such as gland area, gland–stigma distance, and gland–

anther distance; see Armbruster (1988a, 1990)) are evolutionarily highly labile (and homoplastic), and they often vary within species as much as among. This variation apparently reflects adaptation to the local biotic environment by different populations (Armbruster 1985). As a result, pollination ecology and quantitative morphological characters that are related to pollination ecology are not very useful for either analytical or synthetic systematic puposes. An exception is qualitative differences in pollinator rewards (e.g., fragrances vs resins) which are expressed primarily at the species level and can sometimes be systematically informative. By and large, however, the best systematic characters in *Dalechampia* have no *obvious* ecological significance, and some may be selectively neutral. These include traits like number and fusion of bractlets in staminate and pistillate subinflorescences, number of pistillate sepals, and size and shape of stipules (see Webster and Armbruster 1991). Interestingly, Davis (1988) has found a similar pattern in *Puccinellia* (Poaceae), and Davis and Gilmartin (1985) report the pattern from several plant groups: ecologically important characters show high intraspecific variation (among populations) but little consistent interspecific variation. In contrast characters that are useful for distinguishing species have no obvious ecological significance.

Although homoplasy is problematic in phylogeny estimation, it is obviously an important part of the evolutionary process. It remains particularly important to identify the processes that cause homoplasy and to analyze them in conjunction with determining the roles of genetic and developmental constraint, natural selection, and chance in evolution (Wake 1991). Future research that integrates phylogenetic studies with information on the adaptative significance, genetics, and development of character variation will lead to major new advances in our understanding of how organisms evolve and lineages diverge and diversify.

ACKNOWLEDGMENTS

I thank T. P. Clausen, M. E. Edwards, E. Debevec, J. Howard, J. Loquvam, and M. Matsuki for help in the field and the lab, M. E. Edwards, L. Hufford, and M. Sanderson for comments on the manuscript, and the National Science Foundation (BSR-9006607, BSR-9020265, and DEB-9318640) and National Geographic Society for financial support.

REFERENCES

Armbruster, W. S. 1984. The role of resin in angiosperm pollination: Ecological and chemical considerations. American Journal of Botany 71:1149–1160.
Armbruster, W. S. 1985. Patterns of character divergence and the evolution of reproductive ecotypes of *Dalechampia scandens*. Evolution 39:733–752.

Armbruster, W. S. 1986. Reproductive interactions between sympatric *Dalechampia* species: Are natural assemblages "random" or organized? Ecology 67:522–533.

Armbruster, W. S. 1988a. Multilevel comparative analysis of morphology, function, and evolution of *Dalechampia* blossoms. Ecology 69:1746–1761.

Armbruster, W. S. 1988b. Principal pollinators of *Dalechampia*, with locations of study sites and dates of study. Supplementary Publications of the Ecological Society of America 8802: 1–9.

Armbruster, W. S. 1990. Estimating and testing the shapes of adaptive surfaces: the morphology and pollination of *Dalechampia* blossoms. American Naturalist 135:14–31.

Armbruster, W. S. 1991. Multilevel analyses of morphometric data from natural plant populations: Insights into ontogenetic, genetic, and selective correlations in *Dalechampia scandens*. Evolution 45:1229–1244.

Armbruster, W. S. 1993. Evolution of plant pollination systems: Hypotheses and tests with the neotropical vine *Dalechampia*. Evolution 47:1480–1505.

Armbruster, W. S. 1994. Early evolution of *Dalechampia* (Euphorbiaceae): Insights from phylogeny, biogeography, and comparative ecology. Annals of the Missouri Botanical Garden 81:302–306.

Armbruster, W. S., M. E. Edwards, J. F. Hines, R. L. A. Mahunnah, and P. Munyenyembe. 1993. Evolution and pollination of Madagascan and African *Dalechampia* (Euphorbiaceae). National Geographic Research and Exploration 9:430–444.

Armbruster, W. S. and A. L. Herzig. 1984. Partitioning and sharing of pollinators by four sympatric species of *Dalechampia* (Euphorbiaceae) in Panama. Annals of the Missouri Botanical Garden 71:1–16.

Armbruster, W. S., J. Howard, T. Clausen, E. Debevec, J. Loquvam, and M. Matsuki. 1996. Do biochemical exaptations link evolution of defense and pollination systems?—Historical hypotheses and experimental tests with *Dalechampia* vines. American Naturalist (in press).

Armbruster, W. S., C. S. Keller, M. Matsuki, and T. P. Clausen. 1989. Pollination of *Dalechampia magnoliifolia* (Euphorbiaceae) by male euglossine bees (Apidae: Euglossini). American Journal of Botany 76:1279–1285.

Armbruster, W. S., and W. Mziray. 1987. Pollination and herbivore ecology of an African *Dalechampia* (Euphorbiaceae): Comparisons with New World species. Biotropica 19:64–73.

Davis, J. I. 1988. Genetic and environmental contributions to multivariate morphological pattern in *Puccinellia* (Poaceae). Canadian Journal of Botany 66:2436–2444.

Davis, J. I., and A. J. Gilmartin. 1985. Morphological variation and speciation. Systematic Botany 10:417–425.

Goldblatt, P., J. C. Manning, and P. Bernhadt. 1995. Pollination biology of *Lapeirousia* subgenus *Lapeirousia* (Iridaceae) in Southern Africa; floral divergence and adaptation for long-tongued fly pollination. Annals of the Missouri Botanical Garden 82:517–534.

Gould, S. J., and E. S. Vrba. 1982. Exaptation—a missing term in the science of form. Paleobiology 8:4–15.

Johnson, S. D. 1995. Pollination and the evolution of floral traits: Selected studies in the Cape flora. Ph.D. dissertation, University of Cape Town, Cape Town, South Africa..

Maddison, W. P., and D. R. Maddison. 1992. MacClade 3.0. Sinauer, Sunderland, Massachusetts.

Mayr, E. 1960. The emergence of evolutionary novelties. Pp. 349–380 *in* S. Tax, ed. The evolution of life. Univ. of Chicago Press, Chicago.

Sanderson, M. J. 1991. In search of homoplastic tendencies: Statistical inference of topological patterns in homoplasy. Evolution 45:351–358.

Swofford, D. L. 1993. PAUP: Phylogenetic analysis using parsimony, vers. 3.1.1. Illinois Natural History Survey, Champaign.

Wake, D. B. 1991. Homoplasy: The result of natural selection, or evidence of design limitations? American Naturalist 138:543–567.

Webster, G. L., and W. S. Armbruster. 1991. A synopsis of the neotropical species of *Dalechampia* (Euphorbiaceae). Botanical Journal of the Linnean Society 105:137–177.

Webster, G. L., and B. D. Webster. 1972. The morphology and relationships of *Dalechampia scandens* (Euphorbiaceae). American Journal of Botany 59:573–586.

PATTERNS OF HOMOPLASY
IN BEHAVIORAL EVOLUTION

SUSAN A. FOSTER,* WILLIAM A. CRESKO,*
KEVIN P. JOHNSON,† MICHAEL U. TLUSTY,* AND
HARLEIGH E. WILLMOTT‡

*Clark University
Worcester, Massachusetts

†University of Minnesota
St. Paul, Minnesota

‡University of Arkansas
Fayetteville, Arkansas

INTRODUCTION

During the first six decades of this century, ethologists focused attention on the study of pattern and process in the evolution of behavioral phenotypes. Drawing on the seminal works of C. O. Whitman (1899) and O. Heinroth (1910, 1911), these scientists examined both present adaptive value (Tinbergen 1964; Bateson and Hinde 1976) and change in behavior patterns over time (Lorenz 1950; Hinde and Tinbergen 1958). Research examining historical aspects of the evolution of behavior could be subdivided into the use of behavioral phenotypes to reconstruct phylogenies and the overlaying of behavioral phenotypes on phylogenies that had been constructed using other kinds of characters (usually morphological). Both approaches provided surprising insights into the ways in which behavioral characters evolve, and the phylogenies based upon behavioral characters often matched those constructed using morphological characters, suggesting that behavioral phenotypes evolved in much the same way as did other kinds of characters (Lorenz

1950; Hinde and Tinbergen 1958; Mayr 1958; Bateson and Hinde 1976; McLennan et al. 1988).

Optimism about the usefulness of behavioral characters as sources of evolutionary information peaked in the late 1950s at which time Ernst Mayr (1958:545) wrote: "If there is a conflict between the evidence provided by morphological characters and that of behavior the taxonomist is increasingly inclined to give greater weight to the ethological evidence." Unfortunately, this pronouncement immediately preceded a rapid decline in the perceived value of behavioral characters for evolutionary studies in which an historical perspective was taken (for reviews: McLennan et al. 1988; Burghardt and Gittleman 1990). Primary among the concerns cited by those who felt that behavioral phenotypes were unsuitable for evolutionary study was the perceived lability of the characters. If behavioral phenotypes are more evolutionarily labile than other kinds of characters, they might be subject to higher levels of homoplasy or might change too rapidly to reflect historical relationships among taxa (see for discussion, Wenzel 1992; de Queiroz and Wimberger 1993; Proctor, this volume). Recent research on a wide array of taxa and characters suggests, however, that behavioral homoplasy is no more common than is morphological homoplasy (McLennan et al. 1988; Prum 1990; de Queiroz and Wimberger 1993). Thus, behavioral data should be of value in cladistic analyses, and we should be able to reconstruct historical patterns of behavioral evolution.

Despite recent analyses indicating that levels of homoplasy are similar in behavioral, morphological, and even biochemical character sets (Sanderson and Donoghue 1989; de Queiroz and Wimberger 1993), levels of homoplasy may differ at other biologically meaningful levels. All of the studies that have been used to infer levels of homoplasy in behavioral and morphological characters have compared species or higher order taxa (McLennan et al. 1988; Prum 1990; de Queiroz and Wimberger 1993). Studies involving population comparisons, although fewer, and often incorporating less robust phylogenetic hypotheses, suggest a different view of behavior evolution. Adaptive change and homoplasy of behavioral phenotypes may be very common patterns in population divergence (Arnold 1981, 1992; Riechert 1986, in press; Foster 1994a,b, 1995b; Radtke and Singer 1995). If this perception is correct, homoplastic characters must often evolve early in the evolutionary history of a taxon. However, as the lineage evolves through higher levels, the rate of acquisition of homoplastic characters must decrease and existing homoplastic characters themselves be lost or obscured.

Here we ask whether population comparisons reveal higher levels of homoplasy than do higher order comparisons. We first examine patterns of evolutionary diversification of behavior among populations of threespine stickleback, *Gasterosteus aculeatus*, as they invaded freshwater following the recession of the last glaciers. Because marine and/or anadromous threespine stickleback repeatedly and independently colonized isolated freshwater habitats, a large number of populations can be compared to determine the frequency with which similar, derived behavioral phenotypes have evolved independently. Character polarity can be inferred by comparison with the relatively homogeneous and ubiquitous marine/anadromous fish (Bell and Foster 1994b). Our examination of homoplasy revealed through population comparison will be limited to this data set because, as far as we can discern, it is the only one in which a large number of populations have been characterized behaviorally and in which character polarity can be inferred with a high degree of certainty.

We will then contrast the levels of homoplasy observed across freshwater stickleback populations with those observed in comparisons of higher order taxa using some of the data sets analyzed by de Queiroz and Wimberger (1993), two involving Gourami fish (Miller and Robinson 1974; Miller and Jearld 1983) and a partially unpublished data set for dabbling ducks (McKinney 1975; Johnson unpublished). The comparison will be taken further by asking whether most of the homoplasy seen in behavioral characters at the two levels is a consequence of evolutionary loss or whether iterated evolution of novel motor patterns (or their products; see discussion in Proctor, this volume) is a common cause of homoplasy at either level of comparison.

Finally, we will ask whether display behavior shows a different level of homoplasy than does nondisplay behavior in population or higher order comparisons. Differences might be expected because displays often incorporate ritualized elements of behavior patterns that appear to have evolved in other contexts and to have subsequently been coopted for use in the display (Lorenz 1941; Daanje 1950; Morris 1956; Hinde and Tinbergen 1958). This mode of evolution is much less evident in nondisplay behavior patterns (Foster 1995a). The reason for the difference may be that conspicuousness often enhances the effect of a display, and motor patterns performed in unusual contexts are often conspicuous. Thus, there may be many behavior patterns that could be incorporated into displays effectively, yielding many potential display modifications. In contrast, conspicuousness is often disadvantageous in nondisplay behavior, and the modifications of nondisplay motor patterns or their products that can enhance function are probably limited by biomechanical constraints or the need for crypticity (Foster 1995a). Consequently, homoplasy may be more common in nondisplay than in display behavior.

The comparisons we perform in this study are designed to provide insight into the patterns of evolutionary change at two levels of biological organization and in two major classes of behavior patterns. The comparisons must be considered preliminary because of the small number of taxa involved and limitations of the data sets themselves and of the methods used to contrast them. Nevertheless, this is the first effort of which we are aware to explore these issues, and we hope this chapter will stimulate research on the significance of variation in behavioral homoplasy, its prevalence, and its evolutionary causes.

BEHAVIORAL HOMOPLASY AT THE SUBSPECIFIC LEVEL: POPULATION DIVERSIFICATION

Darwin employed comparison extensively in his studies of the evolution of instincts and clearly recognized that behavioral phenotypes were often the products of both phylogeny and natural selection (1859). He, and the ethologists who followed him, focused on comparisons among species and higher taxonomic units in their studies of behavioral evolution (Hinde and Tinbergen 1958; Burghardt and Gittleman 1990; Brooks and McLennan 1991; Harvey and Pagel 1991). Quantitative and population geneticists came to focus on variation within species to provide insight into evolutionary pattern and process, although they rarely studied behavioral characters (Endler 1977, 1986; Foster and Cameron, 1996). Only recently have comparisons among populations been used to explore behavioral evolution. These studies have demonstrated that in a diverse range of taxa geographic (population) differences in behavior are often heritable and that comparisons among populations can provide novel insights into the causes and patterns of behavioral evolution (reviews in Boake 1994; Pomiankowski and Sheridan 1994; Foster and Endler in press).

Perhaps the most compelling insight that has come from population comparisons of behavioral phenotypes is that the behavioral differences appear to be adaptive in a remarkably large number of instances (e.g., Arnold 1981, 1992; Riechert 1986 in press; Foster 1994a,b, 1995b). When gene flow occurs at a high rate among geographic locations or populations, adaptive pattern can become less apparent (Thompson 1990; Riechert 1993), as it occasionally does even when gene flow is restricted (e.g., Goldschmidt et al. 1992). Although these inferences are based primarily on geographic associations between environmental variables and behavior patterns, rather than on experimental tests of adaptation, the pattern is sufficiently common to suggest that natural selection has produced much of the behavioral differentiation observed among populations. If true, a high level of behavioral homoplasy should be expected in certain characters among populations independently exposed to similar, novel selection regimes.

Although the value of population comparison as a source of evolutionary insight is recognized, rarely have behavioral phenotypes been described for more than a few populations (see papers in Foster and Endler in press). In this regard, the threespine stickleback is unusual. Population character states for four behavioral traits that differ markedly among populations have been characterized in 7–13 populations, each of which is likely to have evolved independently after colonization of fresh water. The population data for stickleback provide a unique opportunity to examine patterns and levels of homoplasy, and although we can presently only make comparisons within one taxonomic group, the robustness of our conclusions can be tested by comparison with similar data as they are generated for other taxa.

The Threespine Stickleback Complex

The threespine stickleback fish, *G. aculeatus*, has undergone a remarkable endemic radiation in freshwater habitats in recently deglaciated regions of northwestern North America, Scotland, and other less intensively studied regions in the temperate zone of the Northern Hemisphere. Restricted primarily to coastal freshwater habitats and marine waters in temperate and arctic regions, this small fish must have been absent from much of the northern part of its range at the time of the last glacial maximum (Bell and Foster 1994b, for review). As glaciers began to recede 22,000 years ago (less in some areas; Foster and Bell 1994; McPhail 1994), marine and/or anadromous stickleback colonized the newly formed freshwater habitats, giving rise not only to phenotypically diverse freshwater populations, but also to new species (McPhail 1984, 1994).

Recent phylogenetic analyses of population relationships using allozyme (Haglund et al. 1992; Buth and Haglund 1994) and mtDNA sequence data (Orti et al. 1994) provide evidence that the marine and anadromous forms are ancient. The fossil record shows that the threespine stickleback, now with a discontinuous distribution between the Atlantic and Pacific basins, has been in both oceans for at least 1.9 million years (Bell 1994). Thus, character states shared between Pacific and Atlantic marine populations are presumably primitive character states relative to the postglacial diversification in either basin (Foster and Bell 1994), and character polarity can be inferred within regional radiations.

If populations are selected from geographically disparate sites between which little gene flow has occurred since the formation of freshwater populations, independent origin of similar derived phenotypes can be inferred. This is the approach we will take here in evaluating patterns of homoplasy across populations. Once appropriate molecular data are available, the hypothesis of independent origins can be tested (Withler and McPhail 1985; Bell and Foster 1994b). However, the inference is a strong one even without phylogenetic reconstruction based on molecular data (Bell and Foster 1994b).

In the postglacial radiation of freshwater stickleback, natural selection appears to have played an important role in molding population phenotypes, and much of the population variation has proven adaptive (papers in Bell and Foster 1994a for reviews). Similar freshwater habitats in geographically distant areas, independently colonized by marine stickleback, typically are inhabited by populations with similar, apparently adaptive, morphological and behavioral phenotypes (reviewed in Bell and Foster 1994b). This high degree of ecotypic differentiation is evidence of substantial homoplasy, as the ancestral form seems to be remarkably uniform for many characters.

Six of our lacustrine study populations are in the Strait of Georgia region of British Columbia, Canada. This region was ice covered at the height of the Fraser Glaciation and existing coastal lakes and streams were formed during isostatic rebound over the past 13,000 years (Mathews et al. 1970; Howes 1983). Sproat, Cowichan, and Crystal lakes are located on Vancouver Island in disjunct river drainages. Hotel, Garden Bay, and North lakes are on the Sechelt Peninsula across the Strait of Georgia from Vancouver Island. Hotel and Garden Bay lakes are connected by a stream approximately 0.3 km in length but are of divergent phenotypes. They both are connected to North Lake through creeks and lakes draining a higher elevation lake (Klein Lake) in two directions. North Lake was probably independently colonized through a much shorter direct connection to the Agamemnon Channel to the north. Observations were also made in one anadromous lagoon population on the Sechelt Peninsula (Foster 1994a,b for additional details).

The five Alaskan lakes are located within the MatSu Valley north of Knik Arm of Cook Inlet. Like much of coastal British Columbia, this region was covered by ice during the recent Fraser Glaciation, so that the study populations are only about 22,000 years old (reviewed in Foster and Bell 1994). Willow and Lynne lakes are located near the confluence of Willow Creek and the Susitna River ca. 65 km inland from the river's junction with Cook Inlet. Willow Lake is connected to Willow Creek via a short, direct tributary. Lynne Lake is a relatively isolated lake located in the upper portion of a different Willow Creek tributary. At least 25 stream km and seven intervening lakes separate these populations. Big Beaver Lake is located on a tributary to the Meadow Creek system ca. 22 km (straight-line) southeast of Lynne Lake. Meadow Creek is part of the Big Lake–Fish Creek system, which enters Cook Inlet separately from the Susitna River. Stephan Lake is situated about 10 km southwest of Big Beaver Lake and is near the headwaters of Goose Creek, also a tributary of Fish Creek. However, at least 40 stream km and two intervening lakes separate these populations. Bruce Lake is an isolated water body located at the upper end of the Little Meadow Creek watershed, about 15 straight-line km east of Big Beaver Lake. About 20 stream km and six lakes separate Bruce Lake from Big Beaver Lake, the closest study site. The anadromous population was studied on its spawning grounds, a tidal marsh system just upstream from the mouth of the Anchor River on the Kenai Peninsula. This population is at least 190 straight-line km south southwest of the nearest lacustrine study population.

All of the behavioral data from these populations were collected *in situ*, either while snorkeling or by observation from the shoreline (Foster 1988, 1994a,b, 1995b, for additional details). Methods that differed in studies by others are discussed below when necessary. We will focus on the subset of behavioral phenotypes in which we have detected population variation substantial enough to be studied in the field. Presumably, other behavioral characters will prove to exhibit population differences as well, when studied in more detail under experimental or laboratory conditions. The populations we have chosen for study are, we hope, a representative selection from each region. We have not intentionally selected either populations or behavioral phenotypes in a way that should bias our results, but we stress that we do not yet have a full picture of population variation in behavioral characters in either southern British Columbia or the Cook Inlet region of Alaska.

Our discussion assumes that the phenotypic differences among populations have some genetic basis. As the genetic bases of particular behavioral phenotypes have yet to be examined, the evidence of this is indirect, but will be presented in each instance. Although we find the indirect evidence compelling, the possibility that the differences reflect behavioral lability must be borne in mind until rearing studies are completed.

Nondisplay Behavior: Stickleback Foraging, Antipredator, and Sneaking Behavior

One of the most dramatic examples of ecotypic differentiation in the freshwater radiation of the threespine stickleback in northwestern North America involves trophic divergence between populations in shallow lakes with extensive littoral zones and those in deep, steep-sided, oligotrophic lakes. In southwestern British Columbia, where the divergence has been best studied, many populations in deep lakes are specialized for feeding predominantly on plankton in open water (limnetic populations, *sensu* McPhail 1984). In small, shallow lakes populations are instead specialized for foraging on large benthic invertebrates in the littoral zone (benthic populations; McPhail 1984, 1994; Lavin and McPhail 1985, 1986; Schluter and McPhail 1992). In limnetic fish the mouth is comparatively narrow and the snout long (enhancing sucking action), the eye is large, and the fish have long, closely spaced gill rakers. Benthic fish are deeper bodied, have larger mouths and smaller eyes, and have short, widely spaced gill rakers that prevent escape of prey but do not impede water movement.

These divergent suites of characters are thought to adapt the ecotypes to their respective environments (Hagen and Gilbertson 1972; McPhail 1984; Lavin and McPhail 1985, 1986; Schluter and McPhail 1992), and are clearly related to habitat differences even among lakes within a single drainage (Lavin and McPhail 1985). Each form is more efficient at foraging upon the resource for which it is specialized than is the other (Lavin and Mcphail 1986; Ibrahim and Huntingford 1988; Schluter 1993, 1995). The ecotypes comprise morphological characters which are heritable, and the differences between the ecotypes are based in genetic differences (Hagen 1973; Lavin and McPhail 1987; Day et al. 1994).

Benthic populations differ from limnetic populations with respect to several aspects of behavior as well. Fish from limnetic populations feed high in the water column in large groups, whereas those from benthic populations search substrata in the littoral zone for large benthic invertebrates. The differences in foraging behavior between limnetic and benthic stickleback persist in six small lakes in British Columbia in which are found species pairs of threespine stickleback. In each lake one species is benthic in behavior and morphology, and one limnetic (McPhail 1984, 1994; Schluter and McPhail 1992), despite similarity of available resources. Thus, the behavioral differences between the trophic types are likely to be rooted in genetic differences, but this remains to be tested directly.

Cannibalistic proclivities also differ between the two ecotypes (Foster 1988, 1994a,b, 1995b; Ridgway and McPhail 1988). In both British Columbia and Alaska, stickleback in benthic populations form large groups from which they forage on benthic invertebrates. The groups attack nests guarded by parental males whenever they are detected. If the nest contains embryos or yolk sac fry, they are consumed. Cannibalistic attacks resulted in the failure of 68% of nests under observation during one breeding season in a British Columbia lake (Foster 1988). Although fish in allopatric limnetic populations form planktivorous foraging groups, cannibalism of young in nests has not been observed in any population (Table 1; Foster 1994a,b, 1995b). In at least two lakes containing species pairs, only the benthic species attack nests although both trophic types coexist in the same environment (Ridgway and McPhail 1988), suggesting a genetically based difference in

TABLE 1 *The Relationship between Foraging Mode and Tendency to Group Cannibalism in Lacustrine Populations of Threespine Stickleback in Northwestern North America*

	Noncannibalistic	*Cannibalistic*
Benthic foraging	0	14 (9)
Limnetic foraging	5 (1)	1 (0)

Note. Values are numbers of populations with each combination of characteristics, and those in parentheses are the number in each category in the Alaskan subset. Alaskan populations include the study populations and those from Big, East Sunshine, Stepan, and Trouble lakes (unpublished data). Populations in British Columbia include the study populations plus those from Great Central (Manzer 1976), Kennedy (Hyatt and Ringler 1989a,b), and Trout (unpublished data) lakes (after Foster 1995b).

behavior. Once again, the differences in behavior appear to be adaptive with high levels of homoplasy. The ancestral condition, however, appears to be the presence of cannibalism, with loss being the derived, homoplastic condition (Foster 1995b).

The antipredator responses of threespine stickleback are a second class of nondisplay behavior patterns that vary substantially among freshwater populations (Huntingford 1982; Giles and Huntingford 1984; Foster and Ploch 1990; Huntingford et al. 1994). For example, in lakes where vertebrate predators are abundant, the boldness (a composite behavioral score) of stickleback tends to be less than in lakes where vertebrate predators are rare (see Fig. 10.4 in Huntingford et al. 1994). Again, there is evidence of substantial homoplasy, although directionality of the change, and hence the homoplastic condition, cannot be inferred from data presently available. The differences between populations persist under laboratory conditions and are thought to have a genetic basis, although the behavioral phenotypes may reflect a complex interplay between learning capability and paternal behavior (Huntingford et al. 1994).

The third kind of nondisplay behavior that differs substantially among freshwater populations of threespine stickleback is sneaking behavior. In threespine stickleback, males establish territories, build tubular nests of vegetation and kidney secretions, and court females using a well-described repertoire of courtship behavior patterns (Wootton 1976; Rowland 1994 for reviews). In some populations, all males court females immediately upon completion of their nests regardless of time in the breeding season (Goldschmidt et al. 1992). In others, many males that nest after the first wave of breeders instead build a nest, assume a drab, mottled coloration, and largely ignore their nests as they range widely over the breeding area. When they find a male courting a female, they sink to the bottom and move along it, appearing to move around objects that conceal them from the courting male. If successful, a "sneaker" will enter the nest of a courting male as the female leaves the nest having spawned. Even when the territory holder is prevented from following sneakers into the nest, the eggs are fertilized, indicating that sneakers release sperm while in the nest (Assem 1967).

The differences among populations in the incidence of sneaking behavior is pronounced. In six lacustrine populations in British Columbia, sneaking directed at a courting male was never seen despite long hours of observation (Table 2; Goldschmidt et al. 1992). In contrast, preliminary analyses of data from the Cook Inlet region of Alaska indicate that sneaking may be absent or rare in some populations, but that it is a typical element of male reproductive behavior in other populations. Sneakers intruded upon courting pairs of stickleback in 18.7% of courtships in one lake (Table 2), and the frequency of sneaking may be as high as 50% in others (unpublished data). In one lake, all males that nested after the first wave passed through a period of sneaking before initiating courtship (unpublished data).

TABLE 2   Incidence of Sneaking Intrusion during Courtship in Six Canadian (marked with *) and Five Cook Inlet Lacustrine Populations

Population	No. of males	Sneaking	No. sneaking	% Sneaking
Garden Bay*	21	0	21	0
Sproat*	22	0	22	0
North*	16	0	16	0
Cowichan*	19	0	19	0
Hotel*	16	0	16	0
Crystal*	24	0	24	0
Big Beaver	19	0	19	0
Lynne	16	0	16	0
Willow	15	0	15	0
Bruce	23	1	22	4.3
Stephan	32	6	26	18.7

Sneaking is apparently the ancestral condition in the postglacial radiation of stickleback in northwestern North America. Sneaking occurs in both anadromous populations in the Cook Inlet region of Alaska that we have studied well, and in an anadromous population on the Sechelt Peninsula of southern British Columbia (unpublished data). In the Atlantic, it is also know from at least two freshwater populations that can serve as outgroups for the Pacific North American clade (Goldschmidt et al. 1992). Thus, the homoplastic condition of this character is due to the loss of sneaking behavior, or at least extreme reduction in the frequency of expression of the trait.

Display Behavior: Stickleback Courtship and Diversionary Behavior

The courtship behavior of male threespine stickleback is a classical example of a set of motor patterns considered to be "species-typical," or characteristic of *G. aculeatus* everywhere (Tinbergen 1951). Once they have built nests, males have typically been described as initiating courtship with a stereotyped series of side to side jumps, known as the zig-zag dance. If the female is receptive, she approaches the male in a "head up" posture and he leads her to his nest, displaying it by inserting his nose in the opening and rolling on his side. The female may then move beneath the male and spawn. A second behavior known as dorsal pricking can also be observed during courtship, but is considered characteristic of males who are not fully receptive. Dorsal pricking occurs when the male assumes a position beneath the female and both move forward erratically, the male pausing occasionally while pressing his spines into the female's belly. The behavior has been interpreted as deterring females from approaching the nest, although they may wait nearby (reviewed in Rowland 1994).

Recent field studies have documented adaptive differentiation of court-ship behavior among populations (Foster 1990, 1994b, 1995b). In lakes where adults feed predominantly on benthic invertebrates in the littoral zone during the breeding season (benthic populations), they form large foraging groups that attack nests of conspecific males. Courtship in these lakes is much less conspicuous than in lakes where adults are primarily planktivo-rous and group cannibalism does not occur (limnetic populations). In the limnetic populations, courtship typically proceeds in the manner described in the early literature. Males perform the very conspicuous zig-zag dance to females as soon as they are detected and often lead the females directly to the nest after one or more bouts of dancing (Fig. 1). Dorsal pricking is rarely a component of courtship in these populations (Fig. 2). In benthic populations, however, females typically initiate courtship by positioning themselves on the back of a male and initiating dorsal pricking. In these populations court-ship is often very protracted, the movements are slow, potentially enabling males to survey their surroundings before leading females to their nests (Sargent 1982), and courtship is comparatively inconspicuous. Zig-zag danc-

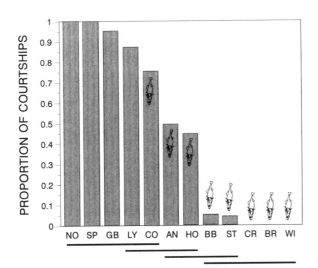

FIGURE 1 The proportion of courtships in each of 12 populations of threespine stickleback that incorporated the zig-zag dance. Shaded bars denote British Columbian populations; un-shaded, Alaskan. Bars with stickleback in or above them represent populations in which group cannibalism was observed. Site (population) names are NO = North; SP = Sproat; GB = Garden Bay; LY = Lynne; CO = Cowichan; AN = Anchor River; HO = Hotel; BB = Big Beaver; ST = Stepan; CR = Crystal; BR = Bruce; WI = Willow. All are lakes unless otherwise designated (from Foster 1995b).

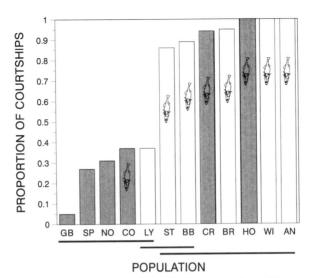

POPULATION

FIGURE 2   The proportion of courtships in each of 12 populations of threespine stickleback that incorporated dorsal pricking. Shaded bars denote British Columbian populations, unshaded, Alaskan. Bars with stickleback in or above them represent populations in which group cannibalism was observed. Site names as in Fig. 1 (from Foster 1995b).

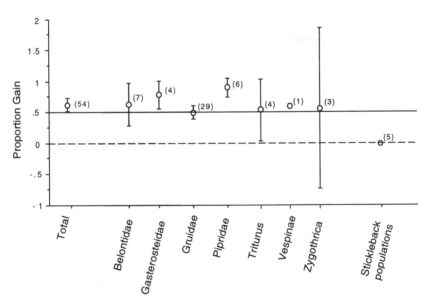

FIGURE 3   The mean proportion of character transitions which were gains, presented with 95% confidence intervals. Area above the solid line indicates a preponderence of character state gains, whereas that below indicates a predominance of losses. The number of homoplastic characters for each data set are in parentheses near the mean value.

ing is also less common in benthic populations (Figs. 2 and 3). The lesser conspicuousness of courtship in benthic populations presumably decreases the frequency with which groups detect the cryptic nests by cueing on nest-directed courtship activities, and is therefore adaptive in cannibalistic populations (Foster 1994a, 1995b).

Observations on anadromous populations, including that in Anchor River (Figs. 2 and 3), indicate that group cannibalism and a preponderance of dorsal pricking during courtship are the ancestral conditions (Whoriskey and FitzGerald 1985; Foster, 1995b) relative to the diversification in northwestern North America. Thus, the homoplastic character states are loss of cannibalism, decreased frequency of dorsal pricking, and increased commonness of the zig-zag dance.

The Anchor River population provides indirect evidence of a genetic difference among populations in courtship behavior. Anchor River fish move into tide pools from the ocean on high tides where they are isolated until the next tidal cycle. When the data presented here (Figs. 1 and 3) were collected in 1992, no benthic foraging groups were present in the pools. The level of zig-zag dancing was intermediate, but dorsal pricking occurred in all courtships. In 1995, foraging groups were common, and the zig-zag dance was rarely displayed (unpublished data). These observations suggest that Anchor River fish are adapted to group cannibalism, but can elevate the frequency of expression of the zig-zag dance to intermediate but not high levels, when cannibalistic groups are missing.

The final display behavior we will discuss here is apparently unique among fishes. Well known in birds, particularly ground nesting birds, diversionary displays serve to deter potential predators from approaching a nest containing vulnerable young (Armstrong 1949a,b; Skutch 1955). In southern British Columbia, Canada, stickleback in benthic populations respond to the approach of cannibalistic, conspecific groups with diversionary displays that vary across populations in complexity and diversity (Table 3). In noncannibalistic, limnetic populations the approach of groups elicits zig-zag dances from courting males rather than diversionary displays. Even when large groups are chased into male territories low to the substratum, limnetic males do not perform diversionary displays, whereas those in benthic populations do so nearly always (Foster 1988, 1994a,b).

Comparison with Atlantic and Pacific anadromous populations (Whoriskey and FitzGerald 1985; Woriskey 1991; Foster 1994b) indicates that the most complex display repertoire, that exhibited in the Hotel Lake and lagoon (anadromous) populations in southern British Columbia (Table 3), is the ancestral character state. Less diverse repertoires seem to reflect partial loss, and the absence of the display, complete loss. Thus, the homoplastic condition is one of loss of a complex display repertoire. Again, the homoplastic condition appears to be adaptive in that diversionary display behavior confers no benefit when there is no risk of group cannibalism, and may cost

TABLE 3   The Number of Diversionary Displays of Different Types Performed by Males at the Approach of Conspecifics in Four Cannibalistic Stickleback Populations

Diversionary behavior	Population[a]			
	Cowichan	Crystal	Hotel	Lagoon
Upright swim-root	10 (0.56)	16 (0.36)	0	0
Erratic swim-root	8 (0.44)	23 (0.51)	8 (0.12)	11 (0.34)
Side swim-root	0	6 (0.13)	8 (0.12)	8 (0.25)
Snout tap	0	0	15 (0.22)	7 (0.22)
Shimmer	0	0	36 (0.54)	6 (0.19)

[a] Numbers in parentheses are the proportions of all displays in a population that were of a given type.

males mating opportunities in that they defend their nests unnecessarily and fail to court females in groups that pose no risk to their nests or embryos within them.

Patterns of Homoplasy in the Stickleback Complex

In the endemic radiation of threespine stickleback in recently deglaciated regions, behavioral characters that display extensive differentiation in freshwater since the last glacial retreat tend to exhibit homoplasy. This holds for display and nondisplay characters, both of which appear to have diverged primarily as a consequence of loss or change in the frequency of expression of ancestral motor patterns. Similar rapid changes in the expression of morphological features of threespine stickleback upon postglacial colonization of freshwater are well documented, and adaptive homoplasy is apparent (Hagen and Gilbertson 1972; Lavin and McPhail 1986; Bell 1988; Reimchen 1989; Bell et al. 1993; Bell and Foster 1994b). Again, none of the novel phenotypes have arisen *de novo*, and most reflect loss of part or all of specific ancestral features.

In the case of behavioral characters, our efforts have uncovered no derived states that are a consequence of the evolution of novel motor patterns. Similarly, we know of no clear cases in which a novel morphological structure has appeared in a postglacial population. Thus, the pattern that emerges in the postglacial radiation of threespine stickleback is one in which adaptive homoplasy is extremely common, and the derived characters are the products of expression shifts or loss of ancestral character states. Presumably, the paucity of true novelty in this adaptive diversification is a consequence of the short time (22,000 years or less, reviewed in Foster and Bell 1984) since the last glacial recession in the regions under study.

BEHAVIORAL HOMOPLASY AT
AND ABOVE THE SPECIES LEVEL

Recent examination of levels of homoplasy in behavioral, morphological and biochemical characters across species (and higher taxonomic units) have failed to reveal unusually high values for behavioral characters (Sanderson and Donoghue 1989; de Quieroz and Wimberger 1993). This finding seems to contrast with our observations on the frequency with which similar character states have evolved independently across freshwater populations of the threespine stickleback. Although the nature of the data used to examine homoplasy at the two levels differs sufficiently that we cannot contrast levels of homoplasy directly, we can ask whether the evolution of homoplasy is due to similar patterns of loss and acquisition in both population and higher order data sets. Before we approach this problem, we will present overall levels of homoplasy for the higher order data sets and examine whether display and nondisplay patterns exhibit similar levels and patterns of homoplasy. Although our efforts must be considered preliminary given limitations of the data sets and analytical methods (below), these issues have not been addressed previously yet are critical to our understanding of the evolution of behavioral homoplasy, and behavioral evolution in general.

General Methods for Species and
Higher Level Comparisons

We used eight data sets (Table 4) to examine levels of homoplasy at the species level and above (hereafter referred to as species level data sets). Our requirements for inclusion were that the character states for each behavioral trait were well described for each taxon involved and that there existed a supported phylogeny. Characters were coded as display behavior if they function to transfer information, and as nondisplay behavior if the primary function is maintenance or locomotion (Foster 1995a). Two of the character sets comprise display behavior only, whereas the remaining six include both types of behavior. Not all data sets could be used in all analyses because of limited sample sizes or character types.

We used two measures to describe the amount of homoplasy in the species level data sets. The first is the consistency index (CI), which is the ratio of the observed number of character state changes over the minimum possible number of character state changes. A value of less than one indicates homoplasy. Consistency indices were calculated using the computer program MacClade (Maddison and Maddison 1993). The second measure we used was the percentage of homoplastic characters. This value was calculated by hand for each data set, and across all data sets. While both measures provide information on levels of homoplasy, each does so in a slightly different way. The percentage provides a measure of the frequency of homoplastic charac-

TABLE 4   The Sources of Species Level Data Sets Used in Our Analyses, Including Citations for Behavioral Characters and Phylogenetic Reconstruction

Data set (n)	Common name	Behavioral characters	Phylogeny hypothesis
Anas (31)	Dabbling ducks	McKinney 1975 Johnson, unpublished data	Johnson, unpublished data
Belontidae (8)	Gourami fishes	Miller and Robinson 1974 Miller and Jearld 1983	Miller and Jearld 1983
Gasterosteidae (7)	Stickleback fishes	McLennan 1993 McLennan et al. 1988	McLennan 1993
Gruidae (13)	Cranes	Archibald 1976 Grier 1984	Krajewski and Fetzner 1994
Pipridae (19)	Manakins	Prum 1990	Prum 1990
Triturus (9)	Newts	Arntzen and Sparreboom 1989	Arntzen and Sparreboom 1989
Vespinae (7)	Wasps	Carpenter 1987	Carpenter 1987
Zygothrica (7)	Fruit Flies	Grimaldi 1987	Grimaldi 1987

Note. The number of taxa for each data set is placed in parentheses (*n*).

ters within or across data sets. In contrast, the CI can be used to calculate the amount of homoplasy within a single character, as well as across all characters in one or more data sets. Whereas the ratio provides an unweighted measure of homoplasy in a character set, the CI provides a measure weighted by the strength of homoplasy in individual characters.

Even when levels of homoplasy are high, character states may be nonrandomly distributed across the phylogeny, indicating some degree of phylogenetic conservation. The probability of obtaining a certain character state pattern can be calculated using a Monte Carlo reshuffling randomization to generate an expected random distribution of character states. This distribution can then serve as a null hypothesis against which to test for phylogenetic information in the character of interest (Archie 1989; Maddison and Slatkin 1991; Peterson and Burt 1992).

We randomly reshuffled characters 500 times for the dabbling ducks and 1000 times for the stickleback, crane, and manakin data sets using MacClade (Maddison and Maddison 1993). To avoid bias, characters were excluded if a state was present in fewer than 15% of the taxa. We calculated the probability of obtaining the same or fewer character state changes by comparison with the null distribution. If this value is less than 0.05, the character can be considered to contain phylogenetic information.

Homoplasy and the Evolution of Display and Nondisplay Behavior

The overall CI for all species level data sets combined was 0.777 ± 0.023 (mean ± 1 SE, Table 5), which indicates the presence of homoplasy, but at low levels. Homoplasy occurred in 36% of the 186 characters surveyed. Data sets differed in their levels of homoplasy as indicated by the CIs (Kruskal–Wallis, $P < 0.0001$, Table 5). Also, variation in the percentage of homoplastic characters differs between the data sets, ranging from 10.8 to 76.3% (Table 5).

In our analysis of display and nondisplay behavior we accounted for variation in levels of homoplasy between data sets by using paired comparisons. Within data sets, levels of homoplasy were greater for display behavior than for nondisplay behavior as measured by the CI and the percentage of characters which were homoplastic (Wilcoxon Sign Rank Test; CI, $P = 0.046$; percentage homoplasy, $P = 0.046$; Table 6).

Although only four data sets could be used because of sample size requirements, the randomization indices show a similar pattern. Display characters appear to be more strongly phylogenetically conserved than do nondisplay characters (Table 6). For the four data sets combined, display behavior again exhibited a greater degree of phylogenetic conservation than did nondisplay behavior (Mann–Whitney U, $P = 0.0023$). Only the dabbling duck and stickleback data sets were large enough that we could calculate randomization indices, and size of the latter was marginal. The dabbling

TABLE 5 *The Number of Characters, Mean Consistency Index (± 1 SE), and Percentage Homoplasy Are Presented for Each Species Level Data Set, as well as for All Data Sets Combined*

Data set	No. of characters	Consistency index	Percentage homoplastic characters
Anas	11	0.405 (0.116)	72.7
Belontidae	14	0.726 (0.078)	50.0
Gasterosteidae	37	0.951 (0.024)	10.8
Gruidae	38	0.557 (0.043)	76.3
Pipridae	44	0.900 (0.037)	15.9
Triturus	16	0.688 (0.101)	43.7
Vespinae	12	0.875 (0.090)	16.7
Zygothrica	14	0.893 (0.057)	21.4
Total	186	0.777 (0.023)	36.0

TABLE 6 *Mean Consistency Index (1 SE), Percentage Homoplasy, and Randomization Index (1 SE) for Each Data Set Subdivided into Display and Nondisplay Behavioral Characters*

Data set	Character group	No. of characters	Mean CI	Percentage homoplastic characters	Randomization index
Anas	Display	6	0.617 (0.172)	50.0	0.002 (0.001)
	Nondisplay	5	0.150 (0.021)	100.0	0.257 (0.086)
Belontidae	Display	8	0.791 (0.104)	37.5	*
	Nondisplay	6	0.638 (0.117)	60.0	*
Gasterosteidae	Display	16	0.948 (0.036)	12.5	0.091 (0.026)
	Nondisplay	21	0.952 (0.033)	9.5	0.076 (0.019)
Gruidae	Display	38	0.557 (0.043)	76.3	0.075 (0.026)
	Nondisplay	0	—	—	—
Pipridae	Display	44	0.900 (0.037)	15.9	0.022 (0.008)
	Nondisplay	0	—	—	—
Triturus	Display	15	0.700 (0.107)	40.0	*
	Nondisplay	1	0.500 (0.0)	100.0	*
Vespinae	Display	1	1.00 (0.0)	0.0	*
	Nondisplay	11	0.864 (0.097)	18.2	*
Zygothrica	Display	9	0.944 (0.056)	11.1	*
	Nondisplay	5	0.800 (0.122)	40.0	*
Total	Display	131	0.773 (0.026)	37.2	0.057 (0.014)
	Nondisplay	44	0.787 (0.047)	32.6	0.126 (0.138)

Note. Statistics for the combined data sets are listed at the bottom of the table. Cells (—) indicate the absence of data, while cells marked (*) indicate values which were not calculated because of insufficient sample sizes.

duck data showed greater conservation of display than nondisplay behavior. There was no difference between classes of behavior in the stickleback data set.

The Nature of Behavioral Homoplasy: Loss or Novel Acquisition?

For each character, we calculated the percentage of transitions that are the consequence of the evolution of a new behavior pattern (gain) as opposed to loss of an ancestral state. The observed number of changes were determined by optimizing the character using parsimony algorithms in MacClade (Maddison and Maddison 1993). Ambiguous states were evenly divided between gains and losses. Percentage gain was calculated for each character as the number of gains divided by the number of transitions. A value greater than 0.5 indicates a preponderance of character state gains. The mean percentage gain was then determined for each data set and for all data sets combined.

Between species level data sets, there is no significant difference in the proportion of character state transitions which are gains, so the data could be pooled (Kruskall–Wallis $P = 0.135$; Fig. 3). The 95% confidence interval for the pooled species level data sets does not overlap the 0.5 level, indicating a preponderance of gains. In stickleback populations, however, nearly all homoplasy in discrete characters is due to loss, or reduction in quantitative characters. Stickleback populations exhibit a significantly smaller proportion of gains than all species level data sets except the Zygothrica. The confidence interval may be large for Zygothrica because of a small number of homoplastic characters. Even given the small sample sizes, homoplasy seems to be a function of different patterns of character state change at different biological levels.

HOMOPLASY AND PATTERN IN BEHAVIORAL EVOLUTION

Although we were unable to compare directly the levels of homoplasy that result from population differentiation and from differentiation at higher taxonomic levels, there appeared to be marked qualitative differences. In particular, iterated evolution of similar phenotypes was extremely common in the stickleback endemic radiation. In most cases the homoplastic characters appeared to be adaptive, in that they were associated with specific foraging environments or predation regimes. In fact, the match between lacustrine environment and phenotype is sufficiently good that one can typically predict the character states in resident stickleback populations from the qualities of a lake. This presents a problem for any comparison between population levels of homoplasy and higher order levels because the frequencies that one detects in population comparisons will depend on the relative representation of lacustrine environment types in the sample. Also, many of the phenotypes are not independent, and have evolved in concert in response to the same selective regimes.

In comparisons of higher order taxa, however, the frequency of appearance of similar homoplastic behavioral states appears to be lower, and the states tend to be more ambiguously associated with environment. The reasons for the latter probably include the longer average time since speciation events and the greater array of potentially confounding correlated characters in more highly differentiated characters (Arnold 1992; Foster et al. 1992).

The apparently lower frequency of homoplastic behavioral states in species level and higher order comparisons could reflect biologically meaningful differences in the rates of retention of different kinds of behavioral characters through lineage sorting, or it could simply reflect the broader array of behavioral phenotypes typically included in these comparative studies. In the stickleback data presented here, we incorporated only those characters known to vary across populations. These may be the characters most likely to display extensive homoplasy because they are subject to strong environmental selection, components of which vary extensively across relatively short distances or periods of time. Some support for the latter explanation is found in our observation of lower levels of homoplasy in nondisplay (maintenance and locomotor characters) than in display behavior. This is not a complete explanation, however, as two forms of display behavior (courtship and diversionary behavior) exhibited high levels of adaptive homoplasy across populations of stickleback.

Although we offer some support for our initial hypothesis that levels of homoplasy should be less in display than nondisplay characters, our findings must be considered preliminary. The dabbling duck data set provided the greatest support for this hypothesis. This is encouraging because it was the only large data set that incorporated good sample sizes of both classes of behavior. The significance of the finding for the total data set must be treated with some suspicion because, in large part, significance could have been due to inclusion of the dabbling duck data. However, both the crane and the manakin data had relatively high consistency indices and low randomization indices, a promising result as only display characters were available for these taxa.

If our tentative finding of higher levels of homoplasy in nondisplay than display behavior holds up as other large data sets become available, the result will suggest that display characters are more appropriate for use in phylogenetic reconstruction than are nondisplay characters. The reason could be, as we suggested initially, that there are more ways to enhance the effectiveness of display behavior than nondisplay because display modifications may, initially at least, primarily serve to increase conspicuousness. In contrast, there may be few possible modifications of, say, foraging or locomotor behavior that enhance function (Foster 1995a,b). The consequence could be a difference in levels of homoplasy in the direction we observed in both the dabbling duck data set and the total data set.

A more robust conclusion from our analyses is that most behavioral homoplasies observed in population comparisons are a consequence of loss of complex ancestral character states, rather than the evolution of novel motor patterns. This is presumably a consequence of the short divergence times separating populations, as behavioral homoplasies observed in higher order comparisons are often the products of the evolution of truly novel behavioral character states.

Although most of our conclusions must be considered tentative, we believe that our results are exciting insofar as they suggest the power of comparative studies which incorporate phylogenetic hypotheses to resolve pattern and process in the evolution of behavioral homoplasy. Furthermore, we believe it likely that some of our findings will apply to other classes of characters as well. Given that an understanding of the evolution of behavioral homoplasy can help us evaluate issues as diverse as the usefulness of particular classes of characters as sources of phylogenetic information, the causes of the breakdown of homogamy, and the nature of constraints on display versus nondisplay behavior, we are delighted with the promise that our results hold. Clearly, as additional large behavioral and phylogenetic data sets are developed, we should be able to test not only the hypotheses advanced here, but also hypotheses that will help us understand the characteristics of different kinds of display and nondisplay characters that are associated with different levels of homoplasy.

ACKNOWLEDGMENTS

We thank L. Hufford and M. Sanderson for inviting this chapter and for their infinite patience. Without the invitation we would not have thought about these fascinating issues. Preparation of the manuscript was supported by a NSF Presidential Faculty Fellowship to S.A.F. Research on the stickleback was supported by this and other awards from NSF and NIMH to S.A.F. Research on dabbling ducks was supported by Dayton and Wilkie Funds for Natural History, the Frank M. Chapman Memorial Fund, and a NSF Doctoral Dissertation to K.P.J.

REFERENCES

Archibald, G. W. 1976. Crane taxonomy as revealed by the unison call. Proc. Int. Crane Workshop 1:225–251.

Archie, J. 1989. A randomization test for phylogenetic information in systematic data. Systematic Zoology 38:239–252.

Armstrong, E. A. 1949a. Diversionary display. Part 1. Connotation and terminology. Ibis 91: 88–97.

Armstrong, E. A. 1949b. Diversionary display. Part 2. The nature and origin of distraction display. Ibis 91:179–188.

Arnold, S. J. 1981. The microevolution of feeding behavior. Pp. 409–453 In A. Kamil and T. Sargent, eds. Foraging behavior: Ecological, ethological and psychological approaches. Garland Press, New York.

Arnold, S. J. 1992. Behavioural variation in natural populations. VI. Prey responses by two species of garter snakes in three regions of sympatry. Animal Behaviour 44:705–719.

Arntzen, J. W., and M. Sparreboom. 1989. A phylogeny for the Old World newts, genus *Triturus:* Biochemical and behavioral data. Journal of Zoology, London 219:645–664.

Assem, J. van den. 1967. Territory in the threespine stickleback Gasterosteus aculeatus L. An experimental study in intraspecific competition. Behavior Supplement 16:1–164.

Bateson, P. P., and R. A. Hinde. 1976. Growing points in ethology. Cambridge University, Cambridge.

Bell, M. A. 1988. Stickleback fishes: Bridging the gap between population biology and paleobiology. Trends in Ecology and Evolution 3:320–325.

Bell, M. A. 1994. Paleobiology and the evolution of threespine stickleback. Pp. 438–471 *In* M. A. Bell and S. A. Foster, eds. The evolutionary biology of the threespine stickleback. Oxford University Press, Oxford.

Bell, M. A., and S. A. Foster, eds. 1994a. The evolutionary biology of the threespine stickleback. Oxford University Press, Oxford.

Bell, M. A., and S. A. Foster. 1994b. Introduction to the evolutionary biology of the threespine stickleback. Pp. 1–27 *In* M. A. Bell and S. A. Foster, eds. The evolutionary biology of the threespine stickleback. Oxford University Press, Oxford.

Bell, M. A., G. Orti, J. A. Walker, and J. R. Koenings. 1993. Evolution of pelvic reduction in threespine stickleback fish: A test of competing hypotheses. Evolution 47:906–914.

Boake, C. R. B. 1994. Quantitative genetic studies of behavioral evolution. University of Chicago Press, Chicago.

Brooks, D. R., and D. A. McLennan. 1991. Phylogeny, ecology, and behavior: A research program in comparative biology. University of Chicago Press, Chicago.

Burghardt, G. M., and J. L. Gittleman. 1990. Comparative behavior and phylogenetic analyses: New wine, old bottles. Pp. 192–225 *In:* M. Bekoff and D. Jamieson, eds. Interpretation and explanation in the study of animal behavior. Vol. II: Explanation, evolution, and adaptation. Westview Press, San Francisco.

Buth, D. G., and T. R. Haglund. 1994. Allozyme variation in the *Gasterosteus aculeatus* complex. Pp. 61–84 *In* M. A. Bell and S. A. Foster, eds. The evolutionary biology of the threespine stickleback. Oxford University Press, Oxford.

Carpenter, J. M. 1987. Phylogenetic relationships and classification of the Vespinae (Hymenoptera: Vespidae). Systematic Entomolgy 12:413–431.

Daanje, A. 1950. On locomotory movements in birds and the intention movements derived from them. Behaviour 3:49–98.

Darwin, C. 1859. The origin of species by means of natural selection. Murray, London.

Day, T., J. Pritchard, and D. Schluter. 1994. Ecology and genetics of phenotypic plasticity: A comparison of two sticklebacks. Evolution 48:1723–1734.

de Quieroz, A., and P. H. Wimberger. 1993. The usefulness of behavior for phylogeny estimation: Levels of homoplasy in behavioral and mophological characters. Evolution 47: 46–60.

Endler, J. A. 1977. Geographic variation, speciation, and clines. Princeton University Press, Princeton, New Jersey.

Endler, J. A. 1986. Natural selection in the wild. Princeton University Press, Princeton, New Jersey.

Foster, S. A. 1988. Diversionary displays of paternal stickleback: Defenses against cannibalistic groups. Behavioral Ecololgy and Sociobiology. 22:335–340.

Foster, S. A. 1990. Courting disaster in cannibal territory. Natural History, November:52–61.

Foster, S. A. 1994a. Evolution of the reproductive behavior of threespine stickleback. Pp. 381–398 *In* M. A. Bell and S. A. Foster, eds. The evolutionary biology of the threespine stickleback. Oxford University Press, Oxford.

Foster, S. A. 1994b. Inference of evolutionary pattern: Diversionary displays of threespine sticklebacks. Behavioral Ecology 5:114–121.

Foster, S. A. 1995a. Constraint, adaptation, and opportunism in the design of behavioral phenotypes. Pp. 61–81 *In* N. S. Thompson, ed. Perspectives in ethology, vol. 11. Plenum Press, New York.

Foster, S. A. 1995b. Understanding the evolution of behaviour in threespine stickleback: The value of geographic variation. Behaviour, 132:1107–1129.

Foster, S. A., and J. A. Endler. in press. The evolution of geographic variation in behavior. Oxford University Press, Oxford.

Foster, S. A., and M. A. Bell. 1994. Evolutionary inference: The value of viewing evolution through stickleback-tinted glasses. Pp. 473–486 *In* M. A. Bell and S. A. Foster, eds. The evolutionary biology of the threespine stickleback. Oxford University Press, Oxford.

Foster, S. A., and S. A. Cameron. 1996. Geographic variation in behavior: A phylogenetic framework for comparative studies. Pp. 138–165 *In* E. Martins, ed. Phylogenies and the comparative method in animal behavior. Oxford University Press, New York.

Foster, S. A., and S. Ploch. 1990. Determinants of variation in antipredator behavior of territorial male threespine stickleback in the wild. Ethology 84:281–294.

Giles, N., and F. A. Huntingford. 1984. Predation risk and inter-population variation in antipredator behaviour in the three-spined stickleback, *Gasterosteus aculeatus* L. Animal Behaviour 32:264–275.

Goldschmidt, T., S. A. Foster, and P. Sevenster. 1992. Internest distance and sneaking in threespined stickleback. Animal Behaviour 44:793–795.

Grier, J. W. 1984. Biology of animal behavior. Mosby College Publishing, St. Louis, MO.

Grimaldi, D. A. 1987. Phylogenetics and taxonomy of Zygothrica (Diptera: Drosophilidae). Bulletin of the American Museum of Natural History 186:103–268.

Hagen, D. W. 1973. Inheritance of numbers of lateral plates and gill rakers in *Gasterosteus aculeatus*. Heredity 30:303–312.

Hagen, D. W., and L. G. Gilbertson. 1972. Geographic variation and environmental selection in *Gasterosteus aculeatus* L. in the Pacific northwest, America. Evolution 26:32–51.

Haglund, T. R., D. G. Buth, and R. Lawson. 1992. Allozyme variation and phylogenetic relationships of Asian, North American, and European populations of the threespine stickleback. Copeia 1992:432–443.

Harvey, P. H., and M. D. Pagel. 1991. The comparative method in evolutionary biology. Oxford University Press, Oxford.

Heinroth, O. 1910. Beobachtungen bei einem Einburgerungsversuch mit der Brautente (*Lampronessa sponsa* (L.)). Journal fur Ornithologie 1:101–156.

Heinroth, O. 1911. Beitrage zur Biologie, namentlich Ethologie und Psychologie der Anatiden. Verhanalung des V Internationalen Ornithologen Kongresses 1911:589–702.

Hinde, R. A., and Tinbergen, N. 1958. The comparative study of species-specific behavior. Pp. 251–268 *In* A. Roe and G. G. Simpson, eds. Behavior and evolution. Yale University Press, New Haven, CT.

Howes, D. E. 1983. Late Quaternary sediments and geomorphic history of northern Vancouver Island, B. C. Canadian Journal of Earth Sciences 20:57–65.

Huntingford, F. A. 1982. Do inter- and intraspecific aggression vary in relation to predation pressure in sticklebacks? Animal Behaviour 30:909–916.

Huntingford, F. A., P. J. Wright, and J. F. Tierney. 1994. Adaptive variation in antipredator behaviour in threespine stickleback. Pp. 277–296 *In* M. A. Bell and S. A. Foster, eds. The evolutionary biology of the threespine stickleback. Oxford University Press, Oxford.

Hyatt, K. D., and N. H. Ringler 1989a. Role of nest raiding and egg predation in regulating population density of threespine sticklebacks (*Gasterosteus aculeatos*) in a coastal British Columbia lake. Canadian Journal of Fisheries and Aquatic Science 46:372–383.

Hyatt, K. D., and N. H. Ringler 1989b. Egg cannibalism and the reproductive strategies of threespine sticklebacks (*Gasterosteus aculeatus*) in a coastal British Columbia lake. Canadian Journal of Zoology 67:2036–2046.

Ibrahim, A. A., and F. A. Huntingford. 1988. Foraging efficiency in relation to within-species variation in morphology in three-spined sticklebacks, *Gasterosteus aculeatus*. Journal of Fish Biology 33:823–824.

Krajewski, C., and J. W. Fetzner, Jr. 1994. Phylogeny of cranes (Gruiformes: Gruidae) based on cytochrome-*B* DNA sequences. Auk 111:351–365.

Lavin, P. A., and J. D. McPhail. 1985. The evolution of freshwater diversity in threespine stickleback (*Gasterosteus aculeatus*): Site-specific differentiation of trophic morphology. Canadian Journal of Zoology 63:2632–2638.

Lavin, P. A., and J. D. McPhail. 1986. Adaptive divergence of trophic phenotype among freshwater populations of threespine stickleback (*Gasterosteus aculeatus*). Canadian Journal of Fisheries and Aquatic Sciences 43:2455–2463.

Lavin, P. A., and J. D. McPhail. 1987. Morphological divergence and the organization of trophic characters among lacustrine populations of the threespine stickleback (*Gasterosteus aculeatus*). Canadian Journal of Fisheries and Aquatic Sciences 44:1820–1829.

Lorenz, K. 1941. Vergleichende Bewegungsstudien an Anatinen. Journal fur Ornithologie 89: 19–29.

Lorenz, K. 1950. The comparative method in studying innate behaviour patterns. Society of Experimental Biology 4:221–268.

Maddison, W. P., and D. R. Maddison. 1993. MacClade Version 3. Sinauer Associates, Sunderland, MA.

Maddison, W. P., and M. Slatkin. 1991. Null models for the number of evolutionary steps in a character on a phylogenetic tree. Evolution 45:1184–1197.

Manzer, J. I. 1976. Distribution, food, and feeding of the threespine stickleback, *Gasterosteus aculeatus,* in Great Central Lake, Vancouver Island, with comments on competition for food with juvenile sockeye salmon, *oncorhynchus nerka.* Fishery Bulletin, U.S. National Marine Fisheries Service 74:647–668.

Mathews, W. H., J. G. Fyles, and H. W. Nasmith. 1970. Postglacial crustal movements in southwestern British Columbia and adjacent Washington state. Canadian Journal Earth Sciences 7:690–702.

Mayr, E. 1958. Behavior and systematics. Pp. 341–362 *In* A. Roe and G. G. Simpson, eds. Behavior and evolution. Yale University Press, New Haven, CT.

McKinney, F. 1975. The evolution of duck displays. Pp. 331–357. *In* G. Baerends, C. Beer, and A. Manning, eds. Function and evolution in behaviour. Clarendon Press, Oxford.

McLennan, D. A. 1993. Phylogenetic realtionships in the Gasterosteidae: An updated tree based on behavioral characters with a discussion of homoplasy. Copeia 2:318–326.

McLennan, D. A., D. R. Brooks, and J. D. McPhail. 1988. The benefits of communication between comparative ethology and phylogenetic systematics: A case study using gasterosteid fishes. Canadian Journal of Zoology 66:2177–2190.

McPhail, J. D. 1984. Ecology and evolution of sympatric sticklebacks (*Gasterosteus*): Morphological and genetic evidence for a species pair in Enos Lake, British Columbia. Canadian Journal of Zoology 62:1402–1408.

McPhail, J. D. 1994. Speciation and the evolution of reproductive isolation in the sticklebacks (*Gasterosteus*) of south-western British Columbia. Pp. 399–437 *In* M. A. Bell and S. A. Foster, eds. The evolutionary biology of the threespine stickleback. Oxford University Press, Oxford.

Miller, R. J., and A. Jearld. 1983. Behavior and phylogeny of fishes of the genus *Colisa* and the family Belontiidae. Behaviour 83:155–185.

Miller, R. J., and H. W. Robinson. 1974. Reproductive behavior and phylogeny in the genus *Trichogaster* (Pisces, Anabantoidei). Zeitschrift fur Tierpsychologie. 34:484–499.

Morris, D. 1956. The function and causation of courtship ceremonies in animals (with special reference to fish). Pp. 261–286 *In* Foundation Signer-Polignac. Colloque International sur l'Instinct, Paris, 1954.

Orti, G., M. A. Bell, T. E. Reimchen, and A. Meyer. 1994. Global survey of mitochondrial DNA sequences in the threespine stickleback: Evidence for recent migrations. Evolution 48: 608–622.

Peterson, A. T., and D. B. Burt. 1992. Phylogenetic history of social evolution and habitat use in the *Aphelocoma* jays. Animal Behaviour 44:859–866.

Pomiankowski, A., and L. Sheridan. 1994. Linked sexiness and choosiness. Trends in Ecology and Evolution 9:242–244.

Prum, R. O. 1990. Phylogenetic analysis of the evolution of display behavior in the neotropical manakins (Aves: Pipridae). Ethology 84:202–231.

Radtkey, R. R., and M. C. Singer. 1995. Repeated reversals of host-preference evolution in a specialist insect herbivore. Evolution 49:351–359.

Reimchen, T. E. 1989. Loss of nuptial color in threespine sticklebacks (*Gasterosteus aculeatus*). Evolution 43:450–460.

Ridgway, M. S., and J. D. McPhail. 1988. Raiding shoal size and a distraction display in male sticklebacks (*Gasterosteus*). Canadian Journal of Zoology 66:201–205.

Riechert, S. E. in press. The use of behavioral ecotypes in the study of evolutionary processes. *In* S. A. Foster and J. A. Endler, eds. The evolution of geographic variation in behavior. Oxford University Press, Oxford.

Riechert, S. E. 1986. Spider fights as a test of evolutionary game theory. American Scientist 74: 604–610.

Riechert, S. E. 1993. Investigation of potential gene flow limitation of behavioral adaptation in an aridlands spider. Behavioral Ecology and Sociobiology 32:355–363.

Rowland, W. J. 1994. Proximate determinants of stickleback behaviour: An evolutionary perspective. Pp. 297–344 *In* M. A. Bell and S. A. Foster, eds. The evolutionary biology of the threespine stickleback. Oxford University Press, Oxford.

Sanderson M. J., and M. J. Donoghue. 1989. Patterns of variation in levels of homoplasy. Evolution 43:1781–1795.

Sargent, R. C. 1982. Territory quality, male quality, courtship intrusions, and female nest choice in the threespine stickleback, *Gasterosteus aculeatus*. Animal Behaviour 30:364–374.

Schluter, D. 1993. Adaptive radiation in sticklebacks: Size, shape, and habitat use efficiency. Ecology 74:699–709.

Schluter, D. 1995. Adaptive radiation in sticklebacks: Trade offs in feeding performance and fitness. Ecology 76:82–90.

Schluter, D., and J. D. McPhail. 1992. Ecological character displacement and speciation in sticklebacks. American Naturalist 140:85–108.

Skutch, A. F. 1955. The parental strategems of birds. Ibis 97:118–142.

Thompson, D. B. 1990. Different spatial scales of adaptation in the climbing behavior of *Peromyscus maniculatus*: Geographic variation, natural selection, and gene flow. Evolution 44: 952–965.

Tinbergen, N. 1951. The study of instinct. Oxford University Press, Oxford.

Tinbergen, N. 1964. On aims and methods of ethology. Zeitschrift fur Tierpsychologie. 20: 410–433.

Wenzel, G. W. 1992. Behavioral homology and phylogeny. Annual Review of Ecology and Systematics 23: 261–281.

Whitman, C. O. 1899. Animal behavior. Woods Hole Marine Biology Lectures, 6:285–338.

Whoriskey, F. G. 1991. Stickleback distraction displays: Sexual or foraging deception against egg cannibalism? Animal Behaviour 41:989–996.

Whoriskey, F. G., and G. J. FitzGerald. 1985. Sex, cannibalism and sticklebacks. Behavioral Ecology and Sociobiology 18:15–18.

Withler, R. E., and J. D. McPhail. 1985. Genetic variability in freshwater and anadromous sticklebacks (*Gasterosteus aculeatus*) of southern British Columbia. Canadian Journal of Zoology 63:528–533.

Wootton, R. J. 1976. The biology of the sticklebacks. Academic Press, London.

ONTOGENETIC EVOLUTION, CLADE DIVERSIFICATION, AND HOMOPLASY

LARRY HUFFORD

Washington State University
Pullman, Washington

ANALOGOUS VARIATION AND EVOLUTIONARY DIVERSITY

Darwin (1859) identified analogous variation as a fundamental aspect of evolutionary diversification in his chapter on "Laws of Variation" in the *Origin of Species*. He recognized that analogous variations included the evolution of similar attributes in parallel in the descendants of a common ancestor as well as the reversion of descendants to attributes that were present in distant ancestors but not close ancestors. Analogous variation was seen not as a single phenomenon but as a set of processes. This insight provided the foundation for our current perception of homoplasy as including convergence, parallelism, and reversal. Other phenomena that Darwin identified as sources of evolutionary diversity included character divergence (i.e., novelty), hybridization, and extinction. In the history of evolutionary biology as a science, the ecological and "mechanistic" bases of various aspects of diversity, including character divergence and hybridization, have been

HOMOPLASY: The Recurrence of Similarity in Evolution, M. J. Sanderson and L. Hufford, eds.

studied repeatedly. The impact of extinction on diversity has also received recent attention (Diamond 1984; Jablonski 1986; Maynard Smith 1989; Norris 1991). In contrast, the evolutionary processes that give rise to homoplasy and its importance as an aspect of organismal diversity have received much less attention.

Homoplasy is a phenomenon associated inherently with clade diversification. Indeed, a clade must consist of at least three taxa before homoplasy can be perceived; but we typically appreciate the significance of homoplastic changes in the evolutionary diversification of groups that are much larger than three taxa. In contrast to character divergence phenomena, which are often investigated in studies focusing on an ancestor and a putative descendant, to study the origin of homoplastic change we need to explore diversification in a much broader monophyletic group. Cladistic phylogeny reconstructions have begun to open avenues along which we can investigate the diversification of broad monophyletic groups.

In this chapter, I consider the historical interplay of different avenues of character diversification via ontogenetic evolution to examine the origins of homoplasies. Ontogenetic evolution creates diversity in morphological characters. In a phylogenetic sense, the distinctions of descendants are established when their developmental patterns are transformed relative to those their ancestors. The evolutionary transformations of ontogenies have been studied to only a limited extent in the context of phylogenetic hypotheses. This approach is essential, however, to show how aspects of clade diversity arise via ontogenetic transformations. This ontogenetic transformation approach to clade diversity helps to address important questions about modes of rapid phenotypic change, correlated changes among characters, novelty and specialization, as well as homoplasy. For example, does homoplasy result because ontogenetic evolution is biased toward particular types of transformations? Data on ontogenetic evolution in clades are far too limited to definitively answer this question, although the scant preliminary data provide some insights that may spur further investigation. In this paper, floral evolution in three groups of angiosperms is used to explore the types of ontogenetic transformations that underlie homoplasies. The goal is to present hypotheses on how the transformations that are a part of ontogenetic evolution and clade diversification create homoplastic patterns.

MODELS OF ONTOGENY AND ONTOGENETIC EVOLUTION

Developmental patterns (ontogenies) can be described in various ways. Two relatively formal models have been used in recent investigations to describe ontogenetic evolution. *Heterochrony models*, derived primarily from Gould

(1977) and Alberch et al. (1979), emphasize temporal aspects of development to characterize how evolutionary transformations may be manifest in the rate of morphological change and in initiation and cessation times of developmental events. *Sequence models* represent a formalization of the traditional narrative description of ontogenies. Sequence models (e.g., O'Grady 1985; Kluge 1988; Langille and Hall 1989; Mabee 1989, 1993; Hufford 1995) emphasize the sequence of forms and their modification during the course of an ontogeny. Heterochrony and sequence models should be viewed as complementary, and each has strengths and weaknesses.

In this paper, I use a sequence model to describe ontogenies and their evolutionary transformation. Flower ontogenies are described in terms of "ontogenetic states." An ontogenetic state is an instantaneous form that can be recognized during the developmental transformations of an ontogeny.

Ontogenetic sequences can be modified in a limited number of ways. O'Grady (1985) identified the following three basic modes of sequence modification: additions, deletions, and substitutions. Hufford (1995) suggested that substitutions can include two different phenomena, distinguished as "novel substitutions" and "reciprocal substitutions." Novel substitutions involve the replacement of an ontogenetic state or set of states in the plesiomorphic sequence with a novel state or set of states in the derived sequence. Reciprocal substitutions involve the interchange of two existing states (or set of states) in the derived ontogenetic sequence relative to their positions in the plesiomorphic sequence. When we compare derived ontogenetic sequences with a plesiomorphic sequence, we recognize that the modes of sequence modification may be instituted at any one of a range of times during development (Fig. 1). In order to characterize the temporal span of development, we can operationally define three phases in an ontogenetic sequence: (1) initial, (2) middle, and (3) terminal.

While some evolutionary transformations are phase-specific, others may disrupt the entire subsequent course of ontogeny (Fig. 1). That is, some evolutionary transformations effectively result in a new ontogenetic sequence from the point at which they are instituted in the descendant. Roth and Wake (1985) characterized these transformations as "ontogenetic repatterning." The recognition of ontogenetic repatterning is important because we might otherwise characterize these modifications as terminal substitutions using the formalization described above since they modify the terminal course of the ontogeny. The significance of ontogenetic repatterning in reshaping the middle and even initial phases is lost if we simply describe these types of change as terminal substitutions. We can incorporate ontogenetic repatterning in the formalization described above by distinguishing between phase-specific transformations and repatterning transformations (Fig. 2). Phase-specific transformations are manifest in the derived ontogeny only during what are recognized as specific phases in the plesiomorphic ontogeny. Repatterning transformations institute a novel sequence in a derived

Plesiomorphic ontogenetic sequence: A→B→C→D

Apomorphic ontogenetic sequence transformations
that are phase specific:

Transformation modes	Transformation times		
	Initial	Middle	Terminal
Addition	X→A→B→C→D	A→B→X→C→D	A→B→C→D→X
Deletion	B→C→D	A→B→D	A→B→C
Novel substitution	X→B→C→D	A→X→C→D	A→B→C→X
Reciprocal substitution	B→A→C→D	A→C→B→D	A→B→D→C

Apomorphic ontogenetic sequence transformations
that are not phase specific:

Transformation modes	Transformation times	
	Initial	Middle
Repatterning	K→L→M→N	A→B→K→L→M

FIGURE 1   Ontogenetic sequences can be described as consisting of a set of ontogenetic states (instantaneous forms) in the course of development. Ontogenies can be characterized operationally as consisting of three temporal phases (initial, middle and terminal) in order to discuss the timing of evolutionary transformations. An ontogenetic transformation that occurs in the course of evolution can be (1) phase-specific and have only a limited temporal impact or (2) a repatterning transformation that alters the entire course of ontogeny from the time that it is instituted in the descendant relative to that of its ancestor. In the example above, individual ontogenies are represented by a sequence of letters, each denoting a different ontogenetic state. The plesiomorphic ontogeny proceeds from an initial state A through middle states B and C to a terminal state D. In the set of phase-specific transformations this plesiomorphic ontogeny can be modified by changes instituted at any of the three temporal phases and by a set of transformation modes that include additions, deletions, novel substitutions, and reciprocal substitutions. Repatterning transformations can be instituted at either the initial or middle phases of any ontogeny.

ontogeny from the time they are instituted in comparison with the plesiomorphic ontogeny. One problem that these definitions present is that terminal additions and substitutions would be regarded as both phase-specific and repatterning transformations. Hence, I will restrict repatterning transformations to those that are instituted in the derived ontogeny at what was either the initial or middle phase of the plesiomorphic ontogeny (and characterize them as initial repatterning transformations and middle repatterning transformations, respectively; Fig. 2).

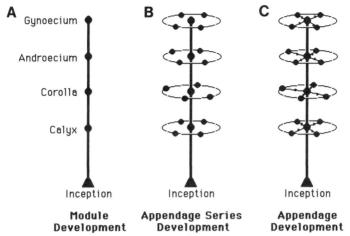

FIGURE 2 Flowers are complex reproductive regions of flowering plant shoot systems. The development of a flower might best be considered in terms of a hierarchy of ontogenies for each of its determinate aspects. This hierarchical pattern of ontogenies for the aspects of a flower are represented here. The most basic (or lowest hierarchical) level of flower development is the ontogeny of the module (A), which proceeds from the inception of the floral apical meristem (or transition from vegetative to floral meristem) through the initiation of the different appendage series. Superimposed on the module ontogenetic sequence, each appendage series (B) has an ontogenetic sequence. The ontogeny of each appendage series consists of the initiation of the appendages and developmental changes that characterize the entire series (e.g., synorganization via zonal growth and size and shape changes manifest in the entire set of appendages of a series). Within an appendage series, each appendage has an individual ontogenetic sequence (C) that begins with its initiation on the floral apical meristem.

FLOWER DEVELOPMENT

Flowers, the reproductive structures of angiosperms, are morphologically complex. They may be perceived simply as a specialized, determinate shoot axis that usually bears both fertile and sterile appendages. The floral axis is usually very short but displays regional specializations along its length. The lowermost region of a conventional flower is the perianth, the central region is the androecium, and the uppermost region is the gynoecium. The perianth of a conventional flower consists of two regions: a lower calyx and upper corolla. Each region of the flower consists of one or more series of appendages. For example, the corolla generally consists of one series of petals. The androecium, in contrast, often consists of more than one series of stamens. The gynoecium also can consist of one or more series of carpels.

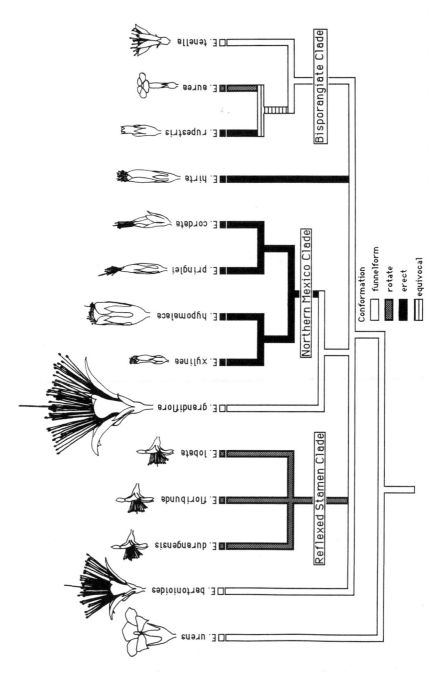

FIGURE 3 Cladogram of *Euenide* (Loasaceae) on which floral forms have been mapped to display the corolla conformations characteristic of the species at the time of pollination.

It is important to recognize that the development of a flower may be described in terms of a hierarchical set of ontogenies (Fig. 2). The basic ontogenetic level of the flower is that of the module. In the sense of a module proposed by White (1984), it is a unit derived from the development of a single apical meristem. The ontogenetic sequence for the module development of a conventional flower might be conceived to consist of its (1) initiation as a floral apical meristem, (2) the formation of the calyx appendage series, (3) the formation of the corolla appendage series, (4) the formation of the androecial appendage series, and (5) the formation of the gynoecial appendage series. Each of the appendages series, including the calyx through the gynoecium, also has an ontogenetic sequence. The ontogenetic sequence of an appendage series consists of the initiation sequence for the individual appendages and any unitary growth processes (e.g., zonal growth subjacent to individual appendage primordia or postgenital unification of primordia) that they display. Each appendage of each appendage series also has its own ontogenetic sequence. The examples in this paper pertain to ontogenetic sequences of appendage series.

HOMOPLASTIC TRANSFORMATIONS IN FLORAL ONTOGENETIC EVOLUTION

Corolla Conformation of Eucnide

The presentation of the flower corolla is an important aspect of reproductive display among most angiosperms. The corolla can serve as a visual display to attract pollinators, as a landing platform, as an aid for locating reproductive appendages and rewards, and even as the source of rewards, such as nectar, oils, or scents.

The small genus *Eucnide* (Loasaceae) has three basic conformations of the petals in corollas. Funnelform corollas, formed by the gentle outward curvature of the upper part of each individual petal in the flower, are hypothesized to be plesiomorphic in the genus since they are shared with putatively primitive members of the closely related *Mentzelia* (placed as the sister-group of *Eucnide* in the phylogenetic analyses of Hempel et al. 1994). Rotate corollas, in which the individual petals of the flower project perpendicularly from the floral axis, arose independently in two clades of *Eucnide* (Fig. 3). Erect corollas, in which the separate petals project upward, parallel to the floral axis, and remain imbricate, also arose two or three times in the genus (Fig. 3). Erect corollas are characteristic of *E. rupestris* of the bisporangiate stamen clade and in four members of the northern Mexico clade. They also characterize *E. hirta*, whose placement is currently equivocal. If *E. hirta* is most closely related to members of the white flowered clade and shares the

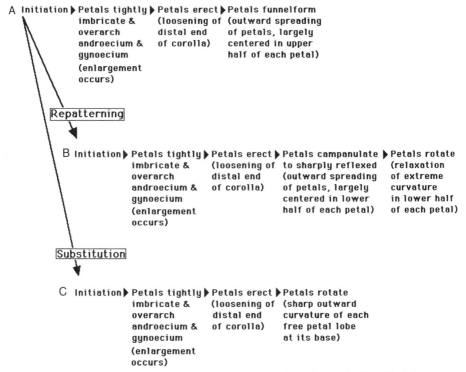

FIGURE 4   Ontogenetic sequences for the development of corolla form in *Eucnide*. A funnelform conformation of the corolla is hypothesized to be plesiomorphic for the genus (sequence A). Rotate corollas have been derived twice in *Eucnide*, and this evolutionary parallelism has involved two different modes of ontogenetic transformation. In the reflexed stamen clade (Fig. 3), the transformation has involved repatterning (sequences A to B). In the *E. aurea* (Fig. 3), this transformation has involved a novel substitution (sequences A to C).

derived corolla conformation of the northern Mexico clade, then erect corollas may have originated only twice in the genus. Alternatively, if *E. hirta* is more closely related to *E. bartonioides*, the reflexed stamen clade, or the clade consisting of both of these groups, then erect corollas originated separately at least three times in the genus.

Erect corollas are paedomorphic in *Eucnide* (Fig. 5). This corolla conformation has arisen with the evolutionary transformation of ontogenies by terminal deletion. The terminal deletion of the outward curvature state present in the more plesiomorphic ontogenies results in derived corolla ontogenies in which erect, loosely imbricate petals are the terminal state. Similar terminal deletions have occurred in the evolution of *E. rupestris*, the northern Mexico clade, and *E. hirta*. Most species with an erect corolla conformation have a loose distal opening. Two members of the northern Mexico

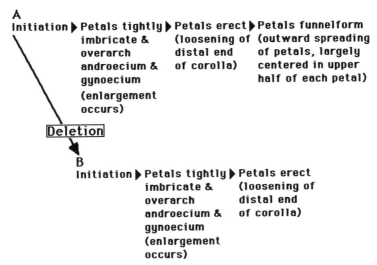

A
Initiation ▶ Petals tightly ▶ Petals erect ▶ Petals funnelform
imbricate &   (loosening of   (outward spreading
overarch   distal end   of petals, largely
androecium &   of corolla)   centered in upper
gynoecium   half of each petal)
(enlargement
occurs)

Deletion

B
Initiation ▶ Petals tightly ▶ Petals erect
imbricate &   (loosening of
overarch   distal end
androecium &   of corolla)
gynoecium
(enlargement
occurs)

FIGURE 5   Ontogenetic sequences for the development of corolla form in *Eucnide*. An erect corolla conformation has been derived in parallel in *Eucnide* at least twice. The derivation of the erect corolla conformation from an ancestor with a funnelform conformation involved paedomorphic terminal deletions as shown in the transformation from sequence A to B.

clade, *E. cordata* and *E. pringlei*, have a tight constriction at the distal end of the corolla. The erect corollas of *E. cordata* and *E. pringlei* display a slightly more paedomorphic form than the other taxa with an erect corolla when compared with the more plesiomorphic forms that have a funnelform corolla. In the northern Mexico clade, it is unclear whether sequential paedomorphic events (evolutionary transformations via terminal deletions) led to the two slightly different erect corolla forms or whether the erect corollas with a tight distal constriction arose first via paedomorphosis and then a slight peramorphic reversal was responsible for creating the erect corolla conformation with a loose distal opening (as found in *E. hypomalaca* and *E. xylinea*) (Fig. 6).

The rotate corolla conformation of *Eucnide* represents a divergence from the phenotypes present in the course of more plesiomorphic corolla ontogenies (Fig. 4). The two clades with rotate corollas differ in the developmental trajectories that result in the parallel conformations (Fig. 4). In the reflexed stamen clade, the rotate corolla originates through an ontogenetic transformation that is a middle repatterning. The corolla ontogeny is transformed at the time of the outward spreading of the petals, following the distal loosening phase of the corolla ontogeny. The outward spreading of the petals in the reflexed stamen clade is centered in the middle to lower half of each petal. This spreading leads to a strong revolute curvature in each petal. Subsequently, the curvature decreases and the upper part of each petal projects

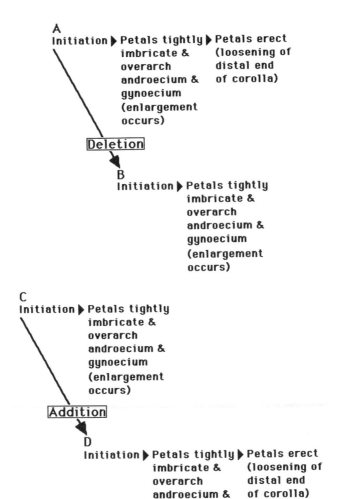

FIGURE 6   Among the species of *Eucnide* with an erect corolla conformation, *E. cordata* and *E. pringlei* differ from the others because the distal end of the corolla in these two species remains tightly imbricate rather than displaying the loosening of the imbrication that is present in all of the other species. It is currently unclear whether this corolla form of *E. cordata* and *E. pringlei* arose from an ancestor with an erect corolla (with a loose distal terminus) via an additional terminal deletion that created a slightly more paedomorphic corolla conformation in these two species (sequence A to B) or whether this conformation arose directly from an ancestor with a funnelform conformation (a slightly more paedomorphic change than shown in Fig. 5). If *E. cordata* and *E. pringlei* arose from an ancestor with a funnelform conformation, then their sister-group which has an erect corolla with a loose distal end may have arisen from an ancestor like them via a terminal addition that restored part of the more plesiomorphic ontogeny (sequence C to D).

perpendicularly from the floral axis, resulting in a rotate conformation. In *E. aurea* of the bisporangiate clade, the rotate corolla conformation appears to have evolved via a novel substitution. The development of the rotate corolla of *E. aurea* of the bisporangiate stamen clade proceeds more directly than in the reflexed stamen clade. Following the distal loosening phase of the imbrication of *E. aurea*, the outward spreading, which is centered at the base of each independent petal lobe, proceeds directly from the erect to perpendicular (rotate) orientation.

Central to the issue of clade diversification is the recognition that both derived corolla conformations in *Eucnide* are homoplastic. This may reflect partly the limited options for variation in corolla conformation among flowering plants as a whole. In flowers with radial symmetry, such as *Eucnide*, the range of possible corolla conformations that function adequately is limited further than for the entire range of corolla forms. It is interesting to look across the Loasaceae clade, outside *Eucnide*, to examine how the patterns within this genus are distributed.

Many members of Loasaceae outside of *Eucnide* possess a rotate corolla. A rotate corolla is found in some derived annual members of *Mentzelia* and also in *Petalonyx thurberi*. In both of these cases the evolution of the rotate corolla form is associated with small flowers that have a tubular basal region. Most members of subfamily Loasoideae have flowers with a rotate corolla conformation. In this subfamily, the nearly pervasive presence of rotate corollas may be associated with the elaboration of staminodes (sterile stamens). The staminodes in this group are not only large and colorful, but some also have a cucullate form that helps them serve as reservoirs for nectar (Brown and Kaul 1981). The large staminodes may help to limit self-pollination by preventing the fertile stamens from coming into contact with the stigma (Brown and Kaul 1981). A rotate corolla associated with another pollination specialization characterizes the monotypic *Schismocarpus*, which has a floral morphological syndrome like that of so-called pollen flowers (Hufford 1989). Pollen flowers offer pollen rather than nectar, oils, or some other material as a reward. Many pollen flowers have buzz pollination and forcibly eject the pollen from the stamens, although this has not yet been demonstrated in *Schismocarpus*. The rotate corolla of pollen flowers may be important in the presentation of a conelike conformation of the anthers and the bilaterally symmetrical gynoecium that is typical of pollen flowers.

The rotate corollas of *Eucnide* may function relatively simply as they do in the species of *Mentzelia* and *Petalonyx* that share this conformation. In these genera that have "generalist" pollination strategies, the rotate corolla may function largely in display and as as landing platform for relatively large insects such as bees and butterflies. Members of the reflexed stamen clade have widely spreading stamens over which bees and other large insects may clamber and may also contact the wings or bodies of resting butterflies. In contrast, the narrow corolla tube subjacent to the rotate corolla lobes of

E. aurea greatly restricts access to the stamens. Outcrossing in this species is most likely accomplished when butterflies or moths (or possibly humming-birds in color morphs that are orange) probe for nectar using a long, thin proboscis. These pollinators may use the rotate corolla as a platform for feeding. The common presence of rotate corollas among Loasaceae outside of *Eucnide* implies that its origin in this genus may not involve significant genomic changes and that it is tied to reproductive schemes that are readily adopted in the family.

The origin of the erect corolla conformation in *Eucnide* is interesting because a similar morphology does not appear to have been adopted as the reproductive form in any other genera of the family. Some South American species of tribe Loasoideae, particularly in *Cajophora*, have a corolla with erect petals that are extremely cucullate and not imbricate; hence, the con-formation in these species differs significantly from that found in *Eucnide*. The erect corolla conformation of *Eucnide* effectively holds the stamens and style of a flower tightly together and occludes the floral throat. One reason why this conformation does not occur among other members of the family may be that most have staminodes. In many species of *Mentzelia* and all those of subfamily Loasoideae, the staminodes are large and sometimes quite elaborate. In flowers with an erect corolla conformation in the which the imbricate petals formed a narrow tube, the display function of these stami-nodes would be nullified, and the staminodes might block access to the nec-tar by further congesting the floral throat. The erect corolla conformation may have evolved repeatedly in *Eucnide* because tight enclosure of the sta-mens against the style and the proximity of anthers to stigma promotes at least some self-pollination (if not self-fertilization) by the time of corolla and stamen abscission (Hufford 1988). Hence, it may help to provide some as-surance of sexual reproduction even if it does not involve out-crossing.

Corolla Size of Besseya

Size change, although not well investigated in angiosperms, is a common avenue of floral evolution. Size change influences how biotic pollinators can interact with flowers (Endress 1994a) and imposes limits on outcrossing when flowers become sufficiently small that anther to stigma distances be-come slight and pollinator entry (or probing) in the flower is restricted. In at least some groups, the size scaling of floral appendages can be an important avenue of evolutionary change (Hufford 1988), whereas other groups dis-play less positive covariation in size change among floral appendages.

Corolla evolution, particularly size change, is dramatic in *Besseya*, a small genus of Scrophulariaceae. The most distinctive aspect of corolla size evolu-tion in *Besseya* involves its extreme diminution in two species. Reproduc-tively mature flowers of *B. rubra* possess only rudimentary lobes of the individual petals, and those of *B. wyomingensis* display no evidence of a corolla (Hufford 1992). Using the most parsimonious cladogram for species relationships in *Besseya* (Hufford 1993), four possible avenues of ontogenetic evolution with different patterns of parallelism and/or reversal are equally possible as explanations for corolla size diversity (Hufford 1995; Figs 7–9).

Two of the alternative avenues of ontogenetic evolution are similar in beginning with an extremely paedomorphic modification of the plesiomorphic corolla ontogeny (Fig. 7). This would have resulted from a deletion of much of the corolla ontogenetic sequence displayed in *B. bullii* and *Synthyris*. In these two scenarios, the rudimentary corolla characteristic of *B. rubra* would have been shared by the group of *Besseya* called the Rocky Mountain clade. The alternative scenarios differ in the course that corolla evolution may have taken within the Rocky Mountain clade. One scenario (alternative 1) calls for further diminution of the corolla with the evolution of the higher elevation clade (*B. wyomingensis* and the Southern Rocky Mountain clade), whereas the other scenario (alternative 2) requires an immediate reversal to a "full size" corolla. The scenario in alternative 1 could be described as displaying sequential paedomorphic events (at the base of both the Rocky Mountain and the higher elevation clades) that significantly diminish corolla size followed by an evolutionary reversal that reestablishes a full size corolla (in the Southern Rocky Mountain clade). This reversal to a full size corolla would be established via an extensive terminal addition of ontogentic states that differ slightly from those in *B. bullii* and *Synthyris* (Hufford 1995). In alternative 2, the second paedomorphic event that resulted in the complete loss of corolla lobe expression in *B. wyomingensis* would not have directly followed the initial paedomorphic event at the base of the Rocky Mountain clade; instead, it would have occurred within the higher elevation clade after the evolutionary reversal reestablished the full size corolla. Hence, the second alternative calls not only for the homoplastic evolutionary reversal to a full size corolla, but also for two paedomorphic events that resulted in parallel origins of extemely diminutive corollas.

The third alternative calls again for the Rocky Mountain clade to be characterized by extreme corolla paedomorphosis (Fig. 8). In this scenario, however, the Rocky Mountain clade is characterized by a transformation that leads to a corolla like that found today in *B. wyomingensis*. This scenario calls for an evolutionary reversal to reestablish the full size corolla in the Southern Rocky Mountain clade. This would have originated by a terminal addition in the corolla ontogenetic sequence that reestablished lobe expression and their subsequent extension growth. This scenario calls for a parallel transformation via terminal addition that would have given rise to the corolla form found today in *B. rubra*. This parallel terminal addition results only in the reestablishment of the corolla lobes but not in extension growth that would result in a full size corolla. In this scenario, the terminal addition is a peramorphic process that is homoplastic in *B. rubra* and the

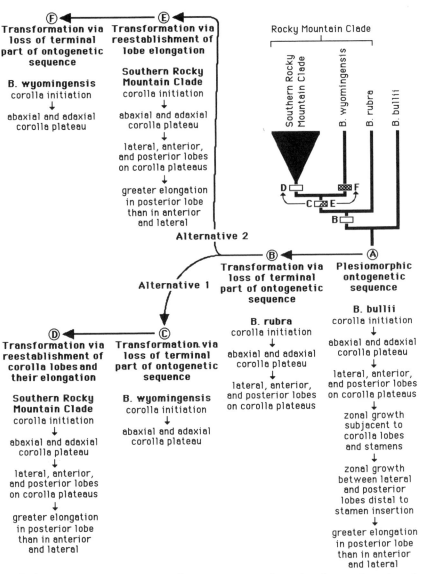

FIGURE 7 Two alternative scenarios for ontogenetic transformations that may have given rise to corolla diversity among species of *Besseya*. Alternative 1 involves successive terminal deletions that generate highly diminutive corollas and a peramorphic reversal to reestablish a full size corolla without a corolla tube. Alternative 1 is mapped on the topology of the most-parsimonious cladogram from Hufford (1993) with the changes marked by open rectangles. Alternative 2 involves parallel paedomorphic changes with intervening peramorphosis (mapped on the cladogram topology using checked rectangles). Two additional scenarios that are consistent with the topology of the most-parsimonious cladogram (Hufford, 1993) are shown in Figs. 8 and 9.

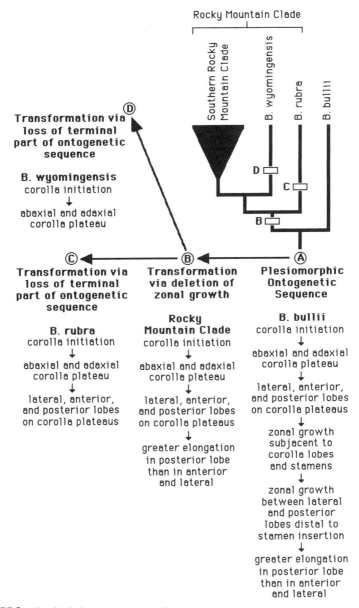

FIGURE 8 The third alternative scenario for ontogenetic transformations that may have given rise to the diminutive corollas of *Besseya rubra* and *B. wyomingensis* and the absence of a corolla tube in the Southern Rocky Mountain clade. This scenario calls for paedomorphosis followed by two peramorphic reversals that reestablish corollas of very different sizes.

Southern Rocky Mountain clade, but note that the corolla forms that result from this parallel change are quite different and would not be regarded as morphologically homoplastic.

The fourth alternative differs from the other scenarios in not requiring a reversal to achieve the corolla characteristic of the Southern Rocky Mountain clade (Fig. 9). In this alternative, the corolla form characteristic of this clade is hypothesized to have originated at the base of the Rocky Mountain clade via a fractionation of the zonal growth that is characteristic of corolla development in flowers of *B. bullii* and *Synthyris*. This alternative calls for parallel paedomorphic events via terminal deletions in an ontogeny like that found today only in the Southern Rocky Mountain clade to have yielded the diminished corollas of *B. rubra* and *B. wyomingensis*.

The range of alternative scenarios that explain equally well the potential avenues of corolla size evolution in *Besseya* precludes formulating a confident hypothesis about a specific pattern of homoplasy. Despite the range of alternatives, some important general observations can be made. First, parallelism and reversal can have complex interrelationships in the diversification of clades. In at least some of the described alternatives, reversals that reestablish a corolla size and form similar to that found in early members of the *Besseya/Synthyris* clade directly follow paedomorphic events that had greatly reduced the size of the corolla. Clade diversification by producing paedomorphs via terminal deletions and then reestablishing forms similar to those in ancestors as peramorphic reversals via terminal additions may be relatively common. This manifestation of homoplastic evolution may give diversification a strong back-and-forth aspect. It also suggests that significant diversity can arise in the absence of novelty.

Gynoecial Merosity of Piperales

Meristic change has been long recognized as one of the key aspects of flower diversification among angiosperms. Merosity has been used to convey at least two ideas in discussions of flower diversity. It may refer to the total number of appendages in any particular appendage series (e.g., five petals or ten stamens) or to the number of appendages in each whorl of an appendage series (e.g., a trimerous flower might have 6 stamens with three in each of 2 whorls). In this discussion, I will use merosity to refer to the total number of appendages in a particular appendage series, although the importance of appendage arrangement in different whorls also has a significant role in the discussion. In order to discuss meristic change we need to focus largely on aspects of phyllotaxy. Phyllotaxy refers to the sequence of appendage initiation at an apical meristem and the positions of these appendages. Hence, meristic evolution has both sequence (temporal) and positional components.

Piperales, an order of dicotyledons consisting of the small families Piperaceae and Saururaceae, has simple flowers that completely lack a perianth.

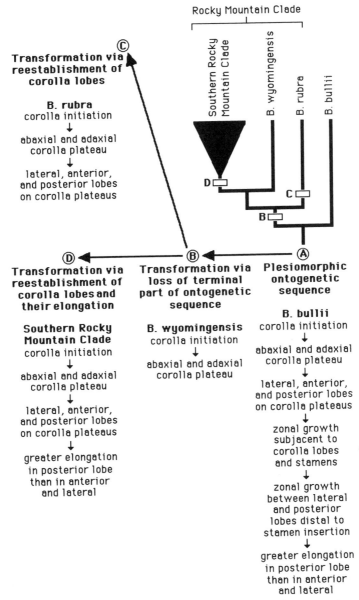

FIGURE 9 The fourth alternative scenario for ontogenetic transformations that may have given rise to corolla diversity among species of *Besseya*. This scenario calls for a direct transformation from corollas with a corolla tube to those without via the loss of zonal growth in the sinus region between the posterior and lateral lobes of the corolla. This is the only one of the four alternative scenarios (Figs. 7 and 8) for loss of the corolla tube that calls for its origin via a direct transformation from an ancestral ontogeny with more extensive sympetaly than that found in the descendant ontogeny. In this scenario, the highly diminutive corollas of *B. rubra* and *B. wyomingensis* result from parallel paedomorphic transformations.

All Saururaceae have bisexual flowers with an androecium and gynoecium. Most Piperaceae are also bisexual, except the genus *Macropiper*. A recent phylogenetic study (Tucker et al. 1993) has resulted in hypotheses of meristic homoplasy in the androecium and gynoecium of Piperales.

The phylogenetic study of Tucker et al. (1993) resulted in two most parsimonious cladograms when outgroup polarization of character states was used. One cladogram resulted in a monophyletic Piperaceae nested within a paraphyletic Saururaceae and the second in a monophyletic Piperaceae as the sister-group of a monophyletic Saururaceae. This second cladogram topology is congruent with results from a phylogenetic analysis of Magnoliidae using *rbc*L sequence data (Qiu et al. 1993) and will serve as the basis for my analysis (although it is important to note that the discussion of evolutionary transformations for Piperales would not differ if the alternative hypothesis of a paraphyletic Saururaceae had been used). The sister-group of Piperales based on the results of Qiu et al. (1993) consists of Aristolochiaceae and Lactoridaceae. They compose a clade that I will call the Aristolochiales. The plesiomorphic conditions for Aristolochiales will be regarded as most similar to those from which Piperales were derived. This basic assumption is necessary since the composition of the sister-group of the Aristolochiales– Piperales clade is uncertain and appears to consist of numerous genera with diverse attributes that could influence our interpretation of the plesiomorphic features of this clade.

Among Piperales, gynoecial merosity is homoplastic because tricarpellate flowers have evolved in parallel in both Piperaceae and Saururaceae (Fig. 10). In both families, the tricarpellate gynoecium evolved from a tetracarpellate state, which was identified on the cladogram as plesiomorphic for Piperales. Among Piperaceae, a clade consisting of *Macropiper*, *Pothomorphe*, *Peperomia*, and *Piper* shares the origin of the tricarpellate gynoecium. *Peperomia* has a derived meristic reduction to a single carpel in this group. In Saururaceae, a clade consisting of *Anemopsis* and *Houttuynia* also shares the derivation of a tricarpellate gynoecium.

Tucker et al. (1993) argued that the tetracarpellate gynoecia of Piperales were homoplastic, although they mapped a single origin for this character state at the base of the Piperales on a summary cladogram (their Fig. 6). They argued that tetracarpellate gynoecia are homoplastic because the sequence of carpel initiation differs among the three genera with this gynoecial merosity. This contrasts with my suggestion above that both sequence (timing) and position are important elements of phyllotaxy and should be used to comprehend meristic evolution. Examples of initiation sequence diversification in the context of conserved positions of appendages have been documented (e.g., the change to unidirectional from helical initiation of floral appendages in some legumes (Tucker 1984)). Hence, I regard the tetracarpellate gynoecia of *Saururus*, *Gymnotheca*, and *Zippelia* as evolutionary homologies despite the diverse initiation sequences that have arisen to produce these appendages. This hypothesis is fundamentally supported by the distribution of character states for gynoecial merosity on the phylogenetic hypothesis (Fig. 10), which suggests that it is congruent with the other character data analyzed by Tucker et al. (1993).

The sequence of carpel initiation is diverse in Saururaceae and Piperaceae (Liang and Tucker 1989; Tucker et al. 1993; it is unknown in the tricarpellate gynoecia of *Macropiper* and *Pothomorphe*). Among the tetracarpellate genera, *Zippelia* of Piperaceae and *Gymnotheca* of Saururaceae share the same sequence of carpel initiation. The lateral pair of carpels initiate first, the adaxial median carpel next, and the abaxial median carpel last. It may be most reasonable to assume that this is the plesiomorphic ontogenetic sequence for the gynoecial appendage series of the Piperales. If this hypothesis of plesiomorphy is true, then it has some interesting implications for understanding the divergent sequence of *Saururus* and the derived meristic reduction that is homoplastic in other genera.

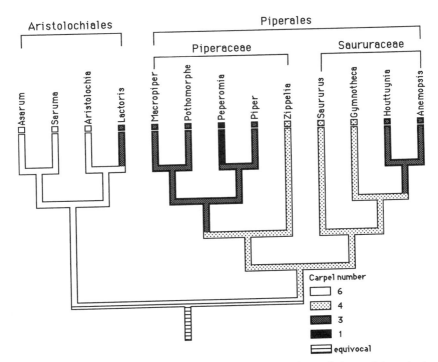

FIGURE 10 Gynoecial and androecial evolution among Piperales. A phylogenetic hypothesis for Piperales showing the optimization of character states for gynoecial merosity (carpel number). Trimerous gynoecia are homoplastic in Piperales. The phylogenetic hypothesis is based on cladistic analyses of Tucker et al. (1993) and Qiu et al. (1993). The sister-group of the Piperales, based on the results of Qiu et al. (1993), is Aristolochiales, which is illustrated here using exemplars from the families Aristolochiaceae and Lactoridaceae.

Saururus differs from *Zippelia* and *Gymnotheca* in its ontogenetic sequence for the gynoecial appendage series in having the adaxial and abaxial median carpels initiated simultaneously and before the lateral carpels. *Saururus* is interesting, however, because the initiation of its lateral carpels is the fifth initiation event in the development of the entire flower. This is the same sequential point in the entire floral ontogeny at which the lateral carpels form in the other tetracarpellate Piperales and in the tricarpellate *Piper*. In regard to the entire development of the flower, this may imply that it is the timing of the abaxial and adaxial median carpel expression that has been modified in *Saururus*.

The tricarpellate gynoecia that are homoplastic among Piperales can be hypothesized to be derived in the course of ontogenetic evolution by the parallel terminal deletion of the last initiated, median abaxial carpel of the more plesiomorphic ontogenies (as displayed by *Zippelia* and *Gymnotheca*). Tucker et al. (1993) previously cited this meristic reduction as an example of Rensch's generalization that the last parts initiated in ontogeny will be the first lost in evolution. Some flowers of *Anemopsis* possess two carpels rather than three, which appears to result from further terminal deletion of the adaxial median carpel (Tucker 1985). One of the most interesting aspects of this shift from a tetracarpellate to tricarpellate gynoecium is a heterotopic change. The heterotopy is associated with the maintenance of equal distances between adjacent carpels. This suggests that the siting of carpel primordia is relatively labile. This lability of carpel positioning is commonly associated with meristic change across the angiosperms.

Aristolochiales display meristic reductions that parallel those of the Piperales. The Aristolochiales have hexamerous gynoecia as the plesiomorphic state (Fig. 10). *Thottea* of Aristolochiaceae has four carpels (Leins et al. 1988) like the flowers of the piperalean *Zippelia*, *Saururus*, and *Gymnotheca*. *Lactoris* of the Lactoridaceae has three carpels (Tucker et al. 1993; Endress 1994b), as occurs in the derived gynoecia of some Piperales. The trimerous gynoecium of *Lactoris* consists of two lateral and one abaxial median carpel in contrast with that of Piperales in which an adaxial median carpel rather than abaxial is present. Floral development in Aristolochiaceae and Lactoridaceae has not been described in sufficient detail to understand how meristic changes in this group have been modified by ontogenetic evolution.

Androecial Merosity of Piperales

Stamen numbers are homoplastic among Piperales because trimerous androecia have evolved in parallel in both Piperaceae (*Macropiper*) and Saururaceae (*Houttuynia*) from the hexamerous androecia that are plesiomorphic in the order (Fig. 11). Tucker et al. (1993) suggested that Piperaceae display a reduction series in their androecial evolution from hex-

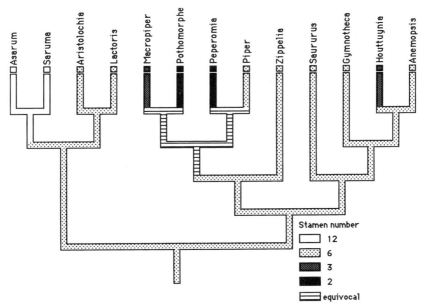

FIGURE 11 Gynoecial and androecial evolution among Piperales. A phylogenetic hypothesis for Piperales showing the optimization of character states for androecial merosity. Trimerous androecia are homoplastic in Piperales. Dimerous androecia may be homoplastic in Piperaceae.

amerous (*Zippelia, Piper*) to trimerous (*Macropiper*) and dimerous (*Peperomia, Pothomorphe*). They mapped the separate origin of dimerous androecia in *Peperomia* and *Pothomorphe* on a summary cladogram (their Fig. 6), although this is only one of three equally parsimonious scenarios for androecial evolution in Piperaceae. For example, it is equally parsimonious on the basis of the Tucker et al. (1993) cladogram to suggest that dimerous androecia evolved once and that flowers with trimerous androecia in *Macropiper* and those with a reversal to hexamerous androecia in *Piper* have been derived from them. This ambiguity may not exist if the cladogram presented by Tucker et al. (1993) is inaccurate because of sampling limitations. They suggest, for example, that *Piper* may be paraphyletic with regard to both *Macropiper* and *Pothomorphe*. Such a revision of the phylogeny, which might occur with more extensive sampling, may indicate that *Piper* retains the hexamerous plesiomorphic androecia (as apparently assumed by Tucker et al.) and that the trimerous and dimerous androecia are derived from this state.

The evolution of the trimerous androecium of *Houttuynia* among Saururaceae is a consequence of ontogenetic transformation via terminal deletion (Fig. 12). This ontogenetic transformation, however, is associated with a number of developmental novelties in the clade that consists of *Gymnotheca*,

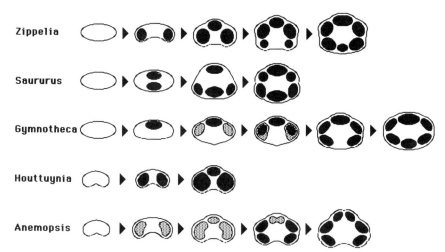

FIGURE 12   Ontogenetic sequences for androecia in Saururaceae and *Zippelia* of Piperaceae (based on Tucker 1975, 1981, 1985, Liang and Tucker 1989, 1995). The diagrams demonstrate the basic shape of the floral apex when viewed from above and the sequence of primordium initiation in the androecium. The ontogenetic sequence for the androecium of each illustrated species proceeds from left to right; and it begins with the floral apex prior to the inception of the androecium and ends with floral apex after the final primordium of the androecium has been established. Solid areas are primordia initiated on the floral apex that develop directly into individual stamens. Stippled areas are common primordia. Common primordia arise initially on an apex, with the general properties of all other appendage primordia, but transform developmentally into two separate stamen primordia in Piperales. Common primordia arise in lateral positions in *Gymnotheca*, and each gives rise to two individual stamen primordia. Common primordia arise in lateral and adaxial positions in *Anemopsis*, and each gives rise to two individual stamen primordia. Diagrams are oriented such that adaxial is toward the top of the page and abaxial toward the bottom.

Anemopsis, and *Houttuynia*. The key to our understanding of the evolution of trimery among Saururaceae lies in a developmental novelty called a "common primordium." A common primordium arises initially on the apical meristem as a single primordium that possesses the general properties of all other appendage primordia. A common primordium soon transforms developmentally into two or more independent primordia. Common primordia have been observed frequently in studies of androecial development, especially in clades of taxa in which polyandry has been instituted secondarily (Ronse Decraene and Smets 1987; Endress 1994a; Leins and Erbar 1994).

In *Gymnotheca* (Fig. 12), androecial development begins with the formation of a median adaxial primordium, followed by the simultaneous formation of a lateral primordium on each extreme lateral end of the ellipsoidal apical meristem. Each lateral primordium is a common primordium that gives rise first to an adaxial lateral primordium and then to an abaxial lateral primordium. The stamen initiation sequence terminates with the formation of an abaxial median primordium. *Anemopsis* has a similar siting of initial androecial primordia to that observed in *Gymnotheca* but its ontogeny is distinctive in three ways (Fig. 12). First, the androecial ontogenetic sequence of *Anemopsis* begins with the simultaneous formation of the lateral primordia at each lateral extreme of the floral apical meristem. The formation of the two lateral primordia in *Anemopsis* is followed by the initiation of an adaxial median primordium. These are the only three androecial primordia formed directly on the floral apical meristem of *Anemopsis*. Each of these three androecial primordia is a common primordium. The second important evolutionary distinction in the ontogeny of *Anemopsis* is that each of these common primordia forms two stamen primordia. The adaxial median common primordium that forms two adaxial median stamen primordia is a novelty in Saururaceae. These two stamens combined with the two that form on each lateral common primordium compose an androecium of six stamens; hence, the hexamerous androecium of *Anemopsis* is homoplastic with that of other Saururaceae. The third important evolutionary transformation of androecial ontogeny displayed by *Anemopsis* relative to the hexamerous *Gymnotheca*, *Saururus*, and *Zippelia* is that the abaxial median primordium does not form (Fig. 12). If the androecial ontogeny of *Anemopsis* is derived from an ancestor with a floral ontogeny like that displayed by *Gymnotheca* and *Zippelia* (Fig. 12), then the loss of its abaxial median primordium represents an ontogenetic transformation via terminal deletion.

The terminal deletion of the abaxial median stamen primordium in *Anemopsis* is central in understanding the trimery of *Houttuynia*. Androecial development begins in *Houttuynia* as it does in *Anemopsis* with the formation of three primordia: two lateral and one adaxial median. Unlike *Anemopsis*, however, each of the three primordia that forms on the floral apical meristem differentiates directly into a stamen rather than serving as a common primordium on which stamen primordia would subsequently form. Thus, a second important aspect of the transformation of the *Houttuynia* androecial ontogeny from that of an ancestor similar to *Gymnotheca* (and expressed also in *Anemopsis*) was the direct development of the lateral primordia as stamens rather than as common primordia.

Although floral development is insufficiently investigated in Piperaceae to trace the ontogenetic sequence transformations in the evolution of the androecial trimery of *Macropiper*, some general hypotheses shed light on this change. *Zippelia* and *Piper*, the Piperaceae with hexamerous androecia, differ in the ontogenetic sequence in which stamens are initiated, but they retain the positions of the secondarily forming stamens that are positioned directly opposite the perianth appendages of *Saruma* and *Asarum* (Leins and Erbar 1985, Dickison 1992, Tucker et al. 1993). *Piper*, for example, displays an androecial development pattern in which three stamens in one whorl are initiated first and then three stamens of the next whorl form. If *Macropiper* is nested within a paraphyletic *Piper* assemblage (Callejas 1986), then its trimery could have evolved via an ontogenetic transformation by terminal deletion of the initiation of the second whorl of three stamens. This scenario of reductive evolution resulting in androecial trimery in *Macropiper* calls for loss of what might be regarded as a "complete" whorl of three stamens.

ONTOGENETIC EVOLUTION, CLADE DIVERSIFICATION, AND HOMOPLASY

We understand little about how diversity arises through ontogenetic evolution. We face this limitation because too few developmental investigations have been conducted in the context of rigorous phylogeny reconstructions. Even when developmental studies have used explicit phylogenies to establish the evolutionary polarities for developmental change, they are often limited by sampling problems. If developmental studies do not approach extensive sampling for the full range of morphological variation in a study group, then estimates of the avenues of ontogenetic transformations and their frequencies may be skewed. Hence, taxon saturation for developmental studies is very important. Developmental investigations often limit sampling only to two or three taxa, which is insufficient to uncover homoplasies and reconstruct the transformations that underlie them. Despite the few investigations, particularly among flowering plants, that may be used to frame generalizations about ontogenetic evolution, it is important to do so in order to present explicit hypotheses that can be tested.

I present hypotheses to help comprehend how homoplasy and other patterns of diversity arise. I will use the metaphor of a railroad to examine ideas about the role of ontogenetic evolution in clade diversification. We can analogize each run made by a train with the ontogeny of an individual. Each run differs slightly in pace and duration. As such, each reaches the stations along the line and the final destination in a distinct way. Like the ontogenies of individuals of the same species, a railroad provides a trajectory that prescribes a similar outcome each trip, but each run is actually unique. Hence, we might see conceptually the railroad as resembling the genetic framework of a species that helps to specify the phenotypes by setting the trajectories and onset and offset signals for the development of each individual.

The stations along a rail line differ in importance, depending upon the schedule of the particular trip. The stations exist whether a particular train stops or not. We might think of stations as akin to special functional states that exist in ontogenies. These special functional states are similar to other ontogenetic states that might be defined during development, but they may have particular importance because they are associated with a suite of functional events or "phase" changes during development. Metamorphosis in amphibians may be a special functional state similar to a major train station where many passengers change and cars may be attached or removed. In shoot systems of flowering plants, special functional states may be akin to the shift from vegetative to reproductive phases. During module development of a flower, these special functional states may correspond to events such as the initiation of a new whorl of appendages. In the development of the module of a conventional flower, we can envision that four special functional states, corresponding to calyx, corolla, androecium, and gynoecium initiation, respectively, would be present. In the evolution of flowers, especially those that are unisexual, some of these functional states become deemphasized or are not made operational (i.e., the train does not stop at a station). For example, in a flower that is functionally and morphologically female, the ontogeny of the floral module may involve initiation of the calyx, corolla, and gynoecium, but the inception of the androecium is by-passed.

Critical for our goals, however, is how species evolve via ontogenetic transformations. If we consider how railroads change, it may be best to look first at the great opportunity provided by regional railroads. They may use the same rail lines of "nonstop," intercity trains but go only a short distance and connect a few small neighboring communities. The diversification of rail systems by exploiting an existing framework may be similar to the way that clades diversify by using the existing genetic and ontogenetic frameworks. A simple way of diversifying a rail system is to convert intermediate stations into final destinations. Such a change would be akin to paedomorphic evolution mediated by heterochrony. These paedomorphic changes may appear most often as ontogenetic transformations via terminal deletions. Kubitzki et al. (1991) suggested that gene silencing may play a role in homoplasy. Gene silencing may be especially associated with the heterochrony that causes terminal deletions. Ontogenetic transformations via terminal deletions led to homoplastic paedomorphic forms in each of the three examples examined above. Notably, parallel paedomorphic events were the most common avenue through which homoplasies arose. Paedomorphic homoplasies have been identified previously in other organisms, including rhipidistian and dipnoan fishes (Bemis 1984), plethodontid salamanders (Hanken 1984), and, perhaps, reptiles (Rieppel 1984). Wake (1991) suggested that paedomorphic homoplasy was one of the significant ways in which heterochrony impacts evolutionary diversification.

Various factors might contribute to the commonality of paedomorphic homoplasy. First, paedomorphic homoplasy may be common if terminal deletions simply occur more frequently (or are more often successful) during diversification than other avenues of ontogenetic evolution. Few data are available on the relative frequencies of the different avenues of ontogenetic evolution. Mabee (1993) estimated that terminal changes in ontogenies account for 76–78% of the modifications in groups of centrarchid fish.

Terminal deletions accounted for 41% of the ontogenetic transformations found by Mabee. Terminal additions were only slightly less common, accounting for 32% of the transformations. Frequency estimates for the avenues of ontogenetic evolution during floral diversification of *Besseya* (Hufford unpublished data) are similar in demonstrating that terminal deletion, middle repatterning, and terminal additions or substitutions are the most common avenues of ontogenetic transformation. These observations may provide some support for the proposal that paedomorphic homoplasy is common simply because of the relatively high frequency of terminal modifications in evolutionary change.

A second factor that may underlie the commonality of paedomorphic homoplasy is that some ontogenetic states have special properties that increase the likelihood that they will become the terminal state following evolutionary truncations via terminal deletion. Wake (1991), for example, suggested that metamorphosis in salamanders acts as a default state to which ontogenies are often truncated by paedomorphosis. The functional, default state presented at metamorphosis may have special properties in some ontogenies that biases heterochrony toward truncation to this point (at least among the successful members of clades). Similar default states have not been hypothesized previously to have a role in the homoplasies observed among plants. Among flowers, however, the initiation points for appendage whorls within a particular series of appendages (e.g., the androecium) might be considered to be "default sets." For example, among dicots, androecia often display homoplastic changes from haplostemony (a single series of stamens: each stamen is opposite a sepal) to diplostemony (two series of stamens: stamens of the outer whorl are opposite sepals, whereas those of the inner whorl are opposite petals) and the reverse. In these cases, ontogenetic evolution tends to proceed by forming an entire whorl or not forming an entire whorl that was present in the ancestor. We can identify those special cases when stamen numbers tend to change by the addition or deletion of incomplete whorls, including polystaminate androecia that consist of more than two whorls (e.g., Hydrangeaceae and Loasaceae), helical rather than whorled androecia phyllotaxy (e.g., among Magnoliidae), bilateral symmetry in the perianth or flower as a whole (e.g., *Penstemon*), and extreme floral simplification or size reduction (e.g., Piperales and tribe Veroniceae of Scrophulariaceae). Among flowers, gynoecia commonly evolve by adding or deleting a single carpel and do not use whorls as default sets. Gynoecia, however, seldom consist of more than one whorl of carpels and tend to be numerous only in flowers with a helical arrangement.

Typically, paedomorphosis has been considered to be an evolutionary dead end (Gould 1977) or associated with clinal trends (McNamara 1982, McKinney 1988, Lammers 1990). We may need a broadened outlook on the role that paedomorphosis plays in diversification. For example, the diversification of some clades may involve a close tracking of paedomorphic and peramorphic transformations affecting the same character suite. Such evolutionary tracking might commonly produce reversals. For example, some of the scenarios for corolla size evolution in *Besseya* demonstrated that homoplasy can arise when peramorphic reversals follow paedomorphosis in clades. Using the Kubitzki et al. (1991) suggestion that gene silencing has a particular role in homoplastic evolution, we could hypothesize that a close tracking of paedomorphic and peramorphic transformations might be observed in a clade because particular genes may be readily silenced at one point and subsequently reactivated.

Ontogenies clearly evolve in more ways than simply shortening existing trajectories (Fig. 1; see Alberch et al. (1979) for an alternative perspective using heterochrony models). Elaborative transformations via both addition and repatterning must be important avenues of clade diversification and essential for the origin of novelties. Consider terminal additions as a simple case of elaborative diversification. Elaborative diversification by terminal addition might seem initially to offer tremendous potential for diversification because of the nearly unlimited directions in which ontogenies could evolve. To see some of the limitations for elaborative diversification more clearly, we can return to the railroad analogy and ask what constrains the choice of a new destination when tracks are added beyond what was formerly its final destination. Two chief constraints for railroads will be the locations of communities or other potential sites for viable new stations and the topography of the landscape. Landscape constrains railroads because of functional limitations imposed by turning and climbing capabilities as well as the expense and technical difficulty of building large bridges and tunnels or surmounting mountain passes. Both of these constraints can be extrapolated to ontogenetic evolution. The locations of new stations for railroads corresponds to the architecture of new special functional states in the ontogeny. Although ontogenies could be transformed by various additions, resulting in novel terminal ontogenetic states, few of these might result in evolutionarily successful individuals because the new terminal states are nonfunctional or insufficient as selection increases. Elaborative evolution by terminal addition may be constrained by the limited number of special functional states that may be assumed by addition. The developmental avenues through which these special functional states may be found is limited by mechanical constraints on organismal structure that are akin to the limitations placed on rail lines by topography. We can think about ontogenetic repatterning in much the same way as we consider terminal addition in terms of these contraints.

Among the examples, the homoplastic rotate corolla conformation of *Eucnide* arose via elaborative transformations. It is particularly interesting that this rotate conformation derived as the terminal ontogenetic state for the corolla of both *E. aurea* and the reflexed stamen clade was achieved by instituting two different trajectories through repatterning and novel substitution. As discussed above, a rotate corolla conformation is common among Loasaceae and has arisen as an evolutionary parallelism at least six times (including twice in *Eucnide*). The occurrence of rotate corollas in *Eucnide* may be controlled by simple genetic changes that occur commonly in small clades. More significant, however, is the fitness of the rotate corolla conformation as a terminal state because of its functional properties in relation to other attributes of floral form and function. For example, it provides a prominent display and a broad landing surface for pollinators (Faegri and van der Pijl 1979) and the petals in this conformation are held away from, and do not interfere with, positional changes among stamens and carpels or pollinator interaction with these appendages. Functional and constructional contraints may tend to channel ontogenetic evolution along a relatively few avenues. The terminal states of some of these avenues may have particularly important functional properties that enhance their evolutionary success under strong selection (cf. Armbruster, this volume). The constraints and patterns of selection under which clades diversify may result in a limited range of successful, adaptive phenotypes that tend to display parallelisms.

CONCLUSIONS

Two phenomena observed in this study are suggested to be particularly important in the generation of homoplasy: (1) biases in the pattern of ontogenetic evolution and (2) a strong back-and-forth aspect in character diversification in small clades when paedomorphosis and peramorphosis follow closely one another. The latter phenomenon may be tied largely to gene silencing and reactivation (Kubitzki et al. 1991). Biased ontogenetic evolution is more complex. The scant data (Mabee 1993; Hufford 1994) on frequencies of different modes of ontogenetic evolution can be used to suggest that biases exist because of the tendency of ontogenies to be modified by terminal transformations, especially deletions. The relative commonality of terminal deletions may generate homoplasies in the form of paedomorphic parallelisms, but also by setting the stage for peramorphic reversals that follow paedomorphosis.

The biased success of different avenues of ontogenetic evolution clearly involves limitations imposed by functional and mechanical constraints. The success of paedomorphs may be high because their evolution does not require experimentation with new constructional arrangements that might create mechanical problems (because the juvenilized architecture of the descendant was already largely present in the ontogeny of the ancestor). Similarly, they may be successful because the elements of the juvenilized architecture may have positional relationships that are to some extent "preadapted" (exapted) because their ontogenetic positions in the ancestor were tied to their eventual functional positions and roles in later ontogenetic states. Elaborative ontogenetic transformations through addition, substitution, or repatterning may be associated less often with homoplastic evolution than is paedomorphosis. An exception may be those elaborative changes that simply reexpress ontogenetic states deleted in a recent ancestor (i.e., elaborative transformations that follow paedomorphosis). Mechanical and functional limitations on elaborative ontogenetic transformations serve to bias those modes of ontogenetic evolution along a few trajectories, and, hence, serve to increase the likelihood that the resulting phenotypes will be homoplastic.

Homoplasy is characteristic of clade diversity. It might be best, however, to see homoplasy as a pattern superposed on the phenomena, such as size scaling, mosaic evolution, origin of novelty (innovation), and juvenilization, that are responsible for the origin of diversity. Thus, one of the most reasonable ways of investigating homoplasy is primarily to approach it through studies that examine the range of avenues through which clades diversify, to ask about the differing roles and significance of the avenues of diversification, and to determine how the trajectory of each of these avenues of diversification may be biased. These more comprehensive investigations of organismal and character diversification may be necessary to understand the complexities that underlie homoplasies.

ACKNOWLEDGMENTS

This work was supported in part by National Science Foundation Grants BSR-8800206 and DEB-9496127. I thank Richard Bateman and Michael Sanderson for comments on the manuscript and the *American Journal of Botany* for permission to reproduce Figs. 7–9.

REFERENCES

Alberch, P., S. J. Gould, G. F. Oster, and D. B. Wake. 1979. Size and shape in ontogeny and phylogeny. Paleobiology 5:2296–317.

Armbruster, W. S. 1996. Exaptation, adaptation, and homoplasy: Evolution of ecological traits in *Dalechampia*. Pp. 225–241 *in* M. J. Sanderson and L. Hufford, eds. Homoplasy: The recurrence of similarity in evolution. Academic Press, San Diego.

Bemis, W. F. 1984. Paedomorphosis and the evolution of the Dipnoi. Paleobiology 10:293–307.

Brown, D. K., and R. B. Kaul. 1981. Floral structure and mechanism in Loasaceae. American Journal of Botany 68: 361–372.

Callejas, R. 1986. Taxonomic revision of *Piper* subgenus *Ottonia* (Piperaceae). Ph.D. dissertation, City University, New York.

Darwin, C. 1859. The origin of species. John Murray, London. (1964 facsimile, Harvard University Press, Cambridge).

Diamond, J. M. 1984. Historic extinctions: a Rosetta Stone for understanding prehistoric extinctions. Pp. 824–862 *in* P. S. Martin and R. G. Klein, eds. Quaternary extinctions: A prehistoric revolution. University of Arizona Press, Tucson.

Dickison, W. C. 1992. Morphology and anatomy of the flower and pollen of *Saruma henryi* Oliv., a phylogenetic relict of the Aristolochiaceae. Bulletin of the Torrey Botanical Club 119:393–400.

Endress, P. K. 1994a. Diversity and evolutionary biology of tropical flowers. Cambridge University Press, Cambridge.

Endress, P. K. 1994b. Floral structure and evolution of primitive angiosperms: Recent advances. Plant Systematics and Evolution 192:79–97.

Fægri, K., and L. van der Pijl. 1979. The principles of pollination biology. Pergamon Press, Oxford.

Gould, S. J. 1977. Ontogeny and phylogeny. Belknap, Cambridge.

Hanken, J. 1984. Miniaturization and its effects on cranial morphology in plethodontid salamanders, genus *Thorius* (Amphibia: Plethodontidae). I. Osteological variation. Biological Journal of the Linnean Society 23:55–75.

Hempel, A., P. A. Reeves, R. G. Olmstead, and R. K. Jansen. 1994. Implications of *rbcL* sequence data for higher order relationships of the Loasaceae and the anomalous aquatic plant *Hydrostachys* (Hydrostachyaceae). Plant Systematics and Evolution 194:25–37.

Hufford, L. 1988. Potential roles of scaling and post-anthesis developmental changes in the evolution of floral forms of *Eucnide* (Loasaceae). Nordic Journal of Botany 8:147–157.

Hufford, L.. 1989. The structure and potential loasaceous affinities of *Schismocarpus*. Nordic Journal of Botany 9:217–227.

Hufford, L. 1992. Floral structure of *Besseya* and *Synthyris* (Scrophulariaceae). International Journal of Plant Sciences 153:217–229.

Hufford, L. 1993. A phylogenetic analysis of *Besseya* (Scrophulariaceae). International Journal of Plant Sciences 154:350–360.

Hufford, L. 1994. Frequencies of modes of ontogenetic evolution in perianth diversification of *Besseya* (Scrophulariaceae). American Journal of Botany 81 (supplement):162

Hufford, L. 1995. Patterns of ontogenetic evolution in perianth diversification of *Besseya* (Scrophulariaceae). American Journal of Botany 82:655–680.

Jablonski, D. 1986. Mass and background extinctions: The alternation of macroevolutionary regimes. Science 231:129–133.

Kluge, A. G. 1988. The characteristics of ontogeny. Pp. 57–82 *in* C. J. Humphries, ed. Ontogeny and systematics. Columbia University Press, New York.

Kubitzki, K., P. v. Sengbusch, and H.-H. Poppendieck. 1991. Parallelism, its evolutionary origin and systematic significance. Aliso 13:191–206.

Lammers, T. G. 1990. Sequential paedomorphosis among the endemic Hawaiian Lobelioideae (Campanulaceae). Taxon 39:206–211.

Langille, R. M., and B. K. Hall. 1989. Developmental processes, developmental sequences and early vertebrate phylogeny. Biological Review 64:73–91.

Leins, P., and C. Erbar. 1985. Ein Beitrag zur Blütenentwicklung der Aristolochiaceen, einer Vermittlergruppe zu den Monokotylen. Botanische Jahrbücher fur Systematik 107:343–368.

Leins, P., and C. Erbar. 1994. Flowers in Magnoliidae and the origin of flowers in other subclasses of the angiosperms. II. The relationships between flowers ofMagnoliidae, Dilleniidae, and Caryophyllidae. Plant Systematics and Evolution 8:209–218.

Leins, P., C. Erbar and W. A. van Heel, 1988. Note on the floral development of *Thottea* (Aristolochiaceae). Blumea 33:357–370.

Liang, H.-X., and S. C. Tucker. 1989. Floral development in *Gymnotheca chinensis* (Saururaceae). American Journal of Botany 76:806–819.

Liang, H.-X., and S. C. Tucker. 1995. Floral ontogeny of *Zippelia begoniaefolia* and its familial affinity: Saururaceae or Piperaceae. American Journal of Botany 82:681–689.

Mabee, P. M. 1989. Assumptions underlying the use of ontogenetic sequences for determining character state order. Transactions of the American Fisheries Society 118:159–166.

Mabee, P. M. 1993. Phylogenetic interpretation of ontogenetic change: Sorting out the actual and artefactual in an empirical case study of centrarchid fishes. Zoological Journal of the Linnean Society 107:175–291.

Maynard Smith, J. 1989. The causes of extinction. Philosophical Transactions of the Royal Society, London B 325:241–252.

McKinney, M. L. 1988. Heterochrony in evolution, an overview. Pp. 327–340 *in* M. L. McKinney, ed. Heterochrony in evolution. Plenum Press, New York.

McNamara, K. J. 1982. Heterochrony and phylogenetic trends. Paleobiology 8:130–142.

Norris, R. D. 1991. Biased extinction and evolutionary trends. Paleobiology 17:388–399.

O'Grady, 1985. Ontogenetic sequences and the phylogenetics of parasitic flatworm life cycles. Cladistics 1:159–170.

Qiu, Y.-I., M. W. Chase, D. H. Les, and C. R. Parks. 1993. Molecular phylogenetics of the Magnoliidae: Cladistic analysis of nucleotide sequences of the plastid gene *rbc*L. Annals of the Missouri Botanical Garden 80:587–606.

Rieppel, O. 1984. Minaturization of the lizard skull: Its functional and evolutionary implications. Symposia of the Zoological Society of London 52:503–520.

Ronse Decraene, L.-P., and E. Smets. 1987. The distribution and the systematic relevance of the androecial characters oligomery and polymery in the Magnoliophytina. Nordic Journal of Botany 7:239–253.

Roth, G., and D. B. Wake. 1985. Trends in the functional morphology and sensorimotor control of feeding behavior in salamanders: An example of the role of internal dynamics in evolution. Acta Biotheoretica 34:175–192.

Tucker, S. C. 1975. Floral development in *Saururus cernuus* (Saururaceae). 1. Floral initiation and stamen development. American Journal of Botany 62:993–1007.

Tucker, S. C. 1981. Inflorescence and floral development in *Houttuynia cordata* (Saururaceae). American Journal of Botany 68:1017–1032.

Tucker, S. C. 1984. Unidirectional organ initiation in leguminous flowers. American Journal of Botany 71:1139–1148.

Tucker, S. C. 1985. Initiation and development of inflorescence and flower in *Anemopsis californica* (Saururaceae). American Journal of Botany 72:20–31.

Tucker, S. C., A. W. Douglas, and H.-X. Liang. 1993. Utility of ontogenetic and conventional characters in determining phylogenetic relationships of Saururaceae and Piperaceae (Piperales). Systematic Botany 18:614–641.

Wake, D. B. 1991. Homoplasy: The result of natural selection, or evidence of design limitations. American Naturalist 138: 543–567.

White, J. 1984. Plant metamerism. Pp. 15–47 *in* R. Dirzo and J. Sarukhán, eds. Perspectives on plant population ecology. Sinauer, Sunderland, MA.

HOMOPLASY IN ANGIOSPERM FLOWERS

PETER K. ENDRESS

University of Zurich
Zurich, Switzerland

INTRODUCTION

Homoplasy, the result of evolution of similar structures in different lineages
from a common ancestor that did not have these structures, is a general
phenomenon in organisms. Evolution proceeds in the direction of the least
resistance (Stebbins 1974). This can be visualized as a landscape, where
travel is easiest along certain pathways, and so certain directions are followed most of the time, while others are hardly ever followed. Speaking in
terms of evolution there is a proximate and an ultimate component: Directions are favored that are easy from a morphogenetical (organizational)
point of view ("epigenetic landscape," Waddington 1940) and where selective forces lead to certain patterns ("adaptive landscape," Wright 1932;
Grant 1963, 1989). These two components may be integrated as an evolutionary landscape that is modeled by both organizational and ecological
constraints (see also Wagner 1988). The most closely related groups of organisms wander in the most similar landscapes. There are certain easy tracks

HOMOPLASY: The Recurrence of Similarity in Evolution, M. J. Sanderson and L. Hufford, eds.

where there is a constant going (parallel evolution) and coming back (evolutionary reversals). Many of the most closely related groups reach a similar goal. This is homoplasy. In contrast, in convergent evolution similar goals are reached from different starting points.

A species has diverse genotypes. The same states of certain genes may be selected repeatedly under special conditions to give rise to several populations that are similar in this respect. A more subtle mechanism is a genetic repertoire within a genotype, whereby different genes may become activated and others remain silent under different conditions, or where a genetically fixed prepattern may give rise to different patterns in one individual (phenotypic plasticity) (see, e.g., Van Valen 1979; Wagner 1982b; Bachmann 1983; West-Eberhard 1986; Kubitzki et al. 1991; Bateman and Dimichele 1994). An "epigenetic" example in flowers is *Distylium* and *Distyliopsis* (Hamamelidaceae), where number of floral organs and sex expression (bisexual or male) is strictly position dependent in the inflorescence: the shorter a branch, the fewer the organs and the more reduced the gynoecium in the flower that terminates this branch (Endress 1978). An "adaptive" example in flowers is *Ipomopsis aggregata* (Polemoniaceae), where the same individuals produce red flowers early in the season, when hummingbirds are available, and pink flowers later in the season, when only hawkmoths as pollinators are present (Paige and Whitham 1985). Here, one of the color morphs may be expected to be apomorphic compared to the other. There would then be homoplasy with other populations of the same polymorphic species showing the same color. Accordingly, in these examples, the analog of homoplasy occurs in one species or even individual.

In their search for the distribution of homoplasies in plants, Donoghue and Sanderson (1992, 1994) point to a dilemma between the feasibility for quantification for an analytical procedure and complexity of morphological features. In an extensive review of recent publications Sanderson and Donoghue (1989) also discuss the difficulties in finding differences in the degree of homoplasies between plants and animals. A sensible quantification is here not possible at least as long as we do not understand the complexity. This is a wide field for morphologists, and it comes now to the fore with more clarity. Maybe there is an inherent impossibility, because organismic complexity remains too multidimensional as to lend itself to a rigorous quantitative analysis. As long as this problem is not solved, the opposite approach, to consider as much of complexity as possible, however without explicit quantification, should not be abandoned. Thus, for the time being, both are complementary approaches with their inherent strengths and weaknesses. Quantification suffers from gross oversimplification. Consideration of complexity suffers from lack of quantification. The second approach is the "connectivity approach," where one attempts to comprehend interrelations between parts of complex systems instead of simply assembling isolated characters (Bock 1989; see also Szalay and Bock 1991; Müller and Wagner

1991; Wake 1991). If morphological features are quantified, at least a careful "prescreening" is important (Donoghue and Sanderson 1994).

AN EXAMPLE—TRIMEROUS FLOWERS IN EUDICOTS

Occurrence and Different Degrees of Expression of Trimery

Trimerous flowers prevail in the monocots, and they are also common in the magnoliids. However, they are quite rare in the eudicots (the group comprising all dicots apart from the magnoliids). From their distribution it seems that floral trimery is autapomorphic for numerous groups, although dimery and trimery may be basal even within eudicots (Drinnan et al. 1994). An extended literature search (of course not complete) resulted in the following results for eudicots, ranunculids excepted (for an earlier review of trimerous flowers see Li 1966).

There are two different aspects for the level of expression of trimery in eudicot flowers: (1) trimery may be complete throughout the organ whorls or only present in part of the whorls, and (2) trimery may be constant or labile in a given taxon. There are all kinds of combinations between the two levels. A discussion on the occurrence of trimery in eudicot flowers makes sense only if the different aspects of expression are considered.

(1) Trimery of the gynoecium alone is quite common among the Hamamelididae/Rosidae/Dilleniidae/Caryophyllidae group (dimery predominant in Asteridae). It is an expression of the general tendency to have a lower number of carpels than organs of other categories, perhaps due to functional–architectural constraints (see Endress 1994). In a number of taxa with trimerous flowers the gynoecium is dimerous or monomerous—an expression of the same trend. On the other hand, there is a tendency of the androecium to have more organs than the perianth whorls (polyandry). This also occurs in some taxa with otherwise trimerous flowers. For unisexual flowers it is easier to become completely trimerous than for bisexual ones, because fewer organ whorls and less organ categories are involved in the change: in some taxa the flowers of only one sex are trimerous.

(2) In many families there is lability in the floral structure at low taxonomic level: genus, species, or even individual. Trimerous flowers may occur among flowers with other organ numbers. In such taxa trimerous flowers may range from common to rare exceptions. Therefore, a clear distinction between families with trimerous flowers and those without is not possible.

In Appendix 1 only families and genera are listed in which completely trimerous flowers occur in at least one species with some regularity. Double positions in the androecium are also allowed. If the flowers are unisexual, only those cases in which this applies for both male and female flowers are listed. In this category 60 genera of 28 families are included.

A second category consists of the families in which groups occur with trimerous flowers except for the gynoecium, the gynoecium having less than three carpels, rarely more than three carpels (and then indicated) (Appendix 2). Here 53 genera of 17 families are included.

A third category contains taxa with the flowers trimerous except for the androecium, which is commonly polymerous (Appendix 3). Here nine genera of six families are included.

A fourth category contains flowers that are trimerous except for the calyx, which is pentamerous (Appendix 4). Here four genera of four families are included.

Developmental Mechanisms in the Evolution of Trimery

Trimery in eudicot flowers has certainly arisen many times, as shown from its distribution in diverse groups of the more basal eudicots (Rosidae, Dilleniidae, Hamamelididae, Caryophyllidae). Different developmental mechanisms suggest that several different evolutionary pathways are obscured under the heading "trimery."

(1) The flowers may start out pentamerous in the calyx, whereby the first two sepals remain small and only the last three sepals become broad and set the framework for a trimerous groundplan of the inner floral whorls. Examples are *Lechea* (Cistaceae), in which the inner whorls are completely trimerous (Janchen 1925), and *Plagiopteron* (Plagiopteraceae), where this is also the case (except for the secondarily? polyandrous androecium) (Baas et al. 1979).

(2) There is also another mechanism to turn a pentamerous groundplan into a trimerous one. Here the calyx is still regularly pentamerous but two pairs of the five primordia of the following whorl have a congenitally united basal region and differentiate into single organs. Thus the second whorl results in three instead of five organs and sets the stage for the following whorls. This is known from *Tripetaleia* (Ericaceae), in which the petal whorl becomes trimerous (Nishino 1988). The same principle operates in male flowers of Cucurbitaceae, where a corolla is lacking and the androecium becomes trimerous, still with two broader ("double") stamens and one narrower (single) stamen (Leins and Galle 1971).

(3) There are families with a constantly trimerous gynoecium but more labile behavior of the outer floral whorls. It is likely that this basic gynoecial trimery may more or less influence the outer whorls. In Hippocrateaceae the androecium is also trimerous, while corolla and calyx are still pentamerous (as also the androecium in the closely related Celastraceae). It has been argued that this hippocrateoid pattern has originated more than once in the

Celastraceae and that, therefore, Hippocrateaceae may be a polyphyletic group (Robson 1965). In Polygonaceae the gynoecium is constantly trimerous but androecium and perianth fluctuate between pentamery and trimery (Galle 1977). In *Hypericum* (Clusiaceae) the gynoecium is trimerous but the number of stamen groups fluctuates between five and three (Ronse Decraene and Smets 1994).

(4) An originally pentamerous pattern may change to a trimerous one by decrease of the circumference of the floral apex (but retention of the size of the floral organ primordia). This was experimentally shown for *Anagallis* (Primulaceae) by Green (1992).

(5) A change from pentamery to trimery in monosymmetric flowers is difficult. However, at least one example is known, in which this occurs in an imperfect way. In *Poroditta* (Scrophulariaceae) three petal lobes (two corresponding to the upper lip, one to the lower lip) and three stamens (position not specified) are present (Molau 1988); the calyx, however, is tetramerous and the gynoecium dimerous. *Besseya* is another example in the Scrophulariaceae where only the corolla is involved in this pattern (Hufford 1992, 1993).

HOMOPLASY AT DIFFERENT EVOLUTIONARY LEVELS IN ASTERID FLOWERS

Floral Organization

Floral structure may be focused at different levels. The number and arrangement of floral organs is referred to as floral organization. The floral organs may be synorganized to a lower or higher degree into structures of a higher level by their spatial coordination, by fusions and differential growth rates, which is referred to as floral architecture. Floral structure is also shaped by ecological factors, especially pollination agents, which is referred to as floral mode (see Endress 1994). These different structural levels are in some way also different evolutionary levels, because evolutionary changes tend to have different frequencies at the different structural levels by exhibiting an increasing sequence from "organization" over "architecture" to "mode" (Endress 1994).

Depending on the degree of synorganization a given floral feature will not be equally prone for change in all subclasses, e.g., if magnoliids and asterids are compared (see Endress 1990). In magnoliids floral organ number and arrangement is highly variable because of often very limited synorganization of organs. As opposed to magnoliids the number and arrangement of floral organs in the asterids is highly constant. However, there is high evolutionary plasticity at other levels, such as floral architecture and floral modes (e.g., by allometric changes of synorganized parts), and this means a high incidence of homoplasy of relevant features (see below).

Asterids are largely characterized by sympetalous flowers. In these the petals are united from the beginning of their development (as opposed to choripetalous flowers with free petals).

Sympetaly constitutes a new organization upon which a new kind of plasticity became possible. It works with the presence of an intercalary meristem. It is extremely flexible in that it allows differential elongation and, in addition, differential expansion of the corolla tube. In this way a wide spectrum of more or less polysymmetric tubular, salverform, or bowl-shaped flowers, or monosymmetric lip flowers of various shapes may be formed (see, e.g., Wiehler 1983; Robbrecht 1988; Sutton 1988; Hilliard 1994). An extreme expression of differential expansion is the formation of a nectar spur at the base of the corolla tube.

Stebbins (1974) had calculated that sympetaly had 10 separate origins at the level of modern angiosperm families. However, sympetaly may not be uniform. As demonstrated by Erbar (1991) there are two developmental ways by which sympetaly originates. The corolla tube arises either before or at the same time as the corolla lobes (early sympetaly) or only after the corolla lobes have become apparent (late sympetaly). These two patterns are not discernible at maturity. However, they are correlated with larger taxonomic groups. The Solanales/Scrophulariales/Lamiales-group is largely characterized by late sympetaly, while the Campanulales and Dipsacales uniformly show early sympetaly (Erbar 1991, 1994; Reidt and Leins 1994). Kim et al. (1994), in addition, suggest that early sympetaly in the asterids had at least two different origins from choripetaly. Further, in Ericales, sympetaly may not be uniform. In the studies by Anderberg (1993, 1994) and Kron and Chase (1993) *Enkianthus* with sympetalous flowers appears as the sister-group of a part of the Ericales, which includes both choripetalous and sympetalous flowers. This implies that (1) sympetaly has evolved more than once in this group, or that (2) reversals from sympetaly to choripetaly occurred, or that (3) the choripetalous flowers within the group may in reality be obscured sympetalous ones that have not been studied carefully enough in their ontogeny.

However, on the whole, reversals from sympetaly to choripetaly seem to be rare. An example, although not completely choripetalous, is the genus *Methysticodendron* that seems to be derived from strongly sympetalous *Brugmansia* (perhaps *B. candida*) (van Steenis 1969; Schultes 1979). Other examples of specific mutants are cited in Hilu (1983). Hufford (1994) pointed to a potential evolutionary pathway by diminution of a sympetalous corolla by paedomorphosis and subsequent reelaboration whereby choripetaly may reappear. There is good evidence for this event in *Besseya* (Scrophulariaceae) (Hufford 1992, 1993).

Floral Architecture and Floral Modes

Rubiaceae

Striking characters of some Rubiaceae are enlarged, often petal-like sepals. This is known from some (but not all) genera of 10 different tribes and, thus, must have evolved several times in parallel (Verdcourt 1958; Schnell 1960; Robbrecht 1988; Bremer and Struwe 1992). It is an example of a homoplastic tendency of a family, i.e., the frequent occurrence of a particular homoplasy (see, e.g., Sanderson 1991). The feature is much rarer in other asterid families (e.g. also in Verbenaceae, where it occurs in *Clerodendrum* p.p., *Holmskjoldia*, and *Petrea*). It may also be noted that in other Rubiaceae (*Cephaelis*, *Psychotria*) the floral bracts may become large and showy (Schnell 1960). This is again paralleled in Verbenaceae, e.g., in *Congea*, *Sphenodesme*, and *Symphorema* (Classen 1986).

Among the Rubiaceae with petal-like sepals all sepals of a flower may be enlarged or only one. If the latter is the case, flowers with an enlarged, optically attractive sepal are sometimes restricted to the periphery of the inflorescence. In fact, there is a strong tendency in rubiaceous flowers to have one sepal slightly enlarged and still inconspicuous. This could be seen as a preadaptation for the evolution of a more enlarged and optically attractive organ. The morphogenetic basis for the originally slightly enlarged sepal has not been studied. Is it the outermost organ that has a special function as a protective organ already in bud (as present, e.g., in *Caesalpinia*, see Tucker et al. 1985)? In some Rubiaceae the trend to enlarge the sepals (again all or only one) appears only after anthesis and then the sepals may have a function in fruit dispersal (Weber 1955; Bremer and Struwe 1992).

Another prominent trait of rubiaceous flowers is secondary pollen presentation on the style, whereby the pollen is transferred from the anthers to the style in an early flowering stage, from where it can be taken up by pollinators. From the systematic distribution within the family and the variations in detail, Yeo (1993) presumes that secondary pollen presentation evolved repeatedly. The same may be the case for the frequent occurrence of heterostyly in the family (Anderson 1973).

Scrophulariales

Personate flowers, in which the floral entrance is closed by upward arching of the lower lip, and which are pollinated by strong bees, have evolved in parallel in a few genera of Scrophulariaceae (tribes Antirrhineae and Gratioleae), Lentibulariaceae, Acanthaceae, Gesneriaceae, and Bignoniaceae (reviewed in Endress 1994). In Gesneriaceae an especially striking example of parallel evolution is the *Hypocyrta*-type corolla shape, whereby the typically red corolla has a prominent pouch below the tightly constricted throat (Wiehler 1983). These flowers are hummingbird-pollinated. This corolla shape occurs in species of at least 10 genera of the family; some of these were originally described as *Hypocyrta* species (Wiehler 1983). In *Pedicularis* (Scrophulariaceae) a particular suite of different corolla shapes has evolved several times in parallel (Li 1951; Macior 1982). There is especially a parallel diversity in Japan and in North America with two different floral biological syndromes: (1) species with nectariferous flowers with long tubes and short upper lip that are early blooming and are nototribically (i.e., by upper side of the body) pollinated by *Bombus* queens; (2) species with nectarless flowers with short tubes but long, rostrate upper lip that are later blooming and are sternotribically (i.e., by lower side of the body) buzz-pollinated by *Bombus* workers or other bees (Macior 1988). In the *Scutellaria angustifolia* complex (a group of nine species) (Lamiaceae) long-tubed flowers have evolved from short-tubed ones at least three times (Olmstead 1989). Vogel (1990) shows the plasticity of corolla tube shapes (and coloration) in addition to Scrophulariaceae also in other sympetalous families, such as Solanaceae, Rubiaceae, and Gentianaceae.

An anther spur to actively open otherwise closed anthers by pollinators (especially bees) has evolved independently at least four times in the Scrophulariales–Lamiales, in *Dicerandra* (Lamiaceae), *Incarvillea* (Bignoniaceae), *Thunbergia* (Acanthaceae), and *Torenia* (Scrophulariaceae) (see Huck 1987; Endress 1994).

The great evolutionary plasticity of corolla shapes and the readiness to change colors makes a transition between bee and hummingbird-pollination syndromes very easy, especially since characteristics of the growth form do not have to change (e.g., Stebbins 1989).

Ornithophily in the four families, Acanthaceae, Gesneriaceae, Lamiaceae, and Scrophulariaceae, has originated at least 80 times among melittophilous taxa (Vogel 1963). The special predisposition of asterids to evolve hummingbird flowers can also be derived from the list of such flowers by Johnsgard (1983): of the 50 genera containing species with hummingbird flowers 32 are in the asterids, while the remaining 18 are scattered among all other subclasses. Even within some single genera this change has occurred several times. In *Penstemon* (Scrophulariaceae) the evolutionary change from bee- to bird-pollination has occurred at least four or five times (Grant and Grant 1968). In *Mimulus* (Scrophulariaceae) ornithophily has arisen in at least two sections from melittophily (see Stebbins 1989). Ornithophily has a peak in Gesneriaceae. Wiehler (1983) estimates that 60% of the neotropical Gesneriaceae are hummingbird-pollinated, in the tribe Episcieae (with over 468 epiphytic species) the frequency among the epiphytic species is at least 80% (Wiehler 1983). According to Vogel (1963) bird-pollinated flowers evolved independently in 12 subtribes among at least 26 genera of Gesneriaceae.

In the *Aphelandra pulcherrima* complex (Acanthaceae) pollination by hermit hummingbirds is primitive, whereas pollination by trochiline birds has evolved twice in distinct lineages, or in other words: short corollas have evolved twice and are not homologous in these two lineages (McDade 1992).

The transition from bee-pollination to bird-pollination has also occurred within several genera of the Lamiaceae, such as *Salvia, Monardella, Trichostema, Stachys, Monarda, Scutellaria,* and perhaps *Satureja* (reviewed in Huck 1992).

Sutton (1988) mentions six genera of the Antirrhineae (Scrophulariaceae) with more or less long nectar spurs (*Chaenorhinum, Cymbalaria, Kickxia, Anarrhinum, Linaria, Nuttallanthus*). In small-flowered species of some of these genera long spurs may be lacking. It is unclear how often a spur has originated in these genera. However, they have been lost several times. A clearly separate origin has to be seen for the double spurs in the oil flowers of *Diascia* (Scrophulariaceae) (e.g., Vogel 1974).

Correlations between anthocyanin pigments and pollination ecology (hummingbirds versus bees) were found in Polemoniaceae (Harborne and Smith 1978) and in Lamiaceae (Saito and Harborne 1992). Multiple color changes were stated in *Columnea* (Gesneriaceae) (Smith and Sytsma 1994).

Campanulales

Within the subfamily Lobelioideae of the Campanulaceae, which seems to be primitively melittophilous, other pollination biological modes have arisen many times in parallel: ornithophilous at least nine times, chiropterophilous at least three times, sphingophilous at least three times, psychophilous at least three times (Vogel 1980; see also Sazima et al. 1994). At a lower systematic level, Stein (1992) discusses the possibility of parallel evolution of sicklebill pollination from pollination by other hermit hummingbirds in different subgroups of *Centropogon* (Lobelioideae). The sicklebill-pollinated flowers are characterized by their strongly curved corolla tube.

Nectar spurs have evolved at least three times in *Lobelia* (*Heterotoma*) (Campanulaceae) (Ayers 1990).

HOMOPLASY BY SIMPLIFICATION

It seems that homoplasy by different kinds of reduction of traits is especially common in flowers at all levels:

Floral Organization and Architecture

In the pentamerous flowers of Scrophulariales the adaxial stamen is almost always reduced concomitant with floral monosymmetry; it is either present as a staminode or completely absent. In addition, sometimes one of the remaining two stamen pairs is also sterile or absent. This reduction of the anticous (lower) or posticous (upper) stamen pair is sometimes labile at low systematic levels, e.g., within a genus, and must have occurred many times in parallel (reviewed in Endress 1994; Hilliard 1994). On the other hand, it is also noteworthy that the fifth stamen has been regained in several instances, when flowers reverted to polysymmetry (Endress 1994; Burtt 1994).

A pappus in the flowers and fruits of the genus *Centaurea* (Asteraceae) has been lost in one or a few species of at least eight sections of the genus (Wagenitz 1974).

In Lobeliaceae the monosymmetric flowers are resupinated, i.e., they are turned 180° during development so that the pollination apparatus comes to lie in the appropriate position for effective pollen transfer to the dorsal surface of a pollinator. Floral resupination has been lost independently in different lineages of the Lobeliaceae where it is not needed due to variant pollen transfer modes (Ayers 1994).

Floral Modes and Breeding Systems

The loss of specialization in an outcrossing pollination system and resulting self-pollination is exceedingly common. Raven (1980) estimates that in Onagraceae "self-pollination has evolved at least 150 times independently, and more likely well over twice that often." Further, in the autogamous groups of the Onagraceae there is a tendency for loss of floral parts (Raven 1980).

At a low systematic level, *Mimulus guttatus* (Scrophulariaceae) is outcrossing. It gave rise to selfing species more than once (*M. micranthus* and *M. laciniatus*) (Fenster and Ritland 1994). In *Turnera ulmifolia* (Turneraceae) homostylous forms arose at least three times from heterostylous forms (together with a loss of self-incompatibility) (Barrett and Shore 1987). If it is considered that within a species there may be outcrossing and selfing populations or genotypes, fluctuations and, therefore, homoplasy, in this trait are particularly ample.

Cleistogamy is generally accompanied by a breakdown or loss of complexity in floral architecture. It has evolved many times from chasmogamy, often as a polymorphism within the same individual, among Scrophulariales, e.g., in various Acanthaceae (Sell 1977; Lord 1981) and Scrophulariaceae (Lord 1981; Hilliard 1994).

The loss of floral colors is also pervasive. As an example, in the *Scutellaria angustifolia* complex (a group of nine species) (Lamiaceae) lack of blue flower color has arisen at least three times separately (Olmstead 1989).

CONCLUSIONS

In their extensive literature review Donoghue and Sanderson (1994) did not find a significant increase in the level of homoplasy from flower characters via leaf characters to pubescence in plants. One of the problems here may be in the fact that in quantitative studies one tends to use superficial floral characters that can easily be quantified and to omit more complex features, as also pointed out by Donoghue and Sanderson (1994).

The same floral feature may be prone to homoplasy to different degrees, as shown here with several examples, dependent on evolutionary plasticity in general of the trait in focus, which differs from group to group. Two prominent examples discussed here are the much lower incidence of floral trimery in asterids as opposed to rosids, and, in contrast, the much higher incidence of a hummingbird-pollination syndrome in asterids as opposed to rosids. In the first case the high level of synorganization of floral organs in asterids constrains meristic changes, while in the second case sympetaly enhances the formation of floral tubes that are an important constituent of the hummingbird-pollination syndrome. In general floral architecture and pollination biological modes seem to be more prone to homoplasy than does floral organization, which is due to the different degree of their evolutionary plasticity (see, e.g., Armbruster 1994; Endress 1994).

The concentration of homoplasy in certain traits is related to what has been called apomorphic tendencies (e.g., Wernham 1912; Cronquist 1963; Cantino 1985; Sanderson 1991). This means that not only synapomorphies but also shared apomorphic tendencies of outstanding characters may be indicative of close relationship.

Although homoplasy is a pervasive phenomenon, part of homoplasy that appears in phylogenetic studies may be only fictitious, due to poor character resolution by too superficial investigation. Examples here discussed include trimerous flowers in eudicots and sympetalous flowers in asterids. Neither trimery nor sympetaly are uniform characters but show an array of different variants in detail as shown above. Also reversals of traits may show subtle variants in detail. Evolution is not reversible, if the focus is on entire organisms. Only characters seem to be reversible because they are described in a coarse-meshed terminology. They are never *exactly* the same, as the organism as a whole has changed and therefore a single character has a slightly different relationship to the other parts of the whole organism (see, e.g., Wagner 1982a; Williams 1987; Sanderson 1993; Donoghue and Sanderson 1994). An example is the apparent superior ovary in *Gaertnera* and perhaps *Pagamea* (both Psychotrieae, Rubiaceae) that has evolved from an inferior one. Its development is different from that of an originally superior one: it begins inferior and becomes superior by pronounced elongation of the uppermost part in later stages (Igersheim et al. 1994).

It is to be expected that by more critical developmental investigations subtle differences between seemingly similar conditions will become more evident. This will help in generating a more refined view of homoplasy.

ACKNOWLEDGMENTS

I thank Larry Hufford, Michael Sanderson, and an anonymous reviewer for most valuable comments on the manuscript.

APPENDIX 1

Families and Genera with Completely Trimerous Flowers

In taxa with unisexual flowers in both sexes; the perianth may be present or absent, the stamens may be in one or in two whorls or—rarely—in double positions; p.p. after a genus indicates that trimery does not occur in every species or is labile even within species.
Achariaceae
  Acharia p.p. (Harms 1925)
Amaranthaceae
  Amaranthus p.p. (Schinz 1933)
Anacardiaceae
  Comocladia (Engler 1892)
Anisophylleaceae (Ding Hou 1958)
  Anisophyllea p.p.
  Combretocarpus
Balanopaceae
  Balanops p.p. (Carlquist 1980)
Burseraceae (Leenhouts 1956)
  Canarium
  Dacryodes
  Haplolobus
  Rosselia (Forman et al. 1994)
  Santiria
Cneoraceae
  Cneorum p.p. (Engler 1931a)
Cornaceae
  Helwingia p.p. (Harms 1897)
Empetraceae
  Corema (Pax 1890)
Euphorbiaceae (Pax and Hoffmann 1931)
  Phyllanthus p.p.
  Sauropus
  Tragia p.p. (Gillespie 1994)

Fagaceae (Fey 1981)
 Castanea p.p.
 Castanopsis p.p. (Okamoto 1983)
 Chrysolepis p.p.
 Fagus p.p.
 Nothofagus p.p.
 Trigonobalanus p.p.
Haloragaceae
 Proserpinaca (Petersen 1893)
Hippocrateaceae
 Pseudocassine (Loesener 1942)
Limnanthaceae
 Floerkea (Reiche 1892)
Lythraceae
 Rotala p.p. (Cook 1979)
Melastomataceae (Krasser 1893)
 Boerlagea
 Lithobium
 Sonerila
Meliaceae
 Amoora p.p. (Harms 1940)
Myrothamnaceae
 Myrothamnus p.p. (Jäger-Zürn 1966)
Myzodendraceae
 Myzodendron (Skottsberg 1913)
Olacaceae
 Olax p.p. (Villiers 1980)
Onagraceae (Raven 1980)
 Camissonia p.p.
 Gaura p.p.
 Ludwigia p.p.
Platanaceae
 Platanus p.p. (Bretzler 1929)
Podostemaceae
 Dalzellia p.p. (Rutishauser, personal communication)
 Indotristicha (Cusset and Cusset 1988)
Polygonaceae (Galle 1977)
 Emex
 Koenigia p.p. (Ronse Decraene and Akeroyd 1988)
 Polygonum p.p.
 Reynoutria p.p.
 Rheum p.p.
 Rumex p.p.
 Symmeria

Rutaceae (Engler 1931b)
  Lunasia
  Fagara p.p.
Santalaceae
  Anthobolus (Stauffer 1959)
  Osyris (Stauffer 1961)
Simaroubaceae
  Soulamea p.p. (Engler 1931c)
Viscaceae (Engler and Krause 1935)
  Antidaphne p.p.
  Arceuthobium p.p.
  Dendrophthora p.p.
  Eremolepis p.p.
  Eubrachion
  Ginalloa p.p.
  Korthalsella p.p.
  Phoradendron p.p.
  Viscum p.p.

APPENDIX 2

Families and Genera with Flowers That Are Trimerous Except for the Gynoecium

The gynoecium commonly has less than three carpels, rarely more than three, and then especially indicated.
Balanophoraceae (Harms 1935)
  Balanophora p.p. (Hansen 1972)
  Corynaea (Engell 1979)
  Helosis
  Langsdorffia p.p.
  Sarcophyte p.p.
  Scybalium p.p.
  Thonningia p.p.
Barbeyaceae (Friis 1993a)
  Barbeya p.p.
Chenopodiaceae (Ulbrich 1934)
  Agriophyllum p.p.
  Atriplex p.p.
  Axyris p.p.
  Chenopodium p.p.
  Corispermum p.p.
  Dysphania (Pax and Hoffmann 1934)

Halimocnemis p.p.
Suckleya p.p.
Teloxys p.p.
Caesalpiniaceae
  Apuleia p.p. (McLean Thompson 1925)
  Gleditsia p.p. (Tucker 1991)
Cunoniaceae
  Vesselovskya p.p. (Engler 1930)
Empetraceae
  Empetrum (6–9 carpels) (Pax 1890)
Hamamelidaceae (Endress 1970)
  Distyliopsis p.p.
  Distylium p.p.
Mimosaceae
  Mimosa p.p. (Barneby 1991)
Moraceae (Rohwer 1993)
  Antiaris p.p.
  Brosimum p.p.
  Dorstenia p.p.
  Ficus p.p.
  Mesogyne p.p.
  Sparattosyce p.p.
  Streblus p.p.
  Treculia p.p.
  Trophis p.p.
  Trymatococcus
Myricaceae
  Canacomyrica (Kubitzki 1993)
Rosaceae
  Potaninia (Focke 1888)
Rubiaceae (Robbrecht 1988)
  Alibertia p.p.
  Asperula p.p.
  Corynanthe p.p.
  Galium p.p.
  Mitchella p.p.
  Opercularia p.p.
  Paraknoxia p.p.
  Pouchetia p.p.
  Scolosanthus p.p.
Salicaceae
  Salix p.p. (Argus 1986)
Scrophulariaceae
  Sibthorpia p.p. (Hedberg 1955)

Scyphostegiaceae
  Scyphostegia (8–12 carpels) (Swamy 1953)
Simaroubaceae
  Amaroria (female flowers 5-merous) (Engler 1931c)
Urticaceae (Friis 1993b)
  Elattostema p.p.
  Lecanthus p.p. (female flowers with three staminodes)
  Pilea p.p.
  Sarcopilea p.p. (female flowers with three staminodes)

APPENDIX 3

Families and Genera with Flowers That Are Trimerous Except for the Androecium, Which Is Polymerous

Begoniaceae
  Begonia p.p. (Irmscher 1925)
Datiscaceae
  Datisca (Leins and Bonnery-Brachtendorf 1977)
Ebenaceae
  Diospyros section (subgenus) *Maba* (=*Forsteria*) (White 1980)
Flacourtiaceae (Gilg 1925)
  Banara p.p.
  Bennettia p.p.
  Phyllobotryum p.p.
  Quadrasia (only male flowers)
Lecythidaceae
  Foetidia p.p. (Tsou 1994)
Rosaceae
  Polylepis p.p. (Focke 1888)

APPENDIX 4

Families and Genera with Flowers That Are Trimerous Except for the Calyx, Which Is Pentamerous

Cistaceae
  Lechea (Janchen 1925)
Ericaceae
  Tripetaleia (Nishino 1988)
Meliaceae
  Aphanamixis p.p. (Harms 1940)

Plagiopteraceae
Plagiopteron (androecium polyandrous) (Baas et al. 1979)

REFERENCES

Anderberg, A. A. 1993. Cladistic interrelationships and major clades of the Ericales. Plant Systematics and Evolution 184:207–231.

Anderberg, A. A. 1994. Phylogeny of the Empetraceae, with special emphasis on character evolution in the genus *Empetrum*. Systematic Botany 19:35–46.

Anderson, W. R. 1973. A morphological hypothesis for the origin of heterostyly in the Rubiaceae. Taxon 22:537–542.

Argus, G. W. 1986. The genus *Salix* (Salicaceae) in the Southeastern United States. Systematic Botany Monographs 9.

Armbruster, W. S. 1994. Evolution of plant pollination systems: Hypotheses and tests with the Neotropical vine *Dalechampia*. Evolution 47:1480–1505.

Ayers, T. J. 1990. Systematics of *Heterotoma* (Campanulaceae) and the evolution of nectar spurs in the New World Lobelioideae. Systematic Botany 15:296–327.

Ayers, T. J. 1994. Floral resupination in the Lobeliaceae: A twist on a twist. American Journal of Botany 81 (6, Suppl.):140.

Baas, P., R. Geesink, W. A. van Heel, and J. Muller. 1979. The affinities of *Plagiopteron suaveolens* Griff. (Plagiopteraceae). Grana 18:69–89.

Bachmann, K. 1983. Evolutionary genetics and the genetic control of morphogenesis in flowering plants. Evolutionary Biology 16:157–208.

Barneby, R. C. 1991. Sensitivae Censitae. A description of the genus *Mimosa* Linnaeus (Mimosaceae) in the New World. Memoirs of the New York Botanical Garden 65:1–835.

Barrett, S. C. H., and J. S. Shore. 1987. Variation and evolution of breeding systems in the *Turnera ulmifolia* L. complex (Turneraceae). Evolution 41:340–354.

Bateman, R. M., and W. A. Dimichele. 1994. Saltational evolution of form in vascular plants: A neoGoldschmidtian synthesis. Pp. 61–100 *in* D. S. Ingram and A. Hudson, eds. Shape and form in plants and fungi. Academic Press, London.

Bock, W. J. 1989. Organisms as functional machines: A connectivity explanation. American Zoologist 29:1119–1132.

Bremer, B., and L. Struwe. 1992. Phylogeny of the Rubiaceae and the Loganiaceae: Congruence or conflict between morphological and molecular data? American Journal of Botany 79:1171–1184.

Bretzler, E. 1929. Ueber den Bau der Platanenblüte und die systematische Stellung der Platanen. Botanische Jahrbücher für Systematik 62:305–309.

Burtt, B. L. 1994. A commentary on some recurrent forms and changes of form in angiosperms. Pp. 143–152 *in* D.S. Ingram and A. Hudson, eds. Shape and form in plants and fungi. Academic Press, London.

Cantino, P. D. 1985. Phylogenetic inference from nonuniversal derived character states. Systematic Botany 10:119–122.

Carlquist, S. 1980. Anatomy and systematics of Balanopaceae. Allertonia 2 (3):191–246.

Classen, R. 1986. Organisation und Funktion der blumenbildenden Hochblatthüllen bei Symphoremoideae (Verbenac.). Beiträge zur Biologie der Pflanzen 60:383–402.

Cook, C. D. K. 1979. A revision of the genus *Rotala* (Lythraceae). Boissiera 29:1–156.

Cronquist, A. 1963. The taxonomic significance of evolutionary parallelism. Sida 1:109–116.

Cusset, G., and C. Cusset. 1988. Etude sur les Podostemales. 10. Structures florales et végétatives des Tristichaceae. Bulletin du Museum National d' Histoire Naturelle, Paris, Sér. 4, B, Adansonia 10:179–218.

Ding Hou 1958. Rhizophoraceae. Pp. 429–493 *in* C. G. G. J. van Steenis, ed. Flora Malesiana I, 5 (4). Noordhoff, Leiden.

Donoghue, M. J., and M. J. Sanderson. 1992. The suitability of molecular and morphological evidence in reconstructing plant phylogeny. Pp. 340–368 *in* P. S. Soltis, D. E. Soltis, and J. J. Doyle, eds. Molecular systematics of plants. Chapman and Hall, New York.

Donoghue, M. J., and M. J. Sanderson. 1994. Complexity and homology in plants. Pp. 393–421 in B. K. Hall, ed. Homology: The hierarchical basis of comparative biology. Academic Press, San Diego.

Drinnan, A. N., P. R. Crane, and S. B. Hoot. 1994. Patterns of floral evolution in the early diversification of non-magnoliid dicotyledons (eudicots). Plant Systematics and Evolution, Supplement, 8:93–122.

Endress, P. K. 1970. Die Infloreszenzen der apetalen Hamamelidaceen, ihre grundsätzliche morphologische und systematische Bedeutung. Botanische Jahrbücher für Systematik 90:1–54.

Endress, P. K. 1978. Blütenontogenese, Blütenabgrenzung und systematische Stellung der perianthlosen Hamamelidoideae. Botanische Jahrbücher für Systematik 100:249–317.

Endress, P. K. 1990. Patterns of floral construction in ontogeny and phylogeny. Biological Journal of the Linnean Society 39:153–175.

Endress, P. K. 1994. Diversity and evolutionary biology of tropical flowers. Cambridge University Press, Cambridge.

Engell, K. 1979. Morphology and embryology of Scybalioideae (Balanophoraceae) I. *Corynaea crassa* Hook. f. var. *sprucei* (Eichl.) B. Hansen. Botanisk Tidsskrift 73:155–166.

Engler, A. 1892. Anacardiaceae. Pp. 138–178 *in* A. Engler and K. Prantl, eds. Die natürlichen Pflanzenfamilien III, 5. Engelmann, Leipzig.

Engler, A. 1930. Cunoniaceae. Pp. 229–262 *in* A. Engler and K. Prantl, K., eds. Die natürlichen Pflanzenfamilien (ed. 2) 18a. Engelmann, Leipzig.

Engler, A. 1931a. Cneoraceae. Pp. 184–187 *in* A. Engler and K. Prantl, K., eds. Die natürlichen Pflanzenfamilien (ed. 2) 19a. Engelmann, Leipzig.

Engler, A. 1931b. Rutaceae. Pp. 187–359 *in* A. Engler and K. Prantl, K., eds. Die natürlichen Pflanzenfamilien (ed. 2) 19a. Engelmann, Leipzig.

Engler, A. 1931c. Simaroubaceae. Pp. 359–405 *in* A. Engler and K. Prantl, eds. Die natürlichen Pflanzenfamilien (ed. 2) 19a. Engelmann, Leipzig.

Engler, A., and K. Krause. 1935. Loranthaceae. Pp. 98–203 *in* A. Engler and K. Prantl, eds. Die natürlichen Pflanzenfamilien (ed. 2) 16b. Engelmann, Leipzig.

Erbar, C. 1991. Sympetaly - a systematic character? Botanische Jahrbücher für Systematik 112:417–451.

Erbar, C. 1994. Contributions to the affinities of *Adoxa* from the viewpoint of floral development. Botanische Jahrbücher für Systematik 116:259–282.

Fenster, C. B., and K. Ritland. 1994. Evidence for natural selection on mating system in *Mimulus* (Scrophulariaceae). International Journal of Plant Science 155:588–596.

Fey, B. 1981. Untersuchungen über Bau und Ontogenese der Cupula, Infloreszenzen und Blüten sowie zur Embryologie bei Vertretern der Fagaceae und ihre Bedeutung für die Systematik. Doctoral Dissertation, University of Zurich. von Dach, Lyss.

Focke, W.O. 1888. Rosaceae. Pp. 1–61 *in* A. Engler, A. and K. Prantl, eds. Die natürlichen Pflanzenfamilien III, 3. Engelmann, Leipzig.

Forman, L. L., R. W. J. M. van der Ham, M. M. Harley, and T. J. Lawrence. 1994. *Rosselia*, a new genus of Burseraceae from the Louisiade Archipelago, Papua New Guinea. Kew Bulletin 49:601–621.

Friis, I. 1993a. Barbeyaceae. Pp. 141–142 *in* K. Kubitzki, J.G. Rohwer, J.G., and V. Bittrich, eds. The families and genera of vascular plants II. Springer, Berlin.

Friis, I. 1993b. Urticaceae. Pp. 612–630 *in* K. Kubitzki, J.G. Rohwer, and V. Bittrich, eds. The families and genera of vascular plants II. Springer, Berlin.

Galle, P. 1977. Untersuchungen zur Blütenentwicklung der Polygonaceen. Botanische Jahrbücher für Systematik 98:449–489.

Gilg, E. 1925. Flacourtiaceae. Pp. 377–457 *in* A. Engler and K. Prantl, eds. Die natürlichen Pflanzenfamilien (ed. 2) 21. Engelmann, Leipzig.

Gillespie, L. J. 1994. A new section and two new species of *Tragia* (Euphorbiaceae) from the Venezuelan Guayana and French Guiana. Novon 4:330–338.

Grant, K., and V. Grant. 1968. Hummingbirds and their flowers. Columbia University Press, New York.

Grant, V. 1963. The origin of adaptations. Columbia University Press, New York.

Grant, V. 1989. The theory of speciational trends. American Naturalist 133: 604–612.

Green, P. B. 1992. Pattern formation in shoots: A likely role for minimal energy configurations of the tunica. International Journal of Plant Science 153:S59–S75.

Hansen, B. 1972. The genus *Balanophora*. Dansk Botanisk Arkiv 28 (1):1–188.

Harborne, J. B., and D. M. Smith. 1978. Correlations between anthocyanin chemistry and pollination ecology in the Polemoniaceae. Biochemical Systematics and Ecology 6: 127–130.

Harms, H. 1897. Cornaceae. Pp. 250–270 *in* A. Engler, A. and K. Prantl, eds. Die natürlichen Pflanzenfamilien III, 8. Engelmann, Leipzig.

Harms, H. 1925. Achariaceae. Pp. 507–510 *in* A. Engler and K. Prantl, eds. Die natürlichen Pflanzenfamilien (ed. 2) 21. Engelmann, Leipzig.

Harms, H. 1935. Balanophoraceae. Pp. 296–339 *in* A. Engler and K. Prantl, eds. Die natürlichen Pflanzenfamilien (ed. 2) 16b. Engelmann, Leipzig.

Harms, H. 1940. Meliaceae. Pp. 1–172 *in* A. Engler and K. Prantl, eds. Die natürlichen Pflanzenfamilien (ed. 2) 19bI. Engelmann, Leipzig.

Hedberg, O. 1955. A taxonomic revision of the genus *Sibthorpia* L. Botaniska Notiser 108: 161–183.

Hilliard, O. M. 1994. The Manuleae. A tribe of Scrophulariaceae. Edinburgh University Press, Edinburgh.

Hilu, K. W. 1983. The role of single-gene mutations in the evolution of flowering plants. Evolutionary Biology 16:97–128.

Huck, R. B. 1987. Systematics and evolution of *Dicerandra* (Labiatae). Cramer, Berlin (Phanerogamarum Monographiae 19).

Huck, R. B. 1992. Overview of pollination biology in the Lamiaceae. Pp. 167–181 *in* R. M. Harley and T. Reynolds, eds. Advances in labiate science. Royal Botanic Gardens, Kew.

Hufford, L. 1992. Floral structure of *Besseya* and *Synthyris* (Scrophulariaceae). International Journal of Plant Science 153:217–229.

Hufford, L. 1993. A phylogenetic analysis of *Besseya* (Scrophulariaceae). International Journal of Plant Science 154:350–360.

Hufford, L. 1994. Frequencies of modes of ontogenetic evolution in perianth diversification of *Besseya* (Scrophulariaceae). American Journal of Botany 81 (6, Suppl.):162.

Igersheim, A., C. Puff, P. Leins, and C. Erbar. 1994. Gynoecial development of *Gaertnera* Lam. and of presumably allied taxa of the Psychotrieae (Rubiaceae): Secondarily 'superior' vs. inferior ovaries. Botanische Jahrbücher für Systematik 116:401–414.

Irmscher, E. 1925. Begoniaceae. Pp. 548–588 *in* A. Engler and K. Prantl, eds. Die natürlichen Pflanzenfamilien (Ed. 2) 21. Engelmann, Leipzig.

Jäger-Zürn, I. 1966. Infloreszenz- und blütenmorphologische, sowie embryologische Untersuchungen an *Myrothamnus* Welw. Beiträge zur Biologie der Pflanzen 42:241–271.

Janchen, E. 1925. Cistaceae. Pp. 289–313 *in* A. Engler and K. Prantl, K., eds. Die natürlichen Pflanzenfamilien (ed. 2) 21. Engelmann, Leipzig.

Johnsgard, P. A. 1983. The hummingbirds of North America. Smithsonian Institution Press, Washington, D.C..

Kim, K.-J., R. K. Jansen, and R. G. Olmstead. 1994. Multiple origins of sympetaly in dicots. American Journal of Botany 81 (6, Suppl.): 165.

Krasser, F. 1893. Melastomataceae. Pp. 130–199 *in* A. Engler and K. Prantl, eds. Die natürlichen Pflanzenfamilien III, 7. Engelmann, Leipzig.

Kron, K. A., and M. W. Chase. 1993. Systematics of the Ericaceae, Empetraceae, Epacridaceae and related taxa based upon rbcL sequence data. Annals of the Missouri Botanical Garden 80:735–741.

Kubitzki, K. 1993. Myricaceae. Pp. 453–457 *in* K. Kubitzki, J.G. Rohwer, and V. Bittrich, eds. The families and genera of vascular plants II. Springer, Berlin.

Kubitzki, K., P. von Sengbusch, and H.-H. Poppendiek. 1991. Parallelism, its evolutionary origin and systematic significance. Aliso 13:191–206.

Leenhouts, P. W. 1956. Burseraceae. Pp. 209–296 *in* C. G. G .J. van Steenis, ed. Flora Malesiana I, 5 (2). Noordhoff, Leiden.

Leins, P., and R. Bonnery-Brachtendorf. 1977. Entwicklungsgeschichtliche Untersuchungen an Blüten von *Datisca cannabina* (Datiscaceae). Beiträge zur Biologie der Pflanzen 53: 143–155.

Leins, P., and P. Galle. 1971. Entwicklungsgeschichtliche Untersuchungen an Cucurbitaceen-Blüten. Oesterreichische Botanische Zeitschrift 119:531–548.

Li, H. 1951. Evolution in the flowers of *Pedicularis*. Evolution 5:158–164.

Li, H.-L. 1966. The phylogeny of trimery in angiosperms flowers. Botanical Bulletin of the Academy of Sciences 7 (1):64–74.

Loesener, T. 1942. Hippocrateaceae. Pp. 198–231 *in* A. Engler and K. Prantl, eds. Die natürlichen Pflanzenfamilien (ed. 2) 20b. Engelmann, Leipzig.

Lord, E. M. 1981. Cleistogamy: A tool for the study of floral morphogenesis, function and evolution. Botanical Review 47:421–449.

Macior, L. W. 1982. Plant community and pollinator dynamics in the evolution of pollination mechanisms in *Pedicularis* (Scrophulariaceae). Pp. 29–45 *in* J.A. Armstrong, J.M. Powell, and A.J. Richards, eds. Pollination and evolution. Royal Botanic Gardens, Sydney.

Macior, L. W. 1988. A preliminary study of the pollination ecology of *Pedicularis* (Scrophulariaceae) in Japan. Plant Species Biology 3:61–66.

McDade, L. A. 1992. Pollinator relationships, biogeography, and phylogenetics. BioScience 42: 21–26.

McLean Thompson, J. 1925. Studies in advancing sterility II. The Cassieae. Publications of the Hartley Botanical Laboratory 2:1–44.

Molau, U. 1988. Scrophulariaceae - Part I. Calceolarieae. Flora Neotropica Monograph 47: 1–326.

Müller, G. B., and G. P. Wagner. 1991. Novelty in evolution: Restructuring the concept. Annual Review of Ecology and Systematics 22:229–256.

Nishino, E. 1988. Early floral organogenesis in *Tripetaleia* (Ericaceae). Pp. 181–190 *in* P. Leins, S.C. Tucker, and P.K. Endress, eds. Aspects of floral development. Cramer, Berlin.

Okamoto, M. 1983. Floral development of *Castanopsis cuspidata* var. *sieboldii*. Acta Phytotaxonomica Geobotanica 34:10–17.

Olmstead, R. 1989. Phylogeny, phenotypic evolution, and biogeography of the *Scutellaria angustifolia* complex (Lamiaceae): Inference from morphological and molecular data. Systematic Botany 14:320–338.

Paige, K. N., and T. G. Whitham. 1985. Individual and population shifts in flower color by scarlet gilia: A mechanism for pollinator tracking. Science 227:315–317.

Pax, F. 1890. Empetraceae. Pp. 123–127 *in* A. Engler and K. Prantl, eds. Die natürlichen Pflanzenfamilien III, 5. Engelmann, Leipzig.

Pax, F., and K. Hoffmann. 1931. Euphorbiaceae. Pp. 11–233 *in* A. Engler and K. Prantl, eds. Die natürlichen Pflanzenfamilien (ed. 2) 19c. Engelmann, Leipzig.

Pax, F., and K. Hoffmann. 1934. Dysphaniaceae. Pp. 272–274 *in* A. Engler and K. Prantl, eds. Die natürlichen Pflanzenfamilien (ed. 2) 16c. Engelmann, Leipzig.

Petersen, O. G. 1893. Halorrhagidaceae. Pp. 226–237 *in* A. Engler and K. Prantl, eds. Die natürlichen Pflanzenfamilien III, 7. Engelmann, Leipzig.

Raven, P. H. 1980. A survey of reproductive biology in Onagraceae. New Zealand Journal of Botany 17:575–593.

Reiche, K. 1892. Limnanthaceae. Pp. 136–137 *in* A. Engler and K. Prantl, eds. Die natürlichen Pflanzenfamilien III, 5. Engelmann, Leipzig.

Reidt, G., and P. Leins. 1994. Das Initialstadium der sympetalen Krone bei *Sambucus racemosa* L. und *Viburnum farreri* Stearn. Botanische Jahrbücher für Systematik 116:1–9.

Robbrecht, E. 1988. Tropical woody Rubiaceae. Opera Botanica Belgica 1:1–272.

Robson, N. 1965. Taxonomic and nomenclatural notes on Celastraceae. Boletim da Sociedade Broteriana 39:5–55.

Rohwer, J. G. 1993. Moraceae. Pp. 438–453 *in* K. Kubitzki, J. G. Rohwer, and V. Bittrich, eds. The families and genera of vascular plants II. Springer, Berlin.

Ronse Decraene, L.-P., and J. R. Akeroyd. 1988. Generic limits in *Polygonum* and related genera (Polygonaceae) on the basis of floral characters. Botanical Journal of the Linnean Society 98:321–371.

Ronse Decraene, L.-P., and E. F. Smets. 1994. Merosity in flowers: Definition, origin, and taxonomic significance. Plant Systematics and Evolution 191:83–104.

Saito, N., and J. B. Harborne. 1992. Correlations between anthocyanin type, pollinator and flower colour in the Labiatae. Phytochemistry 31:3009–3015.

Sanderson, M. J. 1991. In search of homoplastic tendencies: Statistical inference of topological patterns in homoplasy. Evolution 45:351–358.

Sanderson, M. J. 1993. Reversibility in evolution: A maximum likelihood approach to character gain/loss bias in phylogenies. Evolution 47:236–252.

Sanderson, M. J., and M. J. Donoghue. 1989. Patterns of variation in levels of homoplasy. Evolution 43:1781–1795.

Sazima, M., I. Sazima, and S. Buzato. 1994. Nectar by day and night: *Siphocampylus sulfureus* (Lobeliaceae) pollinated by hummingbirds and bats. Plant Systematics and Evolution 191:237–246.

Schinz, H. 1933. Amaranthaceae. Pp. 7–85 *in* A. Engler and K. Prantl, eds. Die natürlichen Pflanzenfamilien (ed. 2) 16c. Engelmann, Leipzig.

Schnell, R. 1960. Note sur le genre *Cephaelis* et le problème de l'évolution parallèle chez les Rubiacées. Bulletin du Jardin Botanique de Bruxelles 30:357–373.

Schultes, R. E. 1979. Solanaceous hallucinogens and their role in the development of New World cultures. Pp. 137–160 *in* I. G. Hawkes, R. N. Lester, and A. D. Skelding, eds. The biology and taxonomy of the Solanaceae. Academic Press, London.

Sell, Y. 1977. La cléistogamie chez *Ruellia lorentziana* Griseb. et quelques autres Acanthacées. Berichte der Deutschen Botanischen Gesellschaft 90:135–147.

Skottsberg, C. 1913. Morphologische und embryologische Studien über die Myzodendraceen. Kungl. Svenska Vetenskapsakademiens Handlingar 51 (4):1–34.

Smith, J. F., and K. J. Sytsma. 1994. Evolution in the Andean epiphytic Genus *Columnea* (Gesneriaceae). II. Chloroplast DNA restriction site variation. Systematic Botany 19:317–336.

Stauffer, H. U. 1959. Revisio Anthobolearum. Aargauer Tagblatt AG, Aarau.

Stauffer, H. U. 1961. Santalales-Studien V-VIII. Vierteljahrsschrift der Naturforschenden Gesellschaft Zürich 106:387–418.

Stebbins, G. L. 1974. Flowering plants. Evolution above the species level. Belknap Press of Harvard University Press, Cambridge, MA.

Stebbins, G. L. 1989. Adaptive shifts toward hummingbird pollination. Pp. 39–60 J. H. Bock and Y. B. Linhart, eds. The evolutionary ecology of plants. Westview Press, Boulder, Colorado.

van Steenis, C. G. G. J. 1969. Plant speciation in Malesia, with special reference to the theory of non-adaptive saltatory evolution. Biological Journal of the Linnean Society 1:97–133.

Stein, B. A. 1992. Sicklebill hummingbirds, ants, and flowers. BioScience 42:27–33.

Sutton, D. A. 1988. A revision of the tribe Antirrhineae. British Museum (Natural History)/ Oxford University Press, London/Oxford.

Swamy, B. G. L. 1953. On the floral structure of *Scyphostegia*. Proceedings of the National Institute of India 19, 2:127–142.

Szalay, F. S., and W. J. Bock. 1991. Evolutionary theory and systematics: Relationships between process and patterns. Zeitschrift für Zoologische Systematik und Evolutions-Forschung 29:1–39.

Tsou, C.-H. 1994. The embryology, reproductive morphology, and systematics of Lecythidaceae. Memoirs of the New York Botanical Garden 71:1–110.

Tucker, S. C. 1991. Helical floral organogenesis in *Gleditsia*, a primitive caesalpinioid legume. Amererican Journal of Botany 78:1130–1149.

Tucker, S. C., O. L. Stein, and K. S. Derstine. 1985. Floral development in *Caesalpinia* (Leguminosae). American Journal of Botany 72:1424–1434.

Ulbrich, E. 1934. Chenopodiaceae. Pp. 379–584 *in* A. Engler and K. Prantl, eds. Die natürlichen Pflanzenfamilien (ed. 2) 16c. Engelmann, Leipzig.

Van Valen, L. 1979. Switchback evolution and photosynthesis in angiosperms. Evolutionary Theory 4:143–146.

Verdcourt, B. 1958. Remarks on the classification of the Rubiaceae. Bulletin du Jardin Botanique de Bruxelles 28:209–281.

Villiers, J.-F. 1980. Olacaceae. Pp. 179–187 *in* A. Aubréville and J.-F. Leroy, eds. Flore de la Nouvelle Calédonie et Dépendances 9. Muséum National d'Histoire Naturelle, Paris.

Vogel, S. 1963. Blütenökotypen und die Gliederung systematischer Einheiten. Berichte der Deutschen Botanischen Gesellschaft 76:(98)–(101).

Vogel, S. 1974. Oelblumen und ölsammelnde Bienen. Tropische und Subtropische Pflanzenwelt 7:1–267.

Vogel, S. 1980. Florengeschichte im Spiegel blütenökologischer Erkenntnisse. Rheinisch-Westfälische Akademie der Wissenschaften, Vorträge N 291:7–48.

Vogel, S. 1990. Radiacion adaptiva del sindrome floral en las familias neotropicales. Boletin de la Academia Nacional de Ciencias, Cordoba, 59:5–30.

Waddington, C.H. 1940. Organisers and genes. Cambridge University Press, Cambridge.

Wagenitz, G. 1974. Parallele Evolution von Merkmalen in der Gattung *Centaurea*. Phyton (Horn) 16:302–312.

Wagner, G. 1982a The logical structure of irreversible systems transformation: A theorem concerning Dollo's Law and chaotic movement. Journal of Theoretical Biology 96:337–346.

Wagner, G. 1982b. On the evolution of multistability in the genome. Progress in Cybernetics and Systems Research 9:283–288.

Wagner, G. 1988. The significance of developmental constraints for phenotypic evolution by natural selection. Pp. 222–229 *in* G. de Jong, ed. Population genetics and evolution. Springer, Berlin.

Wake, D. B. 1991. Homoplasy and parsimony analysis. Systematic Zoology 40:105–109.

Weber, H. 1955. Ueber die Blütenkelche tropischer Rubiaceen. Abhandlungen der Akademie der Wissenschaften und der Literatur Mainz, Mathematisch-Naturwissenschaftliche Klasse, 1955:447–467.

Wernham, H. F. 1912. Floral evolution: With particular reference to the sympetalous dicotyledons. Botany School, Cambridge.

West-Eberhard, M. J. 1986. Alternative adaptations, speciation, and phylogeny (a review). Proceedings of the National Academy of Sciences USA 83:1388–1392.

White, F. 1980. Notes on the Ebenaceae VIII. The African sections of *Diospyros*. Bulletin du Jardin Botanique de Bruxelles 50:445–460.

Wiehler, H. 1983. A synopsis of the Neotropical Gesneriaceae. Selbyana 6:1–219.

Williams, P. L. 1987. Reversible evolution? Nature 328:21–22.

Wright, S. 1932. The roles of mutation, inbreeding, crossbreeding and selection in evolution. Proceedings of the Sixth International Congress of Genetics 1:356–366.

Yeo, P. F. 1993. Secondary pollen presentation. Plant Systematics and Evolution, Supplement, 6:1–268.

HOMOPLASY AND THE EVOLUTIONARY PROCESS: AN AFTERWORD

MICHAEL J. SANDERSON AND LARRY HUFFORD

Evolutionary diversification includes processes that cause species to look different *and* that cause them to look similar once again. The emphasis that is placed on words such as "diversification" in our field is testament to our tendency to view evolution as a process that is synonymous with novelty rather than reiteration. However, a significant component of organismal diversity consists of species that exhibit traits that are similar to those found in near or even distant relatives, despite their absence in a common ancestor. The issue of the relative importance of this factor, homoplasy, in evolutionary diversification has been neglected. This may have been caused in part by controversies over definitions (uncertainties about words like parallelism, convergence, etc.), but more than likely it has also resulted from the difficulty in identifying homoplasy. This is where phylogenetic systematics has played a pivotal role. The contributors to this volume have demonstrated unequivocally that we now have the phylogenetic tools to recognize homoplasy and design investigations to study its pattern of expression in evolutionary history.

However, the study of homoplasy still faces two serious conceptual problems. The first problem is disagreement over the status of homoplasy as "mere" error in phylogenetic analysis. The second is the uncertain relationship of homoplasy to homology, itself a contentious subject.

Regarding the first issue, we note that if homoplasy were nothing more than an error made during homology assessment, then no explanation for the observed patterns of character state distribution among taxa would be needed. Indeed, the perception of homoplasy as mistaken homology suggests that the underlying similarities are irrelevant. This trivialization of homoplasy has hindered progress in understanding its role in diversification. Iterated origins of similarities are biologically significant and have implications for systematics and evolutionary biology. A particular example of iterated similarity may or may not be interesting depending on the kinds of evolutionary processes responsible for that homoplasy. The significance of this similarity is thus dependent on the existence of a relevant underlying process. This is not so different from the observation that the underlying process that lends significance to homologous similarity (or distinctiveness for that matter) is descent with modification.

As to the second problem, it is clear that there is not broad agreement on the relationship and relative significance of homoplasy and homology. Many authors, including several in this volume (e.g., Bateman, Brooks), focus on homoplasy as falsified hypotheses of homology. To many, the identification of homologies and the reconstruction of phylogenetic hypotheses based on them is the first step in the study of homoplasy. The homoplasy that is discovered in a data matrix following phylogenetic analysis has been the subject of much attention in recent years. It is easily quantified by measures such as the consistency index (see Archie, this volume). It is also the kind of homoplasy most likely to hinder accurate phylogeny reconstruction. Interest in it has undoubtedly been motivated in part by this practical issue (see e.g., Proctor, Chang and Kim, Sanderson, and Donoghue, this volume).

Yet this is not the only form that homoplasy takes. As Wake points out in the forward to this volume, homoplasy is *not* just nonhomology. We would go further and say that what is typically meant by nonhomology is really nonhomology within *characters*. A character is a set of alternative mutually exclusive states in an evolutionary transformation series. Homoplasy is not restricted to characters used in phylogenetic analysis although it is often discovered in such characters. Homoplasy is the iterated acquisition of similarity of any kind in phylogenetic history, and therefore it may appear between characters or within characters. "Convergent" evolution is best understood as homoplasy between characters. The morphological and developmental antecedents of insect wings and birds' wings were never transformations of the same character state in some common ancestor. This kind of homoplasy can and must still be studied in a phylogenetic framework, but the characters that provide the raw material for understanding are not the same ones that are used to make the tree.

The study of homoplasy can proceed with or without initial hypotheses of homology but it languishes without an explicit hypothesis about phylogeny. We must see homoplasy as an evolutionary phenomenon distinct from homology. Homology and homoplasy spring from different biological processes and differ in their implications for diversification. At issue in the study of homology is how character states become different despite their common origin. The study of homoplasy, in contrast, concerns any and all processes that produce an iteration of similarity.

Many of the papers in this volume point to future directions in the study of homoplasy. A fundamental avenue is clearly the discovery of nonrandom patterns in the distribution of homoplasy among taxa and characters. These will foster hypotheses about the role of homoplasy in diversification. Armbruster has shown how nonrandom patterns arise when natural selection models exaptations into alternative adaptations. Papers in this volume by Doyle, Foster et al., and Hufford examine how evolutionary biases serve to generate homoplasy. We expect to see many more studies along the lines of McShea's that propose relationships between organismal features such as hierarchical developmental complexity and the likelihood of certain patterns of homoplasy, such as reversal.

Discovery of unexpected patterns in homoplasy may also shed light on its general role in organismal diversity (Endress, this volume). Diversification is generally thought of as a set of processes associated with the generation of species and/or new genotypes and phenotypes in a clade. Homoplasy becomes an issue if characters are nonindependent in some way, because then it is the *combination* of characters that is relevant. Although the recurrence of a character state in one character is not a "novelty," its presence in the genetic background of other characters may be. This recurrence may generate novel epistatic effects on the phenotype, or the interaction of homoplastic phenotypes may have novel effects on how selection acts on the organism and thus on its subsequent evolutionary change.

Studies that appear to be focused on phylogenetic practice may really reveal rather interesting features of the evolutionary process. The clearest example of this is the now diverse set of studies of levels of homoplasy in different classes of characters, including morphological, behavioral (Proctor, this volume), and molecular sequence and RFLP data (Sanderson and Donoghue, this volume). These evidently were motivated mainly by a concern for the relative reliability of different classes of characters for making trees. We would argue that there is a separate, very interesting issue about character evolution that is raised by these studies: Are certain characters more homoplastic than others and why? When we look at the phylogeny of life and see the extensive iteration of states, it seems that the notion that certain characters are *not* homoplastic is unusual and warrants scrutiny. For example, are certain key innovations truly unique or was there homoplasy in taxa derived from near the origin of these traits that may now be extinct (or whose relationships to modern taxa are unclear)?

We believe that homoplasy should take its place alongside such phenomena as rates of diversification, evolutionary trends, and adaptive radiations, as a key area of macroevolutionary inquiry. Homoplasy has been viewed merely as a confounding factor in systematics for too long. Many of the authors in this volume have begun their papers with an acknowledgment of the confounding effects of homoplasy, but in the end what emerges from their chapters is the conclusion that homoplasy is a significant facet of organismal diversification. We hope that these attempts to understand the specific evolutionary processes that generate homoplasy and to evaluate their significance relative to other aspects of diversification will help to set the stage for future investigations of homoplasy's role in the origin and expression of diversity.

INDEX